ESOTERIC ANATOMY

The Body As Consciousness

Bruce Burger

North Atlantic Books

Berkeley, California

Published by
North Atlantic Books
P.O. Box 12327
Berkeley, California 94712

Cover painting by John Ivey
Cover and book design by Andrea DuFlon
Printed in the United States of America

Esoteric Anatomy: The Body As Consciousness is sponsored by the Society for the Study of Native Arts and Sciences, a nonprofit educational corporation whose goals are to develop an educational and cross-cultural perspective linking various scientific, social, and artistic fields; to nurture a holistic view of arts, sciences, humanities, and healing; and to publish and distribute literature on the relationship of mind, body, and nature.

North Atlantic Books' publications are available through most bookstores. For further information, visit our website at www.northatlanticbooks.com or call 800-733-3000.

ISBN 13: 978-1-55643-224-8

Library of Congress Cataloging-in-Publication Data

Burger, Bruce, 1942–
        Esoteric anatomy : the body as consciousness / Bruce Burger.
            p.   cm.
        "Sponsored by the Society for the Study of Native Arts and Sciences."
        Includes bibliographical references and index.
        ISBN 1-55643-224-0
        1. Polarity therapy.   2. Mind and body.   I. Society for the Study of Native Arts and Sciences.   II. Title.
        RZ520.B87   1998
        615.8'52—dc21   96-37686
                                                                                     CIP

Dr. Stone's books were originally published separately as: *Energy, The Wireless Anatomy of Man, Polarity Therapy, The Mysterious Sacrum, Vitality Balance, Evolutionary Energy Charts, Polarity Therapy Principles and Practice, Health Building, A Purifying Diet,* and *Easy Stretching Postures.* In 1985 CRCS Publications brought out an enlarged edition of *Health Building,* which included *A Purifying Diet* and *Easy Stretching Postures.* In 1986 and 1987 CRCS published *Polarity Therapy: The Complete Collected Works of Dr. Randolph Stone* in two volumes consolidating Dr. Stone's writings.

Throughout this work, we refer to Dr. Stone's works variously by title, book number, and abbreviation. Following the order of publication, Book I, then, is *Energy;* Book II, *The Wireless Anatomy of Man;* Book III, *Polarity Therapy;* Book IV, *The Mysterious Sacrum;* Book V, *Vitality Balance;* EEC, *Evolutionary Energy Charts;* ET, *Energy Tracing;* PN, *Private Notes;* HB, *Health Building.* Where *Polarity Therapy: The Complete Collected Works of Dr. Randolph Stone* is referenced by volume, Volume I contains Books I, II, and III. Volume II contains Books IV, V, EEC, ET, and PN. Dr. Stone's publications are readily available from CRCS Publications and the American Polarity Therapy Association.

Quotations from Dr. Stone's works retain the original capitalization and underlining.

7 8 9 10 11 12 SHERIDAN 17 16 15 14 13

Acknowledgments

I am thankful for the support of the contributing editors: Victoria Baker, Buz Blocher, John Chitty, R.P.P., Chandana Becker, Ph.D., R.P.P., Mary Buckley, Michael Drew, Linda Diane Feldt, R.P.P., Roger Gilchrist, M.S., R.P.P., Richard Gordon, Frank Hemingway, Bob Sachs, Jim Said, D.C., R.P.P., Cybele Tomlinson, Don Williams, Aurelia Zoltners, R.P.P., and Elizabeth Shaw (for contributions to the Polarity Diet section). Special thanks to my wife, Chela, who at the eleventh hour shared her editing expertise. Thanks to Krishnakirtana, Pat Neal, Mimi Plouf, Denise Laverty, Chloe Randall, and Mark Allison, R.P.P., who produced the illustrations. Special gratitude to John Ivey for the original airbrush painting on the cover and to Andrea DuFlon for the beauty she brought to the design of the book. I would like to take this opportunity to express my gratitude to the students and staff at Heartwood Institute who have supported the evolution of this work over the last twenty years.

Thanks to CRCS Publications for permission to reproduce selections from the following: *Polarity Therapy: The Complete Collected Works,* Vols. I and II, by Randolph Stone, D.C., D.O., © 1986, 1987; *Health Building* by Randolph Stone, © 1976, 1985, Randolph Stone Trust; and *Astrology, Psychology, and the Four Elements,* by Stephen Arroyo, © 1975. Reprinted by arrangement with CRCS Publications, P.O. Box 1460, Sebastopol, CA 95472. Thanks to Polarity Press, for permission to reproduce Mark Allison's lucid charts from *Energy Exercises,* by John Chitty and Mary Louise Muller, © 1990, and selections from *Relationships and the Human Energy Field* by John and Anna Chitty, © 1991. Reprinted by arrangement with Polarity Press, 2410 Jasper Court, Boulder, CO 80304. Thanks to the American Polarity Therapy Association, for permission to reproduce selections from *Standards For Practice and Code of Ethics,* ©

1996. Reprinted by arrangement with the American Polarity Therapy Association, 2888 Bluff Street, #149, Boulder, CO 80301. Thanks to Shambhala Publications, Inc., for permission to reproduce selections from the following: *No Boundary* by Ken Wilber, © 1979; *Grace and Grit,* by Ken Wilber, © 1991; *The Power of Limits,* by György Doczi, © 1981; *The Tao of Physics,* by Fritjof Capra, © 1975, 1983, 1991. Reprinted by arrangement with Shambhala Publications, Inc., 300 Massachusetts Avenue, Boston, MA 02115.

I would like to express my gratitude to Mathaji Vanamali and Vanamali Ashram, Tapovan P.O. Via Shivananda Nagar, Rishikesh, UP 249192, India for permission to use portions of *Nitya Yoga: The Yoga of Constant Communion* an audio tape commentary on the *Srimad Bhagavad Gita.* (*Nitya Yoga* is also available as a book from Vanamali Ashram and in the U.S.is published as *The Word of God* by Blue Dove Press). Thanks to Blue Dove Press for permission to use material from *The Ultimate Medicine,* Robert Powell, Ed. Reprinted by arrangement with Blue Dove Press, 4204 Sorrento Valley Blvd., Suite K, San Diego, CA 92121.

We would like to thank Sri V.S. Ramanan, President, and the trustees of Sri Ramanasramam for permission to use a photo of Sri Bhagavan Ramana Maharshi and material from the booklet by Sri Ramana, *Who Am I?,* © Sri Ramanasramam, Tiruvnnamalai, India, 1992.

Contents

PART III   *The Practice of Somatic Psychology*

Part IV   Resources

Introduction

When Bruce Burger asked me if I would like to write an introduction to his book, I elected to do so without hesitation. I've never liked reading long introductions, so I'll keep this one short and to the point. I've known Bruce since 1978 and have always been impressed with both his dedication to his work and the ample accessibility of his love. I believe that these qualities clearly manifest through the depth of understanding which he brings to the healing arts. When I hold this text, I feel as though I am holding one of the more important documents of our time—an extremely comprehensive and extraordinary map of our subtle anatomy. *Esoteric Anatomy: The Body as Consciousness* is an ideal book for those who are seeking a deeper understanding of energy in the healing arts. Bruce has mindfully woven insights from the wisdom of ancient India to create a unique approach for understanding and working with life force, the healing presence of the soul.

The first section of this book, Loving Hands Are Healing Hands, contains a series of eleven energy-balancing sessions based upon Polarity Therapy. While these sessions are quite simple and easy to learn, please don't be deceived. These sessions are profoundly effective. Then, through a holistic approach to healing, Bruce presents a system of Somatic Psychology for clearing trauma from the cellular memory of the body.

Bringing in over two decades of exploration and contemplation, Bruce presents us with new models for understanding Esoteric Psychology, Esoteric Anatomy, and Energy Medicine. These models are based upon archetypal understandings of the healing arts, the five elements, and viewing the body as consciousness. This useful, comprehensive, and insightful book is truly a major work of devotion and love. I highly recommend it for students of healing, yoga, and Eastern spirituality. Most especially, I recommend this book to those seeking to bring their most cherished convictions about the sacredness and the unity of all life into their service in the world.

Richard Gordon
Author of *Your Healing Hands*
and
Quantum-Touch—The Power to Heal

Blessing: Hari Aum

Sri Arvind (Bruce) Burger has spent a lifetime of study on the nature of sound vibration and its effect on the physical body. He has been practicing Polarity Therapy for many years. His chanting of the word *Aum* when he uses it for therapy really makes your whole body tingle and vibrate in tune with it. This book, *Esoteric Anatomy,* is the consummation of all his findings and experiments of so many years and will be of enormous benefit to all those who read it. The author is an Indologist as well as a great devotee of Lord Krishna and there is no doubt that the Divine has given him all guidance in bringing out this original and fascinating book.

All manifestation arises from the moment of forces precipitated by the resonance that is *Aumkar.* Aum is the ancient word, the universal sound energy which is pulsating with pure potential. This *shakti* is the very nature of the Absolute. It is the living energy whose vibrations give rise to the whole of the manifest universe. It is the dynamic impulse which reverberates within itself and gives rise to all types of experiences—intellectual, volitional, emotional, and spiritual. The universe arises from sound, as do all things with form. Sound arises in the inner sky of our consciousness—the heart space in the head and the sky space in the heart.

This sound unites with the vital breath and animates the inert body. Sound is like an infinite ocean moving in all directions. It pervades everything both outside and inside. In the form of the Higher Mind it becomes creation, preservation, dissolution—*Brahma, Vishnu, Shiva.* Beyond that it passes into the Thuriya state, which is soundless. This ocean of our potentiality has two aspects—pure potential or Shiva and pure energy or *Shakti,* which is again divided into three aspects—*janana shakti, iccha shakti,* and *kriya shakti*—the energy for knowledge, the energy for desire, and the energy for action. All three are needed for creation to happen. Thus Shakti is the supreme creative aspect of the Absolute—vital and dynamic. It is completely stable yet never still, eternally pulsating within the sea of pure consciousness. This resonance causes movement, waves, and ripples which intersect, mingle, rise, and break.

How exactly this vibrant, self-aware, ever-pulsating yet ever-still ocean of pure consciousness manifests as our familiar material world is a divine mystery which baffles even the scientists of today. We can try to trace a hierarchical development beginning with the one manifesting itself in increasingly differentiated levels. The material world is the grossest and most differentiated level. However all successive levels are contained within their subtle predecessors, so all things basically are the same and share certain basic characteristics. What manifests is always the one potential power—the total energy of sound which is the manifesting power of the Absolute. "Aum vibrates like a storm in the sky, Having neither beginning nor end. It is the stage of the divine drama."—Swami Nityananda.

We as characters in this drama are meant to be directed by this stage manager. Our bodies are made to resonate with this cosmic sound. If we are in harmony with these cosmic vibrations our bodies would never become diseased. Arvind shows how these sounds can be used to cure the body without the use of drugs. May that Divine Energy ever support and guide him to the ocean of Infinite Bliss.

Hari Aum Tat Sat.
Mathaji Devi Vanamali
Vanamali Gita Yogashram
Rishikesh, India

Introduction
Actualizing Your Compassionate Nature

It is a blessing to be called to a life of compassionate service in the natural healing arts. To be an instrument of a caring universe. To actualize one's higher nature through a life of dedicated service. To take personal responsibility for the healing of humanity. To live in integrity with one's most cherished beliefs of the sacredness and unity of life.

Esoteric Anatomy: The Body As Consciousness offers a comprehensive health care system based on an understanding of the body as a field of conscious energy. It shares a spectrum of profoundly effective resources for healing, health building, and self-actualization. We offer a model for working with the client as a person, respecting the depth of his or her life as a conscious, evolving, spiritual being. Our holistic, educational approach honors inner resources and empowers the client in taking responsibility for healing and personal growth.

Esoteric Anatomy offers a prescription for action in the world. It unfolds an effective method for the alleviation of suffering: physical, mental, emotional, and spiritual. The core of our approach is a comprehensive health care system called Polarity Therapy. Polarity Therapy is the fruit of the life's work of Randolph Stone, D.C., D.O., who was almost ninety-two years old when he passed on in 1981. In a professional life that spanned more than a half-century of service as a physician, Stone was a constant student. For decades he burned the midnight oil seeking to understand the mysteries of life and healing. Dr. Stone circled the globe a dozen times in his quest, delving into the ancient wisdom of traditional healing modalities.

In a series of six books published between 1954 and 1957, Dr. Stone offered his synthesis of Ayurveda, Naturopathy, Osteopathy, Craniosacral Therapy, Hermetic philosophy, and modern physics, which he called Polarity Therapy. Stone's consciousness was far ahead of his time and his work was nearly lost to the world. It was only through the renaissance of interest in holistic health and spirituality in the early 1970s that the world caught up with Dr. Stone's profound understanding and Polarity Therapy began to develop a following of earnest students. Robert K. Hall, M.D., founder of the Lomi School, referred to this pioneering physician as "the father of a new field of energetic studies."

Loving Hands Are Healing Hands, Part I of the book, offers a step-by-step initiation to Polarity Therapy. Polarity energy-balancing principles and techniques can be used alone or easily integrated into other bodywork and forms of therapy. We offer an inner approach to healing based on respect, presence, and vulnerability. These attitudes support a safe space in which the client has permission to heal. Loving Hands are Healing Hands offers hands-on professional training in working with a blend of bodywork and counseling techniques that cleanse the body's cellular memory of trauma and promote mental and emotional clarity, physical healing, and personal growth. Our approach is holistic, working with clients on many levels—physical, emotional, spiritual, and lifestyle—in an experiential, educational process for facilitating well-being.

Contemplating Ultimate Truth

Part II, Esoteric Anatomy, offers a series of essays that illuminate an ancient understanding for working with energy in the healing arts. This formulation of universal law has profound implications for our understanding of the healing process. It offers an approach to the practice of natural healing based on Vedanta, Ayurveda, and Yoga philosophy. Western psychology is focused on the persona with little understanding of the deeper Self. In the wisdom of ancient India, we find a psychological model and a spectrum of techniques that have been used effectively for thousands of years to promote healing and self-actualization.

The text is rich in a body of source material that, over the millennia, has been worshiped as ultimate truth. You are invited to join in the contemplation of a vision which sages avow represents the culmination of human knowledge. These essays are challenging. They abandon the assumptions of conventional wisdom and demand that the reader make a leap in consciousness. The teachings invite the student to transcend egocentricity and step into an awareness of Being.

The Culmination of Knowledge

The inspired vision of the ancient *Upanishads,* the revelations of the avatar Sri Krishna in *The Bhagavad Gita,* and the unparalleled clarity of the teachings of the sage Shankara in *The Crest Jewel of Discrimination* offer a formulation of universal law known as the Sanatana Dharma. Sages of the East worship it as the Himalayan summit of the evolution of human consciousness.

Contemplation of the Sanatana Dharma offers an opportunity to consciously accelerate one's evolution through enlightening insight and liberating realization.

The Sanatana Dharma offers a vision of the sacredness and unity of life. All life is an emanation from one transpersonal field of living consciousness. The Sanatana Dharma clearly and simply explains how the living consciousness, or Self, of each being is unified with the one universal Being. Personal realization of the identity of the Supreme Being (Brahman) and the living consciousness of the individual Self (Atman) is the key to this ancient psychology of liberation.

The Ultimate Medicine

The sages explain that our neurosis is rooted in our ignorance of the Self and our identification with the obsessions of the vulnerable persona. "You are not your body … You are not your thoughts … You are not your emotions …" To realize the unity and sacredness of all life is to be released from the pain of ontological insecurity—the self-hatred, shame, alienation, and selfish grasping of the separate ego. Knowledge of the Self is the balm to heal the suffering which is the driving force of our egocentric life.

The Sanatana Dharma invites us to cultivate a way of life that attunes us to the inner Self. It unfolds a spectrum of methods for cultivating integration with the more concentrated levels of consciousness of the Self. This purification process is called yoga, union with the Self. The moment-to-moment experience of union with the Self is the ultimate medicine. Living a life of soul communion frees us from the insatiable insecurity and obsessions of the persona. The experience of union with the Self liberates us from the ignorance of identifying with the suffering ego and delivers us into the peace and bliss of the eternal present.

Truth Can Only Be Lived

The Sanatana Dharma is not empty philosophical speculation. It offers a prescription for action in the world. It avows that all life is sacred. In the Sanatana Dharma knowledge is the personal realization of the identity of the Being of the individual with the Supreme Being. Without this higher knowledge all else is ignorance.

Truth can only be lived. Dharma, which is often translated as righteousness, religion, or law, comes from Sanskrit roots which mean "to uphold." It relates to a way of living that upholds union with the transcendental Self. The practice of the virtuous way of life of the Sanatana Dharma fosters a moment-to-moment experience of the identity of the individual life with the Universal Life. The practice of Dharma harmonizes the individual consciousness in communion with the higher vibrational forces of Universal Consciousness.

The Sanatana Dharma invites us to abandon a life of selfishness and live a life of truth. It maintains that work done with knowledge becomes worship. It invites us to live a life of dedicated service.

How to Use This Book

A newcomer to body-centered therapy should begin with the hands-on material in Part I, Loving Hands Are Healing Hands, which offers a simple introduction to polarity principles. Chapter 4 presents a series of safe, easy to learn energy balancing protocols. Your personal experience of the effectiveness of Polarity Therapy—even in the hands of a neophyte—is the best introduction to this consciousness.

Experienced therapists and the philosophically inclined may choose to focus first on the essays in Part II of the book, which unfold an esoteric perspective for working with energy in the healing arts and expand themes touched on in Part I. Do not merely read this material. Contemplate these revered truths. Take your time with these revelations and join the saints and sages who, over the millennia have enjoyed the rapture of mystical contemplation. Close your eyes, breathe deeply, and savor the nectar of immortality.

Part III of the book, presents a model for the practice of somatic psychology that synthesizes many of the elements of this work. Part IV offers additional resources for health building and the hands-on practice of body-centered therapy.

I honor the Knower of All Hearts who cares for the ancient wisdom over the ages through his devotees, and who offers us this gift of right livelihood through a life of compassionate service.

Namaskar ... Honor to the Divine Within
Bruce Burger
January, 1998
Heartwood Institute, Garberville, California,

PART I

LOVING HANDS ARE

HEALING HANDS

Service is love in action.

—Mother Teresa

CHAPTER 1

What Is Polarity Therapy?

Polarity Therapy is a comprehensive health care system based on an understanding of energy in the healing arts. Created by Randolph Stone, D.C., D.O., Polarity Therapy is the culmination of over sixty years of the study and practice of esoteric healing and natural therapies. Through his studies of the ancient wisdom, Dr. Stone exalted his consciousness and penetrated the deepest mysteries of life and the healing process. He envisioned Polarity Therapy as "Energy Medicine"—the healing science of the future—a science of health rather than disease; a science that works with the person rather than the symptom; a science that understands the body as energy and works with the healing power of life-force energy.

In Polarity Therapy we understand the body as a field of energy. We work with the powerful field of life-force energy emanating from the hands to balance energy, release tension and energy blocks, and bring the body back into harmony with nature and the universe. Polarity Therapy offers techniques for energy balancing and healing through loving touch and caring intention. Polarity offers a dynamic system of movement therapy and dietary regimes for cleansing and health building. It offers a profound body/mind/spiritual psychology and a deep understanding of energy, nature, and the healing process.

*Dr. Randolph Stone,
(1890–1981)*

*Polarity Therapy under-
stands the body as energy.
Ten currents of energy
emanate from the ultra-
sonic core of the
caduceus and chakra
system.[1]*

The Polarity Health Educator

Polarity Therapy offers a holistic, educational approach to healing which empowers the individual with resources for taking responsibility for health and happiness and accelerating spiritual evolution. Polarity Therapists offer education—not treatment. They work with health—not disease. Health is defined as the ability to take responsibility for one' self. Polarity Therapists work with individuals who are seeking to take a higher level of responsibility for their health. Our clients are not individuals who are looking for a doctor to treat disease symptoms. They are individuals who are seeking educational resources for self-understanding and personal growth, and who are willing to make changes in their way of life.

A Polarity Health Educator works with a client over time, sharing techniques that promote enhanced health and happiness. We work with the client as a person, respecting the depth, integrity, and pathos of his or her life as an evolving spiritual being. Polarity is a self-directed therapy in which we support the client in realizing a higher level of inner guidance and integrity in caring for the gift of life. Polarity energy balancing fosters an experience of enhanced aliveness, mental and emotional clarity, and personal integration. This lucid state is an attunement with the Higher Self, a personal experience of exalted consciousness and enhanced well-being. The Polarity Therapist inspires the client to take responsibility for creating a healthier, more conscious lifestyle.

An Evolutionary Approach to the Healing Process

Polarity Therapy is based upon an ancient transpersonal[2] paradigm which offers an evolutionary approach to the healing process. In the evolutionary paradigm, all energy in creation is the energy of an ever-expanding universe. The Polarity Therapist understands the client as the personification of the leading edge of the evolution of the universe. The life of the client is sacred. The healing process of the client is a reflection of the client's evolutionary/actualization issues.

The path between birth and death is a cycle of growth. Life is understood as a learning experience for the Soul. This evolutionary perspective implies a deep respect for the person and the depth of the human condition. Our growth and the challenges we face in the process of maturation are a sacred journey in which we embody a moment of an ever-expanding, ever-evolving universe. Everything we face in life-—every challenge to our growth, all resistance to our actualization—is the turning of the cycles of evolution.

The Polarity Therapist dignifies the disease process by understanding it in an evolutionary context. Because disease is understood as a metaphor for issues in personal evolution, the therapist has an attitude of deep respect for the client. Disease may be understood as a sanctuary, a space where the individual may take refuge from the demands of evolution. The disease is a statement that the individual needs a space to rest, to marshall forces, or to call out for resources and support to meet life's challenges. Accidents and illness may be an unconscious statement of issues that are too threatening for the ego to deal with consciously. In an evolutionary context, the role of the Polarity Therapist is to help bring these issues to consciousness so that clients have an opportunity to take a higher level of responsibility for their healing processes.

In an evolutionary perspective every cell of the body is sentient. The cells contract and energy is blocked when it is too painful to be present in our feelings. Polarity-energy balancing principles facilitate the release of tension, resistance, and body armoring. The practitioner holds a sacred space where it is safe for the client to experience feelings, in order to clear trauma from the cellular memory of the body. The practitioner facilitates a space where the client experiences that it is safe to be present in his or her body—safe to breathe, to be conscious, to be fully alive.

The Sacredness and Unity of All Life

We exist in two dimensions—a dimension of Spirit and a dimension of matter. Though our senses are exclusively tuned into the dimension of matter, Spirit is the source of our aliveness. Spirit exists in a dimension beyond the body and outside the cycle of birth, death, and decay. When Spirit leaves the body, we are left with a corpse that begins to decompose immediately. Without Spirit, the vehicle is now seen as something unclean and decaying—to be disposed of as soon as possible. From moment to moment, Spirit is emanating the energy fields of the body. In Polarity we call this all-pervasive field of Spirit the Soul, the Source, Primary Energy, the Self, or the Life Field.

Oneness is the foundation of this cosmology. In the ancient wisdom, there is an understanding of the identity of all consciousness. Each drop of con-

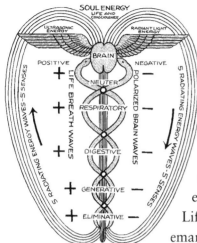

Dr. Stone viewed the soul ... the animating force ... as the nucleus of the energy fields of the body. In chart #1 from Polarity Therapy Dr. Stone portrays the step-down of the soul energy into the physical fields of the body through the caduceus, "a dual force of positive and negative life breaths which are conveyed to all the tissues of the body."

sciousness is one with the ocean of universal consciousness. Each unit of life is one with all life. Each life is a flower on the tree of eternal life. The ancient wisdom is rooted in an understanding of the oneness of all consciousness and the sacredness and unity of all life. The changing world of appearances is an emanation from an unchanging universal field of living intelligence. Himalayan sages call this life essence the Self.

The solar system, nature, and our bodies are part of a single system of Creative Intelligence. All life systems are subsystems of the evolution of the one life of the universe. Spirit is present everywhere. Life force is the presence in the body of Spirit. The material body is an emanation from the spiritual. Our attunement with Spirit is the key to healing and health.

Webster's defines the Soul as: "An animating essence or principle held to be inseparably associated with life or living beings." Living matter is an emanation of the Soul. At the nucleus of every atom, molecule, and cell of the body is an energy field of a higher potential that is in tune with the spiritual realm. The physical field is a radiation from this spiritual nucleus. It is the body's attunement with the Soul that sustains life. The Soul manifests the body as a vehicle for its evolution. The ancient wisdom reveals that the life force in the body is the living presence of the Soul. The living consciousness in the body is nothing less than the Soul. The Soul is the innermost Self—that which is conscious and alive in each being. The Soul is existence itself, in its unity as unbounded awareness.

The First Principle of Polarity Therapy

Dr. Stone understood the Soul to be the animating force in the body. Currents of life-force energy emanating from the Soul sustain the energy fields of the body. In the Summary of Principles in the Introduction to *Wireless Anatomy*, Dr. Stone offers insight into the fundamental basis of healing. He writes:

> The Soul is a unit drop of the ocean of Eternal Spirit which is the dweller in the body as the knower, seer, doer; it experiences all sensation and action. It alone is the power in the body which reacts to any mode of application of therapy or action. Consciousness and intelligence reside within the Soul.[3]

This is the first principle of Polarity Therapy. As Polarity Therapists, our treatments consist of a balancing of the body's attunement with the Soul. Dr. Stone continues:

Research in atomic energy and the corresponding new viewpoint on matter prompts a re-orientation of the healing arts, in conformity with these modern discoveries. In this WIRELESS ENERGY FIELD, ancient and modern science can meet ... Miracles and psychic healing are not dealt with in this course. It describes a rational therapy ... a finer perspective of the actual energies operating in the body as its keynote. The attention is directed to the "dweller" in the body, as the Soul or Nucleus, in relation to the form; and to both of them in relation to Nature or the Universe as a whole. All this brings out a new concept of Health and Disease, according to individual lines of force of attraction, repulsion and relationship ... in the field of energies.[4]

Polarity Therapy is the study of the Wireless Anatomy of the energy fields of the body. It is a healing art whose focus is the Soul as the source of life and healing. Polarity Therapy seeks to understand energy in the healing arts. The understanding of energy offered by Polarity Therapy is drawn from a broad range of esoteric philosophies. The cosmology of ancient India called the Sanatana Dharma (Perennial Philosophy) is the source of many of Polarity Therapy's fundamental concepts. The Sanatana Dharma pictures the universe as one living being called *Brahman*. Brahman exists as a universal field of Ultimate Intelligence. The Sanskrit word Brahman is from the root *brih*, "to expand." Brahman is pictured as an ever-expanding breath of Creative Intelligence. Brahman's nature is boundless creativity, evolution, growth. All creation is a moment-to-moment emanation from this universal field of living consciousness. Dr. Stone referred to the universal field of Creative Intelligence as the Primary Energy, the One Life, or the Soul.

The Three Principles

The universe is the evolution or unfolding of the creative potency of Brahman. In the evolutionary perspective, energy radiates outward from a source creating a field of force; upon reaching the limits of the potency of the energy field, energy returns to the source. All the energy on every level of creation moves through this evolutionary cycle which we call Polarity. The evolutionary cycle expresses the essence of intelligence. It is a cycle of consciousness moving from a source out into experiencing the field and returning to the source. All energy moves in this cycle of intelligence. This cycle is a movement from a Neutral Source to a Positive field, then through a Negative phase, returning to the Source. These three principles, Neutral, Positive, and Negative, are the basis of Polarity Therapy.

The Two Hands of God
A centrifugal spiral, centripetal spiral, and a sunflower. Energy fields in nature are a union of centrifugal and centripetal currents emenating from a center. (Doczi, Power of Limits)

In the Sanatana Dharma, these three principles are called *sattva guna* (neutral), *rajas guna* (positive) and *tamas guna* (negative). The word "guna" is derived from the Indo-European base *gere*, to "twirl" or "wind." The gunas describe spiraling vortices of vibrating energy fields.

The neutral principle, sattva guna, is the force of equilibrium in nature. Sattva guna is an all-pervasive unified field of living consciousness. The Sanskrit word *Sat* refers to the Supreme Being which sustains existence. The life and consciousness of each being is a portion of the life field of sattva guna. The sattvic harmonic reflects truth, light, grace, stillness, and the profundity of the Soul. Sattva guna is lucid, attractive, fascinating, charismatic, peaceful, balanced, and graceful. The charm and attractiveness of babies, the charisma of saints, the glow of health, and the inherent beauty of nature predominates in this harmonic, which is in tune with the neutral source of the life field. In the practice of Polarity Therapy, sattvic contacts promote an attunement and receptivity to the healing presence of the Soul.

Emanating from sattva guna is a positive centrifugal field of force which is called rajas guna. Rajas guna is the creative expression which underlies the evolutionary process. The Sanskrit term *rajas* means "to glow," and rajasic phenomena are passionate, excited, vital, and creative. Rajas guna, the field of force emanating from a center, is the fundamental organizing principle throughout nature. Rajas guna sustains the conscious mind. Rajasic predominant phenomena are creative, kinetic, passionate, directive, and transformational. This positive force rules and regulates the radiant expression of manifestation. The glow of its vision is the light of life.

The negative centripetal field is called tamas guna. Tamas guna sustains form. Tamas rules the structures and foundations which contain and limit energy. The Sanskrit word *tamas* means "to perish." Tamas is a contracting, magnetic field that underlies the crystallization of Spirit into matter. It is the evolutionary return current which can manifest beauty and wisdom or inertia and resistance. Tamas rules the resonance of negativity and the reactive unconscious mind. It is the instinctual motor force. It is the process of elimination and completion. It is the purifying force which eliminates that which is not life sustaining.

The ancient Rishis (yogi-seer-sages) purified their bodies, concentrated their minds, and attained microclairvoyant vision. They directly perceived the fundamental forces of nature and the universe. They witnessed the vibratory basis of the universe which they termed the gunas. They likened the spiraling centrifugal and centripetal undulation of energy fields that they observed on the atomic and molecular levels to the common rope woven of centrifugal

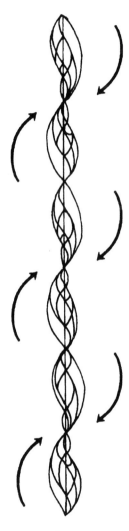

As a simple rope is woven of three strands, all vibration is the spiraling undulation of three forces. Centrifugal ascending, centripetal descending around a stable core. All vibration is the spiraling balance of three forces.

and centripetal strands around a stable core. All vibration is the spiraling undulation of three forces. Centrifugal ascending, centripetal descending around a stable core. All vibration is the spiraling balance of these three forces.

In the Sanatana Dharma, all creation is sustained by its attunement with the three universal forces: sattva, rajas, and tamas. Energy fields in nature are a spiral dance of orbs of force radiating centripetally and centrifugally from a center. Structures in nature are formed by spiral currents that radiate in opposite directions—clockwise and counterclockwise from a center. All life grows through this union of complementary forces. Polarity is a fundamental feature of universal law. The cycle of polarity underlies all the energy fields on every level of macrocosm and microcosm in creation.

The Polarity Cycle

Life Wiggles! Everything in creation is vibrating. The fundamental feature of vibration is that it always exists as an expression of cycles. All of nature is an expression of cycles: of birth and death, summer and winter, night and day, in-breath and out-breath. In nature there is always a dynamic between centrifugal and centripetal, expanding and contracting, pushing and pulling, building and cleansing, active and receptive, electric and magnetic, male and female, acid and alkaline, positive and negative, spirit and matter. This dynamic is best known as "Yin and Yang," and the interaction of these forces in wholeness is known as the "Tao." Orientals call this "the supreme ultimate principle" and "the two hands of God." Dr. Stone called this living cycle of energy that underlies all of nature Polarity.

Energy is always moving. All life is a pulsation, an oscillation, a wave of phases, resonating in attunement with the universal polarity underlying all

Three Modes of Energy

From The Wireless Anatomy of Man *by Dr. Randolph Stone*
The three Gunas as the three universal principles of motion everywhere; positive, Raja; and negative Tamas; spinning around a neuter center, Sattva. These become the polarity principle in the human body, the superior, middle and inferior in importance to consciousness, life and motion. The foundation of the polarity principle is active during the embryonic stages of building the body. This wireless principle carries thru life after the wires and tubes are established. Blocks in this energy field and its circuits are the fine ethereal and material cause of pain and disease. Re-establishment of currents in balance with the center of the neuter field is expressed as the normal stage known as health.

Application of the three modes of energy thru manipulative therapy to balance currents and release blockades in the finer circuits as well as in the gross conductors opens the fields for every known therapy.

A. A neuter principle (Sattva) of any soothing, balancing application is indicated. Here a very light touch thru polarity of the fingers to polarity centers of the body is used.

B. A positive principle (Raja, action motion) of manipulative therapy is expressed in directive force applied to the body.

C. A negative principle (Tamas, darkness, cold, inertia and resistance) which must be aroused and counteracted by therapies which are dispersive, scattering, eliminating, forceful, deep and penetrating, eliciting definite reaction thru the resisting action of the negative pole, thus arousing it and tuning it into the circuit.

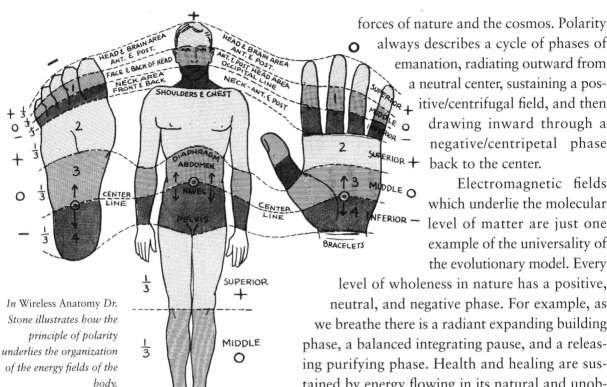

In Wireless Anatomy *Dr. Stone illustrates how the principle of polarity underlies the organization of the energy fields of the body.*

forces of nature and the cosmos. Polarity always describes a cycle of phases of emanation, radiating outward from a neutral center, sustaining a positive/centrifugal field, and then drawing inward through a negative/centripetal phase back to the center.

Electromagnetic fields which underlie the molecular level of matter are just one example of the universality of the evolutionary model. Every level of wholeness in nature has a positive, neutral, and negative phase. For example, as we breathe there is a radiant expanding building phase, a balanced integrating pause, and a releasing purifying phase. Health and healing are sustained by energy flowing in its natural and unobstructed state. In Polarity energy balancing, we work to restore the integrity of the evolutionary cycle by releasing resistance in the energy fields of the body and bringing the flow of energy into harmony with the finer, higher vibrational forces of nature and the cosmos.

The Evolutionary Cycle

All energy moves out from the unitive source through an involutionary phase of expression and then returns through the evolutionary phase back to the source. The involutionary cycle is characterized by a passion for life, experience, expression, and creativity. Attention is directed downward and outward to the world into identity and the ego. It is associated with the five passions: pride, greed, anger, lust, and attachment. In the evolutionary phase, the focus is directed inward and upward. The source of life and action is acknowledged in gratitude. The passions are transformed into the five virtues: detachment, continence, forgiveness, contentment, and humility. In health, the harvest of the evolutionary cycle is equanimity, discrimination, and wisdom.

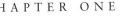

The Five Elements

The ancients understood the world as the product of five archetypal forces: Ether, Air, Fire, Water, and Earth. All energy in nature is attuned with these five harmonics of resonance. These patterns of resonance are traditionally called the elements. Cross-culturally and transhistorically, the science of the ancients is based on these elemental keynotes of energy.

All vibration in creation is sustained by its attunement with these universal forces. The elements sustain five phases of materialization. All energy in nature radiates outward from the center, stepping down from spirit to matter in quantum fashion through the five phases of materialization. The potency of the emanation drops as the energy moves away from the center. The innermost field resonates with the nucleus of Ether, the etheric core. The first field stepping down from Ether in this cycle of phases is Air predominant. The next phase radiates outward as the Fire harmonic, then to Water, a centripetal force. Finally, at the Earth phase, the field is pulled back to the vortex at the center for another cycle.

The Three Principles of Motion

Quality	Rajas Guna	Tamas Guna	Sattva Guna
Polarity	+	-	0
Element	Fire	Water	Air
Pulsation	Centrifugal	Centripetal	Fulcrum
Force	Kinetic	Static	Equilibrium
Direction	Out	In	Neutral
Current	Pushing	Pulling	Stillness
Cycle	Involution	Evolution	Absolute
Manvantara	Nivritti	Pravritti	Pralaya
Ayurveda	Pingala	Ida	Sushumna
Ruling Deity	Brahma	Shiva	Vishnu
Archetypes	Father	Son	Holy-Ghost
Light	Rainbow	Shadow	Light
Gender	Male	Female	Androgynous
Oriental	Yang	Yin	Tao
Embryo	Mesoderm	Endoderm	Ectoderm
Breath	Inhalation	Exhalation	Stillpoint
Consciousness	Conscious	Unconscious	Superconscious
Astrology	Cardinal	Fixed	Mutable
Planet	Mars	Moon	Sun/Mercury
Seasons	Spring, Summer	Fall, Winter	Equinox
Temperature	Warming	Cooling	Unchanging
Plane	Mind	Body	Spirit
Body	Right Side	Left Side	Core
Music	First	Third	Fifth
Polarity	Positive	Negative	Neutral

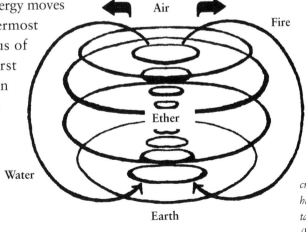

The Soul as the nucleus emanates five fields of force. All creation is attuned to the harmony of these elemental keynotes of vibration. (Purce, The Mystic Spiral)

Creation is woven from the energies of the five elements as: Ether (space), Air (gas), Fire (heat), Water (liquid), and Earth (solid). Our senses weave the fabric of our experience through attunement with these keynotes of nature. Every energy field is made up of boundaries (Earth), a medium for life (Water), directed working energy (Fire), and intelligent feedback (Air), in an overall harmonized space (Ether). The elements underlie our mental, emotional, and physical processes.

Every physical plane phenomenon is made up of all the elements, but one elemental keynote of resonance predominates and gives the form its distinctive qualities. Thus, particular structures and processes in the body are characterized as: Air predominant (nervous system), Fire predominant (muscles and assimilative organs), Water predominant (tendons and urogenital tract), Earth predominant (bones and organs of elimination).

Throughout all of nature the phenomenon of solidity is sustained by a pattern of energy which we call Earth. Earth serves as the predominant resonant frequency in the fields of phenomenal force with a quality of solidity like the bones in our bodies. Earth, which rules at the limits of the energy field, sustains boundaries and predominates in the colon which provides a boundary between that which sustains life and that which is to be eliminated. Earth rules elimination and completion.

Our experience of fluidity predominates in the Water wavelength. The Water element predominates in the endoderm, tendons, and pelvis. Water embodies the liquid medium and the cycles which purify and sustain life, such as blood flowing through veins or rivers flowing to the ocean. Water, which always flows downwards into the Earth carrying off impurities, rules the cleansing processes of the bladder and lymph.

Heat and caloric energy are sustained by a frequency of vibration which we call Fire. This electric pattern of

(Stone, Energy)

Combinations of the Elements in the Human Body
Essential quality of element in bold.

	Ether	Air	Fire	Water	Earth
Ether	**Grief**	Lengthening	Sleep	Saliva	Hair
Air	Desire	**Speed**	Thirst	Sweat	Skin
Fire	Anger	Shaking	**Hunger**	Urine	Blood Vessels
Water	Love/ Attachment	Movement	Luster	**Semen** **Ovum**	Fascia, muscles, viscera
Earth	Fear	Contraction	Laziness	Blood	**Bones**

energy predominates in the fields of force which manifest warmth. Fire directs energy in our vital organs and musculature.

The quality of lawfulness and the profound harmony and integration of all movement in creation is expressed in the field of force called the Air element. Our nervous system resonates with the keynote of Air.

Ether is the space where energy flows freely without obstruction. Ether provides a container which limits and defines the overall space in which the elements cycle. When Ether is in balance, there is a sense of freedom and psychic space.

All phenomena are made up of these five archetypes of vibration: the boundary of Earth, the medium of Water, the directive energy of Fire, and the organizing lawful movement of Air in the space of Ether. Like a television set that is tuned into a particular channel of vibration, our senses and mind are tuned into the elemental frequencies of vibration and build our experience through this attunement.

The elements are the archetypal beings whose Creative Intelligence and will underlie manifestation. Working with these five transpersonal forces is a foundation of Polarity Therapy. It is valuable for the therapist to have a reverent and intelligent rapport with these ancestors.

Air-Predominant Fields

Air is the first aura of emanation from the etheric core, stepping down from Ether through the lengthening phase of the in-breath. Air sustains the fundamental rhythms that attune the body to the forces of the cosmos. Air rules movement as an expression of the harmony of nature and the universe. The transverse currents of energy, which spiral out of the crown of the head and move from east to west around the surface of the body acting as sensory and feedback mechanisms, are Airpredominant. Air defines the body from side to side uniting the periphery of the energy fields with the core.

Air governs movement, and sustains the harmony of the cycles which attune all vibration in the body with the one source. Air rules the beat of the heart, the lungs, respiration, and peristalsis throughout the body. The thoracic cavity, home of the fundamental life rhythm of breath, is the center of the Air harmonic in the body.

From an astrological perspective the element Air rules the spine, shoulders, and arms (positive, Gemini) the diaphragm and kidneys (neutral, Libra) and the ankles (negative, Aquarius).

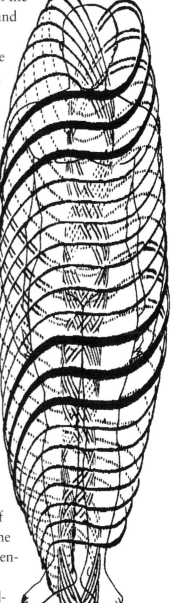

Transverse currents.

The Air element rules our sense of movement through touch, the desires and aversions that move us in life, the agility of the mind, flexibility of the joints, and rhythms of the breath of life. Air is a neutral balancing force between the centrifugal and centripetal polarity of Fire and Water, whose living intelligence attunes the body to the Source. A gentle touch of stillness activates this keynote of sattvic vibration.

Fire-Predominant Fields

Fire rules directed energy in the body. It expresses the purpose, intelligence, willpower, motivation, desire, and excitement of creation. Fire is the driving force of vitality in the body. Fire radiates throughout the body from its center at the umbilicus to define the body from front to back, within to without. Fire is the centrifugal energy centered in our solar plexus, located between the diaphragm and navel. Fire rules our mesoderm, internal organs, the processes of metabolism, and our musculature.

From an astrological perspective Fire, directed energy, rules the head as vision, intention, focus, and concentration (positive, Aries). Fire rules the solar plexus and heart as warmth, drive, vitality, will power, and enthusiasm (Neutral, Leo). Fire rules the thighs as purpose, the power to move in the world, to care, and to take responsibility (negative, Sagittarius).

Any stimulating contact resonates with this rajasic keynote and promotes its attunement.

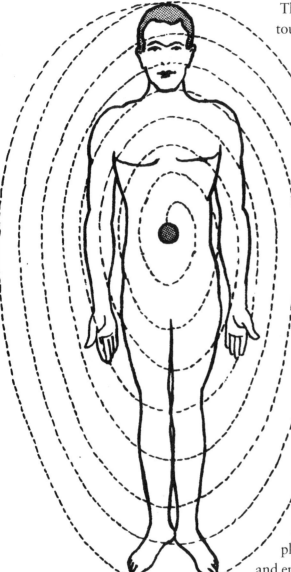

Spiral currents. (Stone, Evolutionary Energy Charts)

Water-Predominant Fields

Water is the phase in which consciousness precipitates into substantial solidity. Water is the medium of life. Water is characterized by the centripetal spectrum of receptivity, materialization, resistance, inertia, and crystallization into form. Water resonates with the tamasic, contracting, descending arc which is always flowing downward as the negative fields of the body.

The magnetic force of Water rules attraction, cohesion, and love. Water rules the process of giving and receiving love and nourishment, connecting us to the life around us.

The water element predominates in the pelvic cavity. As water rules form, reproduction and gender issues are centered on this level of vibration. Water rules the cycles of renewal, purification, and elimination in the body. In the body as a whole, the positive field of Water is the breast (Cancer), the neutral field the pelvis (Scorpio), and the negative field the feet (Pisces).

In Dr. Stone's work, there is a major focus on the negative field. Energy always moves in cycles. It is in the negative field of the cycle that problems of resistance, tension, and elimination dam the flow emanating from the center. A key to Polarity Therapy is working with physical and emotional elimination in the body/mind. The Water-predominant negative fields are a key to restoring healthy elimination. Working with the perineum, pelvis, and feet as negative fields releases chronic tension and unexpressed emotions and restores healthy elimination. The work of the Polarity Therapist is to reconnect energy with the Source.

Long line currents. Energy steps down into physical manifestation through the chakras which are the nuclei or the physical energy fields. This lucid chart by Mark Allison illustrates how each chakra sustains a long line current of elemental resonance. (Chitty and Muller, Energy Exercises)

The Three Currents

The three principles, sattva, rajas, and tamas, govern three important energy currents in the body. The transverse or east-west current relates to the sattvic quality of neutrality and defines the body from side to side. The transverse current predominates in the Air harmonic of resonance and spirals around the surface of the body, emanating from the long axis of the core at the top of the head. The transverse current functions in balance, communication, and coordination, integrating the periphery of the body with the ultrasonic core and carrying sensory input from without to within. It is thus associated with the parasympathetic nervous system which performs these functions in the body.

The fiery current, which Dr. Stone called the spiral current, relates to the rajasic quality of directed energy. It spirals from the umbilicus to energize the whole of the body for warmth, vitality, and movement. The spiral current defines the body from within to without, back to front. Fire predominates in the mesoderm, musculature, and assimilative organs of the body. Fire relates to the sympathetic nervous system.

Tamas guna, the Water principle, rules the long line currents which pulsate from the core to sustain five currents of life force on each side of the body. At the core of the body are a system of chakras which are vortices stepping energy down into physical manifestation. The chakras serve as nuclei of the elemental energy fields. Each chakra sustains a current flowing out to function in the body and then returning to the chakra. Energy flows through the body as the long line currents which govern the five senses and rule physiological processes. The ten long line currents define the body from side to side and from top to bottom. The long line currents are associated with the central nervous system and the energy of the cranial rhythm.

The three currents define the energy field of the body along three axes: side-to-side transverse currents; within-to-without/front-to-back spiral currents; and top-to-bottom long line currents.

CHAPTER 2

The Practice of Polarity Therapy

The Polarity of the Fingers and Hands

The right hand resonates predominately with the yang, pushing, radiant, positive, electric, centrifugal, vibrations. The left hand resonates predominantly with the yin, pulling, receptive, negative, magnetic, centripetal vibrations. Each finger resonates with a predominant element and corresponding current of life force. The thumb is neutral and resonates with the Etheric core vibrations. The index fingers are negative or receptive and relate to the Air element. The middle fingers are positive and relate to the Fire current. The ring fingers are negative or receptive and relate to the Water harmonic. The little fingers are positive and relate to the Earth keynote of vibration. It is important to use these polarities in sattvic contacts.

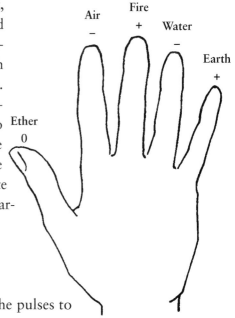

Polarity Pulse Assessment

Assessment in Polarity Therapy often begins with checking the pulses to ascertain the attunement of the body's harmonics with nature. Taking the pulses helps us to assess whether the body is in a degenerative phase of disharmony or in a health-building attunement with nature and the universe.

Sattvic Pulse

We feel for the sattvic pulse, which is a reflection of the attunement of the Air and Ether elements and the functioning of the nervous system, near the carotid artery on each side of the neck and at the flexor hallucis longus behind the ankles.

If the pulses are synchronized and exhibit the same rhythm on both sides of the neck and feet, the sattvic pulse of the Air element is in balance. An Air imbalance manifests as a crooked, snakelike, zigzag, fast pulse. If the pulses are out of sync or in any way asymmetrical, sattvic energy balancing is needed to harmonize the body with nature and the cosmos and to initiate a health building attunement.

Rajasic Pulse

If the pulses are weak at both the carotid artery and the flexor hallucis longus, the Fire principle is blocked. Vitality is not radiating from the Fire center at the navel and the rajasic currents are not in harmony with their keynote in nature. Rajasic energy balancing is needed to promote strengthening of the energies that sustain vitality.

Tamasic Pulse

If any of the pulses are weak while other pulses are strong, tamasic vibrations that sustain form may be too crystallized and resistant. The more subtle vibrations are blocked by this tension. A tamasic focus is indicated to bring resistance to the surface, to facilitate awareness, and to promote elimination. Water imbalances manifest soft, weak, waddling, slow, heavy pulses.

Dr. Stone writes:

> The excess of the AIRY force in the blood makes a fast and zig-zag or crooked pattern of the pulse beat. It resembles the movement of a snake. When the heat element of FIRE is in excess in the pulse, it is restless and jerky or jumpy like a frog pattern. When the WATER . . . element predominates in the circulation, the pulse is slow, weak, restricted and soft."[1]

The Air-predominant sensory currents emanate from the core of the energy system and move from east to west on the surface of the body. Sattvic contacts on the surface of the body are carried to the deepest core of the being by these forces. Sattvic contacts touch the Soul!

Making Contact

In Polarity Therapy we always use two hands in making contact. In this way we work with the powerful electromagnetic field of force resonating between the active and receptive hands. Polarity therapists do not do adjustments, manipulations, or strokes. Polarity Therapists hold contacts. Contact is very special. In making contact you are fully present with another being. The more deeply present you are for another—the more powerful your work as a therapist. To be present is a very powerful state of being. The present is a conscious intersection with eternity and a potent energy nexus. Sages tell us to Be Here Now. The present is the eminence of the Supreme Being. An energy block is an inability to be present in the body, an inability to be feeling, conscious, alive, evolving. As Polarity Therapists we maintain contact by keeping our presence in our fingers and hands. We are constantly feeling for pulsation, heat, energy, attunement, and focusing our attention into the contact. Your care for the client's well-being and intention to contribute to his or her healing is a powerful expression of presence.

Sattvic Contacts

In Polarity Therapy, gentle contacts of stillness with caring intention on the surface of the body are called sattvic contacts. Sattvic contacts facilitate a receptivity to the healing presence of the Soul. The Sanskrit term *sattvic* describes energies in which the spiraling waves of electromagnetic vibration are in a profound state of harmony with the life field and vibrate sympathetically with the finer, higher vibrational forces of the Soul. Thus, sattvic qualities are characterized by their ability to reflect physically the lucidity, peace, harmony, wisdom, beauty, well-being, and inspiration of the life-giving forces of higher intelligence.

Both the Ether and Air elements are ruled by sattva guna which rules the stepdown from the Source into manifestation. Sattvic contacts facilitate a receptivity to the inner core of life essence, they touch the heart of Being. Sattvic contacts resonate with the nervous system and the vibrations which carry sensory messages from the surface of the body to the core. Sattvic contacts promote inner peace and balance throughout the nervous system. Ayurveda, the ancient Indian science of healing, teaches that seventy to eighty percent of disease has its basis in imbalances of the nervous system. Since pain is carried by the sensory surface currents, sattvic contacts ameliorate pain.

Sattvic contacts promote a profound state of stillness and receptivity to inner guidance and attunement with our higher Self.

Rajasic Contacts

Rajasic touch is stimulating, directive, and penetrating. In rajasic contacts, we use our muscles to direct energy through stimulation, movement, and strength. Rajasic contacts balance the Fire principle and resonate with the fiery keynote of positive, electric, vital energy throughout the body. Rajas centrifugally radiates throughout the body from the fire center at the navel. The rajasic resonance predominates in the mesoderm, muscles, and organs of the body. Rajasic contacts facilitate the release of tension and trauma from the musculature and promotes well-being. Rajasic contacts enhance vitality and the functioning of the internal organs and eyes.

Tamasic Contacts

Tamasic touch is forceful and penetrating in order to disperse blocked energy. Tamasic touch is intense in order to evoke the client's resistance. Always ask the client's permission before initiating a tamasic contact. Tamasic contacts should be held for only one or two seconds, just long enough to evoke resistance. This reaction stimulates the polarity cycle at the negative field which rules elimination and facilitates the clearing of disharmonious patterns of emotional and structural resistance. Tamasic touch can be effectively used to release resistance at the negative fields of the body at the feet and the soleus/achilles. Stimulating the evolutionary return current is a key to the facilitation of structural integration and grounding. Stimulating the evolutionary return current gets a person unstuck and often facilitates a turning point in his or her evolution.

Balancing both the Water and the Earth elements, tamasic contacts may evoke a process of elimination which can be dramatic and cathartic. It can be frightening to the client and the practitioner as the worst moments

Stone, *Evolutionary Energy Charts*

of a person's life are released from the cellular memory and surface to be experienced consciously and expressed in the safety and support of the bodywork session. Do not identify with the elimination. It is only energy being released. Hold a safe and sacred space. Pray for the client's well-being. Always hold sattvically at the end of the contact. Later in the book, contacts will be introduced that prevent retraumatization during emotional release.

Pain Relief Through One's Own Polarity Currents

In *Health Building* Dr. Stone offers these simple guidelines for hands-on healing:

> An excess amount of the positive current produces irritation, pain, swelling and heat in the tissues, organs and areas of the body, due to excess amount of blood in that area, and the opposite or negative pole energy is required to balance it. The positive current is the Sun energy of fire and radiant warmth in normal amounts. The right hand is the conductor of this energy. For negative tension, congestion, spasm and stasis, the right hand contains the antidote, the positive polarity current. Place it over the negative symptoms for relief.
>
> The left hand is the conductor of the negative or moon current, which is cooling, soothing, refreshing and toning. Place it over the seat of pain, where the positive currents are in excess, giving the symptoms mentioned above. Wherever the pain is, that excess calls for release of the irritation, heat, and swelling, which the negative current can provide.
>
> The positive or right hand is placed opposite the negative. If the pain is in front, place the left hand over it and the right hand on the same area of the back, to get the current through the congestion. If the pain and congestion is on the side, place the left hand over the area, and the right hand opposite on the other side. If the pain is on top of the head, place the left hand there and the right hand below it, on the back of the head, or under the jaw if the pain is nearer to the front than the back.
>
> If the pain is in the spine, you can use either the flat of the left hand, or double the hand up into a fist, and lie on it, wherever the pain may be located, so as to get the pressure along with the application, as both are indicated. This is done easily by lying on the back in bed or even on the floor. Then the right hand is used over the abdomen, directly opposite to the left hand in the back. Hold this position for ten or fifteen minutes, and changes will take place in the electromagnetic fields of that area and bring relief. This is the most potent remedy, ever present in the hour of need, which can be used by every person on himself or herself.

Gentle stroking with the left hand has a very soothing, sedative effect. With the right hand it is stimulating. For the same reason downward stroking is soothing, while upward stroking—from the feet up to the head—is stimulating. This can be done with both hands because the direction determines the polarity here.[2]

Energy Medicine—Releasing Energy Blocks

A key to effective therapy is working with the principles rajas and tamas in therapy. Rajas guna is the positive, motor, pushing, centrifugal current. The shoulders, solar plexus, and Fire center near the navel are centers of rajas guna. Tamas guna is the negative, sensory, pulling, centripetal current. The pelvis and feet are centers of tamas guna.

Centrifugal Treatments: Rajas Guna in Distress (+)

The centrifugal yang energy flows from the center outward. Its nature is expansion. Dr. Stone writes:

> There are definite signs of the plus principle in distress. They are based on the centrifugal force being checked and not being able to move or flow outward. Heat, fever, friction, pain, swelling, inflammation, stagnation, are all plus (+) factors that must be inhibited or drained off to make room for the output and direct it into proper channels. For these conditions, cooling, soothing, inhibitory treatments are indicated. Relaxation and balancing the Cerebrospinal, the Sympathetic and the Parasympathetic Systems by means of deep PERINEAL TECHNIQUE are always indicated, to balance the body with nature, Prana and the pulse beat—the inner and outer forces—that is essential . . .
>
> Contacts in the CENTRIFUGAL APPLICATION OF POLARITY THERAPY, are from the head down and from within outward. In acute and painful conditions, the outgoing currents run into blocks and obstacles with back-pressure registering as pain and acute aggravation. Remove the energy blocks by stimulating the current flow from above downward and from within outward, with the flow of the life currents.[3]

Dr. Stone continues, emphasizing the importance of working with the negative pole:

> Centrifugal or efferent energy currents flow from the brain to the extremities and return, to complete the circuit. Moderate resistance is met in all

tissue, but more so at the negative pole because of the distance from the center of energy, and by sedimentation at the lowest level. Therefore, THE NEGATIVE POLE USUALLY NEEDS STIMULATION in order to remove the obstructions and to activate the current flow back to the source, to complete the cycle. Stagnation of currents at the negative pole have much to do with mental conditions.[4]

Dr. Stone offers these specific instructions to release energy blocks in the centrifugal fields:

TREATMENTS: The position of the hands in the application of CEN-TRIFUGAL POLARITY TREATMENT: The positive hand or finger of the operator is on the superior region which is the positive pole, to strengthen the current impulse by the indicated manner of stimulation in order to speed the current from its accumulated back pressure location. The negative hand or finger is placed on any of the anteriorly located regions, negative to the superior or positive pole, to draw the current from the positive contact. Then the left hand descends further down to the neuter pole and finally down to the negative pole, to complete the circuit … The application of the hands is in a positive manner, a firm grip or pressure type manipulation of physical energy expenditure rather than a mere magnetic contact. Like poles repel and like applications repel.[5]

Centripetal Treatments: Tamas Guna in Distress (–)

The centripetal, space-building force of Nature is yin. Its nature is contraction. The centripetal application of Polarity Therapy is for subnormal or negative conditions that need stimulation of the return energy currents.

Dr. Stone explains the process of crystallization:

The minus (–) factor is bound up with the centripetal or inward flow of energy, from the circumference to the center. This we find in chemistry as surface tension in the molecule. In muscles it is also surface tension of the cells, and causes severe contraction, like in lumbago, which is usually due to acid crystals, sediments and other precipitates that have not been eliminated. When the muscles contract upon millions of these sharp, crystalline, needle-like particles, naturally a spasm is the result, which is a severe sharp, stabbing pain—with limitation of motion due to contraction, spasms, hardness and tension that cannot let go. The muscle tissue is very sore and stands out. Later it becomes ropy, if it is not drained of waste AND REPOLARIZED WITH ITS CIRCUIT FUNCTION.[6]

The key to chronic degeneration and blocked energy can be often be found working at the feet. Dr. Stone writes:

> The feet and the toes are the extremities of the negative pole in the current flow where most of the blocks are formed by precipitation. Releasing the blocks usually present at the junction of the positive, outgoing motor and the negative, inflowing sensory return currents is therefore a very important factor in POLARITY THERAPY BALANCING of these two forces. Precipitation occurs most often here because of the great distance from its invigorating center.[7]

Acute and painful conditions are treated with the current flow typically from above downward. Subnormal and chronic conditions are treated beginning at the extremities, moving in an upward and inward direction, to stimulate the weakened currents. Dr. Stone points out, "the real object of any treatment is to BALANCE THE ENERGIES AND RELEASE THE EXCESS SURFACE TENSION on any system, so the center can radiate again, and energy can flow where it is destined."[8]

In any condition it is valuable to work with the extremities as a foundation for more subtle work nearer the core.

Contraindications to Polarity Energy Balancing

Polarity Therapy is a *health resource*. Health is defined as the ability to take responsibility for one's self. Polarity Therapy is for individuals who are willing and able to take responsibility for their health. If a client is being treated by a physician or psychotherapist for a disease condition, you should consult with his or her doctor before practicing Polarity. Polarity is a valuable and effective complementary health-building modality. Exercise care in the following situations:

Pregnancy: Tamasic work, deep abdominal work, and work on the pelvic and reproductive reflex points at the ankles is contraindicated and may trigger miscarriage or labor. Care should be taken to support the comfort of the client using pillows.

Osteoporosis: Osteoporosis is a weakness in the bones that is quite common in postmenopausal women. Bones may become paper thin and easily broken. Work only sattvically if this is a concern. Polio may also leave bones brittle.

Spine: Do not put pressure directly on the spinous processes. Use special postures to work deeply with the spine.

Crura: Crura or connective tissue can be injured at the diaphragm. Use care in rajasic contacts at the rib cage.

Xiphoid Process: Be careful not to injure the xiphoid process at the tip of the sternum.

Varicose Veins: If raised, bumpy, or clotty, do not work, due to the danger of clots breaking off and circulating to the brain or heart.

Skin problems: Skin problems like herpes, cold sores, staph, poison oak, rashes, burns, cuts, abrasions, athletes foot, and fungi should be avoided. The practitioner must be careful to avoid contact if there is a risk of spread or irritation. Eczema and psoriasis are not contagious.

Menstruation: Avoid rajasic work in general. Only work sattvically on the abdominal area.

Hepatitis: Hepatitis, which is highly contagious, can be transmitted through contact with blood or fecal matter. Care must be taken in bodywork to avoid contact. Gloves or other appropriate barriers should be used. The practitioner must be careful to avoid contact with fecal matter and to wash hands thoroughly after perineal therapy.

Diabetes: Diabetes will sometimes cause fragile skin or blood vessels at the legs or feet. Avoid rajasic stimulation to the left of the umbilical center where a major artery may be irritated.

HIV Positive: If the client has open lesions or if you are working in an area where you may contact bodily fluids you should wear examination gloves. If you have any fears about working with HIV positive individuals, wearing gloves may help you feel more comfortable. You protect the client with a weakened immune system when you use gloves.

If you have questions as to the potential adverse effects of Polarity Therapy in any situation, only use sattvic contacts.

Polarity Health Education Client Information Form

Name _____ Sex _____

Address_____ Sun sign _____Asc _____Moon_____

City _____ State _____Zip _____

Telephone (home) _____ (work) _____

Age _____ Birth date _____

Occupation _____ Employer_____

Physician _____ Referred by _____

Primary reason for appointment _____

_____

Areas of complaint, pain, or tension _____

_____

Health goals _____

Are you presently under the care of a physician? (If yes, describe conditions) _____

List any medications you are currently using_____

Have you ever had any serious disease conditions? _____

Are you pregnant?_____Have you ever had polio? _____Are you HIV positive? ____

Do you wear contact lenses or dentures? _____

Do you have any skin problems or allergies?_____

Do you have any heart problems? _____

Do you have any blood pressure problems?_____

Do you have varicose veins or blood clots? _____

Have you suffered an acute injury recently? _____

Do you have any spine problems? _____

Do you have arthritis or osteoporosis? _____

Do you have any conditions that you would like to bring to my attention? _____

_____

_____

Polarity Health Education Client Release Form

I_____, understand that the Polarity Energy Balancing Therapy given here is for the purpose of promoting relaxation, clarity, increased energy flow, and health-building balanced energy.

I understand that Polarity Health Education is for healthy individuals who are taking responsibility for health maintenance and health enhancement.

I understand that the Polarity Therapist does not diagnose illness, disease or any physical or mental disorder.

I understand that the Polarity Therapist does not prescribe medical treatment or perform spinal manipulations.

It has been made clear to me that Polarity Therapy is not a substitute for medical examinations and/or diagnosis and that it is recommended that I see an acupuncturist, chiropractor, physician, or psychotherapist for any physical or mental illness, condition, or disease.

Client Signature _____Date _____

Practitioner Signature _____Date _____

CHAPTER 3

Transformational Body Therapy

In an evolutionary approach to healing we work with the person as an evolving spiritual being. We work with the person, not merely with the body. The bodywork session serves as a space for working with health and disease as issues in the person's self-actualization process. Each person represents a unique moment in creation, a unique facet or reflection of the Infinite Self. Growth is a microcosm of the expression and experience of this Beingness.

The body reflects a process of steps in which Spirit materializes into matter. The body is understood as consciousness, a crystallized expression of mental and emotional patterns. In Polarity Therapy, the body is regarded as the negative pole, the mind as the positive pole, the soul as the neutral core of the energetic field of the body. As we move energy in the body, energy moves on the more subtle levels of emotions and mind. By moving energy in the body as the negative pole of the mind, we facilitate mental and emotional integration. This is an important aspect of the body/mind connection that forms the basis of contemporary somatic therapy. Working with the body is indeed working with the mind.

Somatic Emotional Clearing

There are times in life when the ego, the executive of the personality, inadequately processes our experiences and self-expression. These may be moments when the personality is overwhelmed by trauma (for example, loss of a loved one), or periods of time when expression and integration are threatened by outside circumstances (for example, abuse in childhood), or incidents of fear that the ego is unwilling to confront (for example, almost drowning). These issues may be suppressed by the ego, hidden from conscious experience and expression. Though hidden, they are points of inner focus and fixation in that they represent a threat to the ego's sense of security. The ego may invest life energy in suppression of these threats and in distorting perception and emotions in a fabric of defense mechanisms.

These unconscious fixations of energy often have a powerful bodily focus. The ego fixates on threats and obsesses on issues. Obsession sends a myriad of excess, confused, and contradictory impulses through the body into the musculature. Because the ego was not successful in expressing these impulses or integrating these experiences, these excessive mental and emotional impulses lodge in the body as tension, resistance, and armoring. This armoring keeps the body from vibrating in harmony with its source of aliveness in Spirit. We become ruled by unconscious negative forces. Dealing with this resistance and negativity is a focus of somatic emotional clearing.

The Essence of Somatic Therapy

The essence of somatic therapy is facilitating the experience for the client that it is safe to be feeling, safe to be present, safe to be fully alive in his or her body. An energy block is the inability to have and express feelings, to be present in the body, to be fully alive. Release may involve consciously processing and releasing past issues and fears. The goal of body-centered therapy is to support the client's experience that it is safe to be present in his or her body. Experiencing that it is safe to be present with our feelings releases resistance and blocked energy.

In every moment the body is sustained by its attunement with Spirit. Our work as body therapists is to facilitate the release of resistance in the body so that the energy fields can resonate with the finer forces of spiritual attunement. Because the goal of the work is to bring the physical body back into harmony with Spirit, the work is deeply integrating. The person abides in the source of joy, inspiration, and aliveness. In the peace of this attunement, the individual experiences a sacred communion with life. There is a great-fullness and the experience of the natural spirituality which is inherent in good health. Being in harmony with the forces of Creative Intelligence brings inspiration and guidance, clarity of purpose, and meaning into our moment-to-moment living. This is sound health.

Intuition

In a spiritual approach to bodywork, we work with realms of being that are beyond the five physical senses. As a practitioner, day by day we foster a relationship with higher guidance. We open ourselves to forces that are intuitively sensed rather then physically experienced. This process involves cultivating a deep trust of our intuition, or nonlinear mind, and cultivating a sensitivity to unseen forces. This process begins with faith. But as we witness the healing process, there grows a trust in the power of Spirit to heal and to make whole.

Grounding and Centering

The more grounded and centered you are as a practitioner, the finer the forces of vibration with which you resonate. The limiting factor in the energy flowing through you is your grounding to the Earth. By grounding to the Earth, you open to the energy of the heavens. Feel your feet on the ground. Let your legs be like roots going down into the Earth. Feel your connection with the Earth through these roots. Picture a cord from your coccyx connected to the ground. Experience your legs and the grounding cord from your coccyx as a stable base. Picture energy flowing from Heaven to Earth through your spine. Tune into your spine as the center of your energy field. Energy fields emanate from a dynamic center called the fulcrum. When the fulcrum is not centered, the field is off- center and eccentric. The more deeply centered the fulcrum, the more subtle the convergence of the energy fields and the more profound the attunement to the higher-vibrational healing forces. As you begin the Polarity session, take a few deep centering breaths. With each out -breath, relax your shoulders, arms, wrists, and hands. Feel a deep alignment with your center. Invite clients to breathe deeply into the umbilical center. With each deep breath, letting the breath out slowly and naturally, clients will feel themselves relaxing more and more.

Presence

Being totally present for the client is a potent healing force. Being a witness for another's process may facilitate clearing through "being heard." It may offer him or her a safe and supportive space to get something "off his chest," or to get in touch with her issues. Many practitioners speak about sharing a vulnerable space with the client as the foundation of a healing rapport.

It is very important that you be present in the polarity contact. Your presence may be your most potent healing resource. Focusing your attention on the point of contact by keeping your awareness in your hands and fingers is essential. Keep your attention in your fingers and hands by feeling for sensitivity, resistance, energy, pulses, unwinding, vibration, and warmth. Be fully present in your hands.

Intention

The intention of the Polarity Therapist is to address the healing presence of the Soul, the animating force on the body. The clarity of this intention to address sacred life-force energy defines Polarity Therapy. Intention is a very potent resource in a transpersonal approach to bodywork. The intention of the practitioner and the client are a foundation for the healing process created in the bodywork session. Clarifying the goals of the session at the outset focuses this creative potency of intention.

Effortless Effort

As a practitioner it is very important that you cultivate soft, relaxed hands. Energy radiates into manifestation through your hands. The more deeply relaxed you are as a practitioner, the more profound your attunement. As a Polarity Therapist, it is important that you stay in touch with the feeling of the Polarity contact. If a contact feels tense, awkward or uncomfortable, move your hands, arms, body, etc., until the contact feels relaxed, easy, centered, and effortless. A contact should give the practitioner energy and not drain energy. You can always tell that you are doing a polarity contact correctly— when it is easy and effortless.

Psychic Protection

Bodywork can be a very intimate psychic space. Psychic etiquette begins with staying in your own space. Some individuals have trouble doing this in the psychic intimacy of a bodywork session. Healthy boundaries need to be established. The practitioner may need to begin the session by creating boundaries between practitioner and client by mentally acknowledging differences in hair color, skin, body type, etc. Sometimes during a session the practitioner will experience energy, thoughts, or feelings that belong to the client. By knowing yourself, you can witness these sensations without identifying with them.

If you do emotional release work with your clients, you may choose to keep a candle or incense burning in your treatment room. You might also picture the treatment room filled with diamond white light as you begin your session. Or you may burn sacred sage or sacred cedar at the end of the session and smudge your clients, blessing them with these purifying essences. If possible, work in a treatment room that gets direct sunlight and fresh air. It is useful to flick your fingers toward the Earth between Polarity contacts to facilitate elimination and to reestablish the integrity of your energy fields.

Water always flows downward carrying off impurities. It is important to begin and end each bodywork session by washing your hands, wrists, and forearms in cold water, aligning yourself with this purifying and grounding force. Cold water is contracting. It resonates with the Water element and the centripetal harmonics of elimination. Some practitioners keep a glass of water in the treatment room as an energy sink. They change the water after each client.

It is also very important to shower at the end of a day's bodywork as a form of grounding. Many bodyworkers have clothing that they only wear during work. They shower and change to street clothes after their last session of the day. White clothing offers the best psychic protection.

By cultivating psychic protection on a day-to-day basis, it will be available to you when it is needed during stressful periods. In a deeply relaxed state, visualize and affirm that your body is emanating a field of diamond white light that is impenetrable to any negativity or disharmony. Doing this over a period of time fosters psychic protection in your subconscious. Be aware of the elements of your lifestyle that support you in feeling clear and those that leave you spaced out, low in energy and vulnerable. For many sensitive individuals, careful attention must be paid to dietary choices and to a commitment to exercise and a healthy lifestyle.

Establishing Rapport

In beginning a session, it is important to take time to establish rapport with the client. We stand facing the client or we hold hands with the client on the table. We connect with the client by looking into her eyes, the windows of the Soul, and asking about her health and well-being. Through the warmth of your voice and the kindness of your approach, let the client know that this is a safe space and that you are present to serve him, in a deep and sensitive way.

Invocation

Many practitioners find it valuable during the initial polarity contact to silently pray or share a prayer with their client. This acknowledges the presence of a higher power and invites higher guidance to manifest during the session as a healing force. Many practitioners fill the psychic space with diamond white light and invoke healing guides and spiritual masters. An invocation defines a sacred space for the healing process. The more exalted your state of consciousness as a practitioner, the more profound the work you will do with your clients.

Defining The Situation

It is useful to define the situation for the client, letting him or her know that this is a safe and appropriate place to process issues. You may say:

> Take a few deep breaths deep into your belly, feeling yourself relaxing more and more with each breath. Deep breaths, deep into your belly. Relaxing breaths, feeling yourself letting go more and more with each breath. This is a time for you to relax and allow yourself to be nurtured and cared for. A safe space where you don't have to be anyone or do anything, in which you can just open and let go.
>
> During this session we will be balancing your energy. Energy balancing facilitates natural processes of emotional release and elimination. Feelings and images from your past may come to the surface of your awareness. I simply invite you to be as open as possible to experiencing or expressing whatever feelings or images surface for you. This a safe and appropriate place for you to work with your feelings. I encourage you to experience or express as fully as possible whatever feelings may surface for you. Understand that this is a safe space and you are in control in this situation. If at any time you feel that anything is too intense or inappropriate, simply say stop. You are in control and this is a safe space.

Listening

During the initial cradles and in the first half of the bodywork session while working on the feet, I encourage clients to express themselves. As I hold the Quiet Mind Cradle I ask, " What is an issue you are working with in your heart and mind?"

As the client speaks, I listen for emotional charge in the voice. I encourage him to get in touch with his feelings as issues emerge with emotional investment. I encourage him to breathe into his feelings and to focus his attention and breath on areas of pain or resistance in his body. If he cries or emotes, I say "good," again encouraging the client to express himself and letting him know that this is a safe and appropriate space for emotional expression. Bring the client's awareness to his or her body. Constantly ask the client, "What are you experiencing in your body?"

The client will always express what is going on if you are truly present. Listen for verbal clues, for example, words with an emotional charge, repeated words, or phrases taken literally (pain in the neck). Watch for words accompanied by a nonverbal underscore, for example, gestures, change in posture, change in eye contact. Look out for repeated gestures and movement, for example, scratching, holding part of the body. Notice tense and unmoving places in the body, irregular or erratic breathing, holding the breath, or going unconscious. Be alert to body sensations presented by the client, for example, nausea, headache, pressure, pain, tension. Constantly encourage the client to take a deep breath and experience her feelings and bodily sensations.

Pay attention to body sensations resonating in the therapist physically and psychically.

Often the client will clear the past by simply having her experience witnessed. Profound healing takes place through the practitioner's willingness to be present to receive the communication of the client's pain. Acknowledge her self-expression by saying "good." Employ active listening by repeating her statement back to her.

You may invite the client to get off the table and do woodchoppers or beat a pillow as a way of moving energy through the musculature and clearing the cellular memories. You may invite the client to use *etheric plane communication*. This is a powerful psychic tool in which the client, picturing the individual with whom he feels incomplete, gets things off his chest by expressing his feelings as if he were speaking to the other person. You may encourage the client to take sensations into kinetic expression, for example, dancing the feeling on a stage, or painting the emotion on a canvas.

You are not a psychotherapist. You do not provide theories or answers. You are working with fundamentally healthy individuals who are willing and able to take responsibility for actualizing a more integrated state of being. Encourage clients to listen to their own inner guidance and to find inspiration and integration within. The body is a fabric of conscious expression.

The woodchopper is a powerful tool for moving Fiery energy and safely venting anger.

When we work with the body in communicative ways respecting its intelligence, it can be profoundly cooperative.

Breathing

Energy moves in spirals. Respiration is the most physical process through which our bodies are attuned with the spiraling harmonics of nature. Each breath radiates through the atoms, molecules, and cells of the body as a wave moves through a body of water. The cycle of the breath sustains the fundamental rhythms that build and cleanse the body. With each in-breath, we attune ourselves with the radiant, electric, building forces of nature. With each out-breath, we align ourselves with the releasing, magnetic, eliminating forces of the cosmos. Consciously working with the reality of these forces can foster profound results in the release of energy blocks and body armoring. Resistance can be defined as resistance to the breath, the fundamental physical expression of our aliveness, feeling, and consciousness. It is valuable to have the client breathe into the practitioner's fingertips constantly throughout the session. Aliveness and awareness will return as the resistance is released and the area of the body begins to resonate harmoniously again.

Visualization

The body is woven of a fabric of consciousness. Armoring is resistance to conscious experience. Release can often be facilitated by focusing consciousness toward the resistance through visualization. Asking the client to picture the point of sensitivity, pain, or tension as melting ice or expanding light can facilitate release.

Toning

In Polarity Therapy we use a mantric technique called toning. Toning is an amazingly powerful technique which melts tension, dissolves energy blocks, and balances the elemental pulses. Client and practitioner tone together using the sound Ah, held for as long as the breath will allow. It is useful to blend visualization, focused sound/breath, and conscious intention in toning. Visualize light emanating from the nucleus of every atom of the client's body, flooding the body with light and releasing any resistance or negativity in the physical body. Picture the physical body aligning itself with this glow of awareness in divine perfection.

The Brush Off

At the end of the session, it is valuable to "brush off" the client's aura as a ritual for breaking the physical and psychic connection at the end of the body-work session. This ritual clearly defines the end of the session and lets the client know that it is time for him or her to be moving on. It is good to hug the client at this time to balance the energy cycle by giving the client the opportunity to express gratitude. Thankfulness is the consciousness of the return current of energy to the heart which completes the cycle of involution and evolution. It is often deeply rewarding to look into the client's eyes, the windows of the Soul, and experience the profound gratitude of the spiritual being for your support in its evolutionary journey.

Love

Working with compassionate touch is working with the essence of the healing force. Our touch resonates with divine love and is the living presence of Spirit in the world. Cherish your clients. Touch your clients with warmth and safety. Touch your clients with tenderness as you have always wanted to be touched. Manifest unconditional love, impersonal love, universal love. We are instruments through which love is expressed in the world. Let yourself be an instrument of a compassionate universe. Love heals.

CHAPTER 4

Hands-on Energy Balancing Protocols

Sattvic Session[1]

Balancing the Air Principle

Sattva guna, the Air principle, governs movement in nature. Air sustains the fundamental rhythms which fill the body with attunement to the harmonies of the cosmos. The beat of the heart is the neutral source, the cycle of in-breath and out-breath of the lungs is the positive field, and the peristalsis of the colon is the negative field of the Air-predominant torso. Air rules the desires which move us in life: its neutral expression is in the heart as attachments and aversions, its positive field is in the chest as feelings and human nature, and its negative field is in the colon as the crystallization of emotions (fear, resentment, etc.).

Pain is often related to Air imbalances. Excess Air from poor digestion manifests as painful muscle tissue, gas pressure in the bowels and rises up to the head to ache.

Energy tends to block at the periphery far from the vital centers. Air reflexes in the calves or forearms are often a key to balancing the Air element.

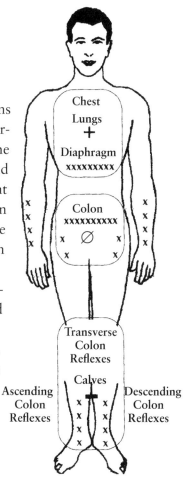

Anterior Air Balancing

Introduction: Sattvic Presence to Evoke Life Essence

In Polarity Therapy light contacts, predominating in stillness and presence, are called sattvic contacts. The word sattvic comes from the Sanskrit root *Sat*, which refers to the Supreme Being in its quality of Existence. Sattvic contacts resonate with the life essence of a living universe and the higher intelligence that is the force of equilibrium in nature. The sattvic resonance is a super-conscious presence of living cosmic Intelligence. The energy fields of the body emanate from this neutral field of mind energy. The thoracic cavity and diaphragm, home of the life breath, predominate in sattva guna.

In Polarity Therapy we translate sattva guna as the Air principle. The Sanskrit word *guna* describes spiraling fields of resonance which are the highest essence out of which manifestation arises. The Air principle predominates in the east-west/transverse currents which arise from the ultrasonic core and unite the periphery of energy fields with the nucleus. The transverse currents integrate sensory imput from the surface of the body with the central nervous system and brain. The Air principle relates to the parasympathetic nervous system, which unites the periphery of the body with the healing equilibrium at the core, fostering deep relaxation, meditative, and ecstatic states. Sattvic intention attunes the body to the healing presence of life essence spiraling through every level of vibration in the energy fields of the body.

Sattvic touch is a tender touch of stillness, presence, and caring. Sattvic touch resonates with the heart as the center of the Air *oval field*. Sattvic resonance touches the deepest core of the body—the heart of Being. Sattvic contacts resonate with the Soul, the portion of divine intelligence that animates the energy fields of the body. Through Sattvic contacts the practitioner's intention reaches the higher intelligence which is the medium of creation. Stillness, a gentle touch, soothing movements, and balancing contacts activate the life essence of Sattvic vibration. Sattvic touch balances the Air-predominant nervous system. Ayurveda teaches the seventy percent of disease has its basis in imbalances of the nervous system. Sattvic contacts ameliorate pain which is carried by the transverse currents. Sattvic contacts resonate with the Air principle to balance both the Ether and Air elements.

Etheric Atomic Lines of Force[2]

Sattvic contacts balance the Air Principle, the currents that emanate from the neutral sushumna of the core energy system and spiral transversely, east to west on the surface of the body. The function of these fields of resonance is intercommunication and binding carrying sensory input and motor feedback from the surface to the core. Sattvic contacts touch the heart of Being and facilitate a receptivity to the healing presence of the Soul.

The Healing Power of Stillness

The profound Cosmic Intelligence of sattva guna rules the electromagnetic balance, which is the foundation of matter on the molecular level. Rajas guna, the centrifugal positive field, and tamas guna, the centripetal negative field, arise out of the neutral ground state of sattva guna. Sattvic contacts evoke profound inner resources of living Cosmic Intelligence to promote healing and well-being.

Cradling the body in stillness and presence evokes sattva guna. Sattvic contacts facilitate a receptivity to the healing presence of the Soul—the profound intelligence that animates the body. The transverse currents arise from the core of the body—the heart of Being. Sattvic contacts are neutral in polarity and resonate with the nucleus of every atom of the body. Tender contacts with a compassionate intention touch the Soul. Sattvic contacts facilitate a superconscious state of lucidity on the cellular level, which invites healing.

Being present in our hands, fully concentrated in our caring, supports a sacred space where the client has permission to release the past and open to the gifts of the present. In Polarity Therapy we always use two hands in making contact. In this way we work with the powerful field of life force resonating between the hands. Focus your awareness into your hands and finger tips, feeling for life-force energy, pulsation, and heat. Sattvic contacts are soothing and balancing; Dr. Stone explains: "neuter or gentle ... (using) ... the lightest touch of a polarized finger to an opposite polarity region ... bathing the body in the finest waves of electromagnetic force. This must be so fine that it is imperceptible to the tissue and does not arouse a reaction. It must be a magnetic likeness and soothing in action ... Sympathy and polarity is its keynote."[3]

Hold contacts for three minutes or until you feel heat, pulsation, or until you intuitively feel a sense of completion. You are holding the contact correctly when it feels effortless and easy. If the contact feels awkward, breathe deeply, center yourself, and move your body, hands, and fingers until you feel comfortable and the contact feels effortless. It is valuable to end every polarity contact with a few seconds of sattvic stillness and presence.

Posterior
Air Balancing
Triune Function

Gemini
+
Shoulders

Kidneys
Adrenals
O
Libra

Calves
Ankles
Aquarius

The Sattvic Session

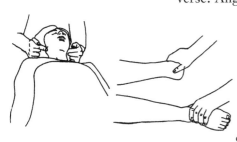

1. **Making Contact.** Squat or stand holding hands with the client. Make eye contact, holding a heartfelt intention to do your best to support the client's healing process. Ask about the client's physical and emotional well-being. Invite the client to set a healing intention for the session.

The eyes are the windows of the Soul. Making eye contact deepens the rapport. We are creating our healing. Intention is the seed of our manifestation.

2. **Position Client.** Have the client lie on his or her back with head facing toward the north or northeast with left hand on his or her heart.

The body is a microcosm of all the energy fields of nature and the universe. Align the body with the core energy fields of the Earth. The heart is the center of the Air oval field, the body cavity predominating in a sattvic resonance. The hands are the most physical and outgoing expression of the neutral sattvic resonance.

3. **Take Pulses.** Contact carotid artery on each side of the neck and contact the flexor longus hallucis behind the ankle. See Polarity Pulse Assessment, Chapter 2.

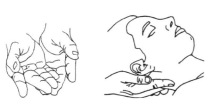

4. **Quiet Heart Cradle.** Right hand above left, hands crossed under neck at occiput. Skull cradled in palm, Air fingers along side of neck paralleling tenth cranial nerve.

The Air fingers contact the vagus nerve to release the sympathetic nervous system. The contact also balances the energy field of the neck, which is a powerful neutral nexus where mental and emotional impulses move toward motor expression in the body and where motor and sensory feedback returns to the brain.

5. **Quiet Mind Cradle.** Left hand above forehead, Fire and Water fingers contacting third eye, Earth and Air fingers over supraorbital ridge, Ether (thumb) at fontanel. Right hand securely cradling head, spine between Air and Fire fingers at occiput.

Promotes balance between the anterior positive sensory, posterior negative motor, and neutral brain fields.

6. **Headache Cradle.** Hold left palm one-half-inch above forehead and securely cradle head at acciput with right hand, spine between Air and Fire fingers at occiput.

    Balances Fire principle at its origin at the brow center. Excellent for headache due to indigestion.

7. **Balance Fire Principle.** Left palm on forehead with thumb at brow center, right palm over navel rocking satvically and holding in stillness.

    Balances the positive and neutral fields of the Fire principle.

8. **Stimulate Core Reflexes at F.L.H.** Stimulate for a minute and then hold sattvically. Continue this pattern until the client feels warmth or energy.

    Stimulates the most negative fields of the etheric core reflexes to the brain, spine, womb, bladder, prostate, and rectum.

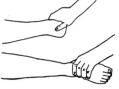

9. **Hold Ten Fingers on Ten Toes.** Tips of fingers on tips of toes.

    Balances long currents at the negative field.

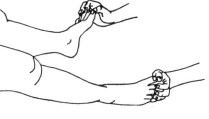

10. **Chakra Balancing through Toning and Visualization.** Picture red/brown light radiating from the Earth chakra at the sacrococcygeal junction, as both client and practitioner tone using the seed mantra Lam. Picture crystal clear light, the color of the water in a bubbling brook, radiating from the lumbosacral junction while toning with the mantra Vam to balance the Water element. Picture a brilliant golden luster radiating from the Fire chakra in the spine behind the navel while toning with the seed mantra Ram. Picture green, the color of spring grass, emanating from the spine behind the heart as you tone Yam sweetly. Picture sapphire blue light radiating from the spine behind the throat as you tone Ham. Picture violet light radiating from the third eye as you tone Ksham. Picture diamond white light radiating from the crown as you tone Aum. Invoke/invite blessings from transcendental intelligence by honoring Mother Earth, the Angel of Water, Grandfather Fire, the Angel of Air, Father Sky, at each appropriate chakra.

11. **Five-Pointed Star.** Gentle rocking with left hand on client's right shoulder, right hand on opposite hip . . . pause . . . then move left hand to opposite shoulder. Continue rocking . . . pause . . . then right hand to opposite hip . . . continue rocking . . . pause . . . then hold left Air finger at foramen magnum sattvically. Balances the functional and sensory energy fields from above to below and within to without.

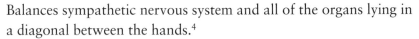

Balances sympathetic nervous system and all of the organs lying in a diagonal between the hands.[4]

12. **Balancing the Long Currents at the Positive Pole.** [5]

13. **Toning To Balance the Etheric Core.** Ask the client if there is an area of the body where she or he would like to focus healing intention. With tips of thumbs at the ear opening, practitioner and client picture the light of the Soul at third eye, filling the cranium and spine and radiating outward to fill the whole body. Picture this diamond white light of such spiritual power that resistance and disharmony in the energy fields of the body dissolve. Both client and practitioner tone, blending their voices, using the tone Ah.

Voice and visualization are powerful tools for manifesting healing intention. The Ether-predominant thumbs resonate with the Ether predominant ears and the sonic medium of the Etheric body. Toning manifests healing intention, stimulating the *manomayakosa* (mental body) and *pranamayakosa* (life force).[6]

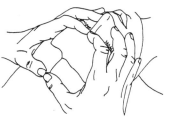

14. **Cradle of the Sages of Old.** Hands on face with Earth fingers contacting jaw, Water fingers contacting lips, Fire fingers contacting base of nose, Air fingers at third eye, and thumbs aligned with the Etheric core. Balances the Primordial mind pattern at the positive field.[7]

15. **Recheck Pulses.**

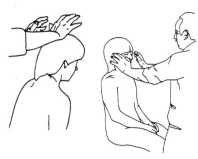

16. **Sit Client Up and Brush Off.** With sensitivity and presence caress the aura, blessing the client. On the back of the body hands cross over at neck and sacrum so that positive hand is at receptive side of the body, active hand is in polarity with the receptive side of the torso. No cross over on the front of the body. Breaks physical and psychic contact.

17. **Hug.**

18. **Wash Hands and Forearms in Cold Water.**

Dr. Stone writes: "Air expands and locks up function in tissues, causing spasm and pain ... it acts as a brake to the motive energy, like an air gap in conduction. Most pains are due to expansion and gas formation in tissues.... Great relief can be given by removing these energy blocks."[8]

Summary of the Sattvic Session

1. Making contact.
2. Position client.
3. Take pulses.
4. Quiet heart cradle.
5. Quiet mind cradle
6. Headache cradle.
7. Balance Fire principle.
8. Stimulate core reflexes at F.L.H.
9. Hold ten fingers on ten toes.
10. Chakra balancing through toning and visualization.
11. Five-pointed star.
12. Balancing the long currents at the positive pole.
13. Toning to balance the etheric core.
14. Cradle of the sages of old.
15. Recheck pulses.
16. Sit client up and brush off.
17. Hug.
18. Wash hands and forearms in cold water.

CHART NO. 10 — THE FIVE POINTED STAR IN THE HUMAN BODY AS NATURE'S GEOMETRIC KEYBOARD OF LINES OF FORCE AND THEIR REFLEXES.

ANTERIOR TENSION IS MOSTLY FUNCTIONAL AND SENSORY REFLEX ACTION FROM THE VISCERA, PLUS EMOTIONAL DISTURBANCE FROM PELVIC ORGANS. POSTERIOR TENSION IS USUALLY CAUSED BY RESISTANCE OR BLOCKS IN THE MOTOR IMPULSES WHICH ARE CONTROLLED BY SENSORY IMPULSES.

SOFT TISSUE WORK IS DONE WITH DEFINITE DIRECTIONS OF LINES OF FORCE CARRIED THROUGH THE BLOCKANCE BY A STEADY PRESSURE OR STIMULI FROM ANY POLE. THE IMPACT OF THE DIRECTED FORCE ACTS UPON THE MOLECULES AND ATOMS IN THE WIRELESS CIRCUIT. IT DOES NOT HAVE TO BE AN ADJUSTMENT.

ALL TENSION SEEMS TO ACCUMULATE IN THE NECK BECAUSE IT IS THE MAIN ETHERIC NEUTER FIELD AND THE SUM TOTAL OF THE RETURN FLOW OF ALL ENERGY FROM BELOW UPWARD TOWARD THE HEAD. TENSION CAUSES ARE MOSTLY REFLEXES WHICH MUST BE FOUND AND RELEASED. LIGHT PRESSURE SOOTHES AND RELAXES. HEAVY PRESSURE STIMULATES AND RELEASES STRONGER RESISTANCE IN THE CURRENT.

THAT WHICH IS ABOVE IS REFLECTED IN THE AREA BELOW, AND THAT WHICH IS BELOW HAS A REPRESENTATION ABOVE IN ITS GOVERNMENT.

ARROWS ON LINES OR ALONG SIDES OF THEM INDICATE THE DIRECTIONAL FORCE IN THE CONTACT. THIS IS ACCOMPLISHED BY THE LINE UP OF THE FINGERS, THE FOREARM AND THE HUMERUS IN THE EXACT ANGLE OF THE IMPACT OF ENERGY AND HELD LONG ENOUGH TO CREATE A MOLECULAR PUSH IN THE CURRENT FLOW.

ALL LINES WHICH CROSS OVER FROM ONE SIDE TO THE OTHER BELONG TO THE BIPOLAR BRAIN REFLEX CHAIN OF THE CADUCEUS.

THIS ARROW FOR DIRECTIONAL CONTACT IS POINTING DOWNWARD OVER THE SIGMOID WHICH IS INVARIABLY HIGH IN ITS POSITION.

SIGMOID VALVE AS A HEART REFLEX. TREAT TOWARD INFERIOR AND POSTERIOR POSITION AGAINST THE CURRENT TO RELEASE HEART SPASMS.

HIP REACTIONS TO RIGHT SHOULDER.

RESPIRATION IS THE FIRST ACT OF LIFE. THE DIAPHRAGM HAS A DEFINITE SUPERIOR POSITIVE POLE—A NEUTER POLE AND AN INFERIOR OR NEGATIVE POLE WHERE IT CAN BE INFLUENCED.

SORENESS AND TENSION OVER ANY ORGAN IS A PROTECTIVE MEASURE OF THE DEEPER TISSUES AND AN EXTENSION OF THE SORENESS AND STAGNATION TO THE SURFACE AREA. RELEASE ENERGY BLOCKS BY DIRECTIONAL CONTACTS IN THE LINE OF FORCE.

ENERGY IMPULSES FLOW DOWNWARD. THE RETURN CURRENTS ARE THE REFLEX IMPULSES FLOWING UPWARD. THESE MAY CROSS OVER AT NEUTER POINTS AND CENTERS OR FLOW STRAIGHT AS ELECTRO-MAGNETIC LINES OF FORCE AND AS GRAVITY IMPULSES. CURRENT RESISTANCE ANYWHERE BECOMES REFLEX PAIN.

ALL THE AREAS AND ORGANS WHICH THE LINES OF FORCE PASS OVER IN THEIR ASCENDING FLOW, HAVE THEIR REPRESENTATIVE REFLEXES ABOVE THE DIAPHRAGM IN EXACTLY THE SAME ORDER AS THEY WERE PASSED. SO THE OVARIES AND PELVIC ORGANS HAVE THEIR REFLEX IN THE BREASTS ON THE OPPOSITE POLARITY SIDE. OTHER ORGANS FOLLOW IN LINE AND CAN BE TRACED. BY THE SAME TOKEN THE DIAPHRAGM HAS A DEFINITE REFLEX BELOW POUPART'S LIGAMENT ON THE MUSCLES OF THE THIGHS ON EACH SIDE. THIS IS A VALUABLE REFLEX TO RELEASE IN SPASMS OF THE DIAPHRAGM.

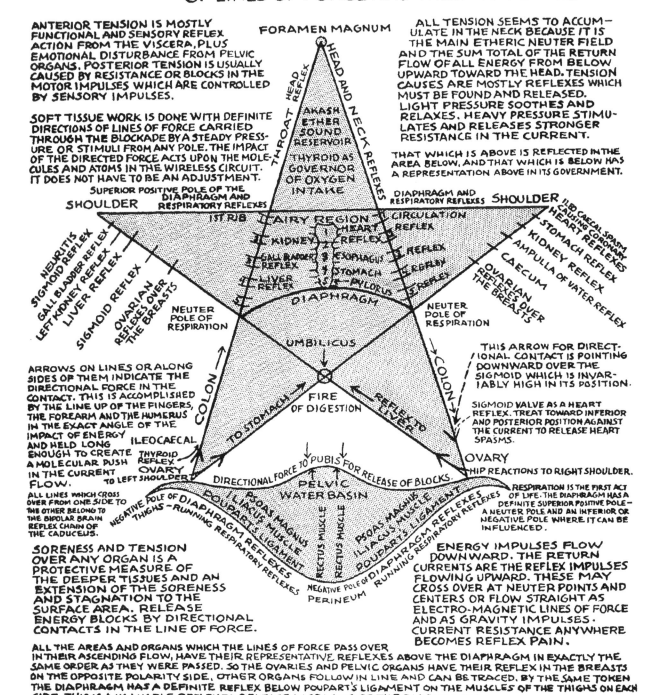

FORAMEN MAGNUM — HEAD AND NECK REFLEXES — THROAT — HEAD REFLEX — AKASH ETHER SOUND RESERVOIR — THYROID AS GOVERNOR OF OXYGEN INTAKE — SUPERIOR POSITIVE POLE OF THE DIAPHRAGM AND RESPIRATORY REFLEXES — SHOULDER — DIAPHRAGM AND RESPIRATORY REFLEXES — SHOULDER — 1ST RIB — FAIRY REGION — CIRCULATION REFLEX — HEART REFLEX — KIDNEY — GALL BLADDER REFLEX — ESOPHAGUS — STOMACH — LIVER REFLEX — PYLORUS — DIAPHRAGM — NEUTER POLE OF RESPIRATION — NEUTER POLE OF RESPIRATION — NEURITIS — SIGMOID REFLEX — GALL BLADDER REFLEX — LEFT KIDNEY REFLEX — LIVER REFLEX — SIGMOID REFLEX — OVARIAN REFLEXES OVER THE BREASTS — ILEO CAECAL CAUSING CORONARY HEART REFLEXES — STOMACH REFLEX — KIDNEY REFLEX — AMPULLA OF VATER REFLEX — CAECUM — OVARIAN REFLEXES OVER THE BREASTS — UMBILICUS — COLON — COLON — FIRE OF DIGESTION — TO STOMACH — REFLEX TO LIVER — ILEOCAECAL — THYROID REFLEX — OVARY TO LEFT SHOULDER — OVARY — DIRECTIONAL FORCE TO PUBIS FOR RELEASE OF BLOCKS — PELVIC WATER BASIN — PSOAS MAGNUS ILIACUS MUSCLE — POUPART'S LIGAMENT — DIAPHRAGM REFLEXES RUNNING RESPIRATORY REFLEXES — NEGATIVE POLE OF DIAPHRAGM THIGHS—RUNNING RESPIRATORY REFLEXES — RECTUS MUSCLE — RECTUS MUSCLE — NEGATIVE POLE OF DIAPHRAGM — PERINEUM

(Stone, Wireless Anatomy)

The Rajasic Session

Introduction: Rajas-Directed Energy

Rajas is the instrument of the conscious or active mind. Rajas,

> the positive pole of energy is an active, energetic manifestation of life . . . It must flow, act, express its potential energy or there is pain through repression, disappointment, frustration, and inhibition . . . This type of energy flows mostly over muscle tissue to act and express itself. Any therapy designed to assist this must be active, vibratory, and positive in nature. All stimulating applications are in this field . . . [using] vibratory directional force over the muscles involved . . . blending into the energy current in purpose, speed, and direction like two rivers joining into one. This has the greatest therapeutic value . . . The impulse must have momentum, but no compulsory force which causes a reaction."[9]

Rajas moves from the center outward and "is a positive centrifugal energy of heat and expansion, flying upward and to the right, repelling in its red, fiery nature . . ."[10] Rajasic contacts are characterized by movement and focused energy. By using our muscles we focus Fiery resonance into the Fire-predominant musculature and vital organs. We can entrain with the radiant tone of rajas through contacts that resonate with the *elasticity* of the tissue. In making contact spiral the tissue clockwise to cultivate a quality of elasticity. Spiraling the umbilical contact into elasticity entrains the fascial sheath to radiate energy throughout the body, through the connective tissue. The connective tissue, where the cerebrospinal fluid steps down energy, is a major avenue of vital force. Directing energy through this elastic tone radiates vitality throughout the body.

Dr. Stone points out that concentrated intention, expressed through the practitioner's focused body mechanics in the contact, sets up a vortex of energy in the electromagnetic fields on the molecular level. Rajasic contacts enhance vitality, the functioning of the organs of assimilation, and the sympathetic nervous system. They facilitate the release of tension and trauma from the musculature to allow emotional clearing and structural integration. Balancing Fire enhances the functioning of the Fire-predominant eyes.

Evolutionary Radiating Vital Energy[11]
The spiral of radiance from the Fire center defines the body from within to without it "radiates like atomic warmth from the chakra at the umbilicus . . . it works through the sympathetic nervous system as autonomic function to sustain the body."[12] Fire predominates in the red tissue of the muscles and organs of the body. The Fire element predominates in the solar plexus. Rajasic contacts resonate with the Fire element, the keynote of vital, electric, positive energy throughout the body.

The Rajasic Session

1. **Making Contact.** Squat or hold hands with Client. Make eye contact and ask about the client's physical and emotional well-being. Invite the client to set an intention for the session.

2. **Position Client.** Have the client lie on his or her back, with head facing toward the north or northeast with left hand on his or her heart.

3. **Take Pulses.** Contact carotid artery on each side of the neck and contact the flexor longus hallucis behind the ankle.

4. **Quiet Heart Cradle.** Right hand above left, hands crossed under neck at occiput.

5. **Quiet Mind Cradle.** Left hand above forehead, Fire and Water fingers contacting third eye, Earth and Air fingers over supraorbital ridges. Right hand securely cradling head with spine between Air and Fire fingers at occiput.

6. **Balance Fire Principle.** Left palm on forehead, thumb at brow center, right palm on solar plexus below navel, spiral palm clockwise to entrain with an elasticity in the tissue, rocking rajasically with energy rippling through the body.

   The solar plexus is the center of the rajasic radiance of the Fire of vitality in the physical body; the brow center is the center of the Fire of the intellect, the Fire of the conscious mind, and the source of the Fire principle.[13]

7. **Rotate and Pull Legs.** Measure legs at ankles. Begin on long, relaxed leg.

8. **Flex Foot and Rotate Ankle.** Rotate rajasically, first in one direction, stillness, then reverse direction.

   The ankles are a diaphragm, ileocecal, ovaries/testes bladder, kidney, and hip reflex. The long currents reverse polarity at the ankles.[14]

9. **Stimulate Air Toes Tamasically.** Quickly pinch tips of Air toes, evoking a reaction. Then work Air toes rajasically for one minute, hold sattvically.

   The Air toes are the negative field contact for the Fire principle and reflex to the gallbladder and duodenum.[15]

Balancing Fire

Fire rules vitality, organ functioning, and the musculature of the body. The Fire center radiates like a whirling centrifugal fountain spray from deep in the umbilical area to define the body from within to without. Fire rules directed force and the positive motor fields of the shoulders and motor fields of the hips and thighs. The Fire Principle has its positive pole at the eyebrow center, neutral field at the umbilicus, and negative pole at the first phalange of the Air toe. The Fire element triad is a harmonic relationship between the vision of Aries (+), the drama of Leo (0) and the intention of Sagittarius (–) in which the seed of individuating vision matures into the principles that guide action in the community. The Fire Principle has its positive pole at the Eyes and brow center, neutral at the umbilicus and first phalange of the Air finger and negative at the first phalange of the Air toe. The Fire Element Triad has its positive pole at Aries the head, neutral at Leo the Diaphragm and Solar Plexus and negative at Sagittarius the thighs.

The Fire center has its neutral pole at the umbilicus, positive at the Diaphragm and Solar Plexus and negative at the thighs.

The Fire principle has its positive pole at the Eyes and brow center, neutral at the umbilicus and first phalange of the Air finger, and negative at the first phalange of the Air toe.

The Fire element triad has its positive pole at Aries the head, neutral at Leo the diaphragm and Solar Plexus, and negative at Sattitarius the thighs.

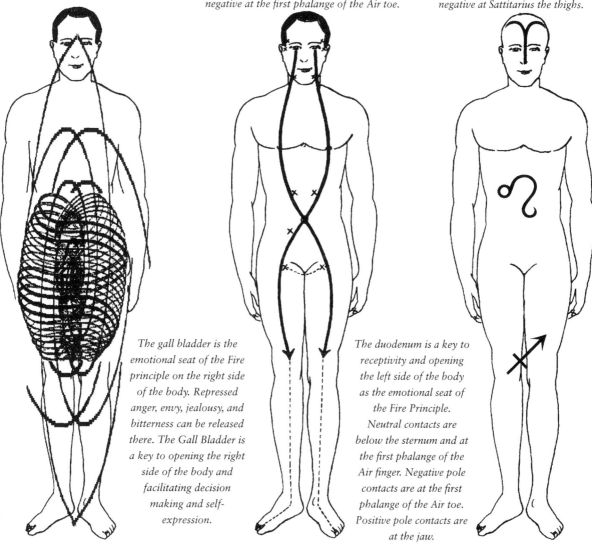

The gall bladder is the emotional seat of the Fire principle on the right side of the body. Repressed anger, envy, jealousy, and bitterness can be released there. The Gall Bladder is a key to opening the right side of the body and facilitating decision making and self-expression.

The duodenum is a key to receptivity and opening the left side of the body as the emotional seat of the Fire Principle. Neutral contacts are below the sternum and at the first phalange of the Air finger. Negative pole contacts are at the first phalange of the Air toe. Positive pole contacts are at the jaw.

Fire Center   **The Fire Principle**   **Fire Element Triad**

10. **Work All Toes Rajasically.** Stimulate each toe rajasically working from outside in, from Earth toward Ether. Pay particular attention to the Fire and Air toes, the area between the Ether and Air toe and the entire metatarsal area.

The long/bipolar currents emanating from the chakras respond best to stimulation at the feet and head. The Ether toe is a reflex to the head, brain, and neck. Webbing between Ether and Air is a neck, diaphragm, liver reflex. The Air toe is a chest, lung and heart reflex. Fire relates to the abdomen and organs of assimilation. Between Fire and Water are kidney and colon—the Water toe resonates with the pelvis, prostate, uterus, and urinary; Earth to rectum, bladder, and elimination.[16]

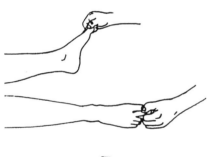

11. **Stimulate Umbilical Reflexes on Bottom of Foot Rajasically.**[17]

12. **Work Ball of Foot.** Reflex to chest, diaphragm and brachial plexus.[18]

13. **Stimulate Air Toes Tamasically.** Quickly pinch Air toes alongside toenail tamasically, then work rajasically for one minute and hold sattvically.

The Air toe is the negative field reflex to the Fire principle, gallbladder, and duodenum.[19]

14. **Hold Ten Fingers on Ten Toes.**
Stimulates long currents at the negative pole to enhance the functioning of internal organs and vitality.[20]

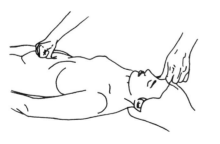

15. **Stimulate Fire Principle.** Move to client's right side. Gently palpate solar plexus around navel. Right thumb 1½ inches below navel, spiral contact clockwise to resonate with an elastic quality in the tissue. The left thumb is on third eye with the thumbs pointing at each other. Alternate rajasic stimulation (one minute) and sattvic holding (twenty seconds). Work contact for two to three minutes to stimulate vitality, or three to five minutes to stimulate emotional release. In any case work contact until client feels warmth or energy in solar plexus.

"The umbilicus is the center in the body through which the life energy can be influenced."[21]

16. **Stimulate Fire Principle to Negative Field.** Move to client's left side. Client bends knees and puts soles of feet together. Right thumb pointing toward heels, continue stimulating contact below navel. Left thumb at heel pointing toward navel, feel for pulse. Work contact for two to three minutes or until you feel a pulse at heels.

Excellent therapy for heart, kidneys, digestion, and sciatica.[22]

17. **Leo Sagittarius Rock.** Practitioner on client's left. Left hand works tense and armored muscles on thighs, right hand on solar plexus, thumb on umbilical center. Spiral umbilical contact to resonate with an elastic quality in the tissue.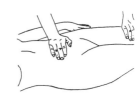

The solar plexus, ruled by Leo, is the neutral field and the thighs, ruled by Sagittarius, are the negative field of the Fire triad (Aries positive). The inner thighs are a seat of reserve energy and vitality. Releasing armoring in the outer thighs facilitates self expression.[23]

18. **Woodchoppers.** If necessary have the client do woodchoppers to facilitate emotional release.[24]

Caution!
Ask the client if he or she has had back problems before doing woodchoppers. Modify woodchoppers if necessary to care for the client's back.

19. **Six-Pointed Star.** Client turns over and lies on abdomen.

   a. Fingers of left hand consciously contact trapezius tension. Right forearm stimulates sacroiliac in a circular motion. Tenderly, with presence, stimulate trapezius, while client breathes out through the muscle telling it to let go.

   b. Stimulating buttock with an energetic circular motion using the forearm, fingers of the left hand are present with the tension at the trapezius on opposite shoulder inviting client to breathe into the muscle, releasing tension with the breath.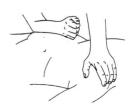

   c. Wide rocking from hip to trapezius, tenderly focusing on trapezius tension until trapezius tension is released.

   Balances posterior structural fields and motor currents. The sacrum reflexes to the brain, buttocks to the scapula, and hips to the shoulders.[25] Trapezius to buttock contacts balance parasympathetic and sympathetic nervous systems.[26]

20. **Bloodless Surgery to Stimulate Fire Principle.** Move to client's right side. Thumbs point at each other. Right thumb rajasically stimulates below navel. Left thumb feels for pulse at diaphragm, clavicle, jaw, third eye. First on right side about five inches from center, then on left side.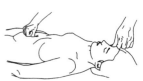

Clears gallbladder and duodenum as emotional seats of the Fire principle.[27]

21. *Optional:* **Bloodless Surgery.** Bloodless Surgery is appropriate for all chronic, cold, blue, tense condition *(inappropriate for inflamed, red, hot, acute conditions).* Thumbs pointing at each other make rajasic contact from navel to any appropriate problem area.[28]

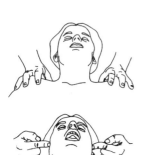

22. **Trapezius Rock.** Work the trapezius muscles rajasically. Spiral thumbs clockwise to entrain with an elasticity in the tissue.

    The shoulders are the positive field (Fire predominant) of the torso.[29]

23. **T.M.J. Release.** Hold Air finger sattvically at T.M.J. for one minute. Rajasically stimulate T.M.J. area until the client feels energy.

    The T.M.J. reflexes to the hip sockets and the vital reserve of Fire at the buttocks, hips, and thighs.[30]

24. **Cradle of the Sages of Old.** Hands on face with Earth fingers contacting jaw, Water finger contacting lips, Fire finger contacting base of nose, Air finger at third eye, and thumbs aligned with the Etheric core.

    Balances the primordial mind pattern at the positive field.[31]

25. **Recheck Pulses.**

26. **Sit Client Up and Brush Off.** To break physical and psychic contact.

27. **Hug Client.** Opens heart and balances polarity of receiving and giving.

28. **Wash Hands and Forearms in Cold Water to Ground.** Good hygiene, grounding, and psychic protection.

More on the Six-Pointed Star

Dr. Stone writes: "The posterior part of the shoulder is covered by the trapezius muscle which is supplied by the spinal accessory nerve and is one of the cranial parasympathetic reflexes. The upper portion of the glutei is also supplied by the three sacral parasympathetic nerves. These must be balanced with the upper parasympathetics … of the neck region and the occipital area … The sympathetic current must be coordinated with this, as the parasympathetic, emotional outlets can only be expressed when nature, the sympathetic nervous system, responds. This is the key to mental-emotional and functional treatment for frustration."[32]

Summary of the Rajasic Session

1. Making contact.
2. Position Client.
3. Take pulses.
4. Quiet heart cradle.
5. Quiet mind cradle.
6. Balance Fire principle.
7. Rotate and pull legs.
8. Flex foot and rotate ankle.
9. Stimulate Air toes tamasically.
10. Work all the toes rajasically.
11. Stimulate umbilical reflexes on sole.
12. Work ball of foot.
13. Stimulate Air toes tamasically.
14. Hold ten fingers on ten toes.
15. Balance Fire principle.
16. Balance Fire principle to negative field.
17. Leo Sagittarius rock.
18. Woodchopper
19. Six-pointed star.
20. Bloodless surgery to stimulate Fire principle.
21. *Optional:* Bloodless surgery.
22. Trapezius rock.
23. T.M.J. Release.
24. Cradle of the sages of old.
25. Recheck pulses.
26. Sit client up and brush off.
27. Hug client.
28. Wash hands and forearms in cold water.

Tamasic Session

Introduction

Tamasic contacts facilitate the release of structural, vital, and emotional energy blocks. Tamas rules resistance and crystallization into matter as the negative pole of energy fields. Tamas rules elimination in the body as a centripetal, downward, contracting force which facilitates expression, completion, and release. The word tamas comes from a Sanskrit root which means "to perish." Writing about tamas guna, Dr. Stone explains: "It is the crystallizing effect of the subconscious mind as rigidity and fixation."[33]

"A negative principle (Tamas, darkness, cold, inertia, and resistance) which must be aroused and counteracted by therapies which are dispersive, scattering, eliminating, forceful, deep, and penetrating, eliciting a definite reaction through the resisting action of the negative pole, thus arousing it and turning it into a circuit."[34] He continues: "All heavy pressure and forceful technique is dispersing in its effect. It is of the tamas quality of energy as a potential compelling force scattering accumulations . . ."[35]

Tamasic contacts promote elimination. The key to breaking through physical and emotional energy blocks is often found in working with the resistance in the negative field. Dr. Stone writes: "The negative pole has slowed down its eliminative, scattering force of energy distribution. It ceases to respond to the positive pole of the life force above. The inertia effect of the negative pole rules by itself, unpolarized, unresponsive to the other currents. This inertia is the most negative block and must be overcome by positive steady force to counteract the resistance and reaction to impulse."[36]

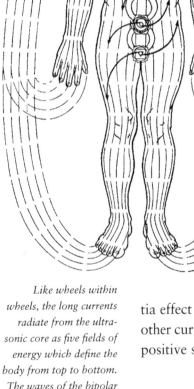

Like wheels within wheels, the long currents radiate from the ultrasonic core as five fields of energy which define the body from top to bottom. The waves of the bipolar currents emanating from the chakras respond best to stimulation at the extremities.[44]

The Five States of Energetic Pathophysiology

As energy becomes blocked in its pathway of motion back to its energy center in the body, the body experiences varying states of ill health, or disease.

These states, varying from mental and emotional disturbance, to acute inflammation, to chronic illness, and finally to full degeneration, are characterized by energetic states corresponding to the step-down of the five elements.

Tamasic contacts may be painful for an instant to evoke the client's resistance. This reaction resonates with the resistance (tamas) and stimulates the eliminating, dispersive, scattering force at the negative field. Stimulating the tamasic centripetal resonance facilitates elimination of emotional and structural trauma and energy blocks. Stimulating the return currents will usually get a person unstuck emotionally, reestablish healthy elimination, and often facilitate a turning point in the person's evolution.

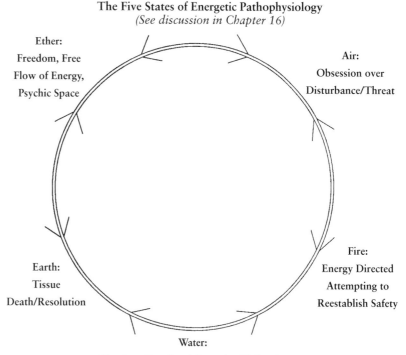

The Five States of Energetic Pathophysiology
(See discussion in Chapter 16)

Ether:
Freedom, Free
Flow of Energy,
Psychic Space

Air:
Obsession over
Disturbance/Threat

Fire:
Energy Directed
Attempting to
Reestablish Safety

Earth:
Tissue
Death/Resolution

Water:
Fixation, Emotional Armoring/Catharsis

Assess Resources

Before doing tamasic somatic emotional clearing assess the client's resources. This work is for "healthy people." Health is defined as the ability to take responsibility for one's self. If the client is in emotional crisis, tamasic release may not be appropriate. For example, if the client is confused and unstable or dissociated and not present in his or her body you would not want to do tamasic work. Only use the tamasic approach with a client who has the resources to take responsibility for integrating the process.

Somatic Emotional Clearing

In threatening or traumatic situations the mind fixates on the threat, sending a myriad of contradictory and confused mental and emotional impulses into the musculature. These impulses to the musculature sustain tension and resistance in the energy fields; this resistance is called emotional armoring. This hardening drops our level of aliveness and feeling. Health is a capacity to be alive and feeling. (See Chapter 16, Somatic Psychology, for additional discussion.)

Moving energy in the negative field eliminates armoring from the cellular memory to release fixation and reestablish healthy elimination. Tamasic contacts may provoke a process of elimination which can be dramatic and cathartic. It can be frightening to the client and to the practitioner as the worst emotional moments of a person's life are released from the cellular memory and surface, to be experienced consciously and expressed in the safety of the bodywork session.

Do not identify with the elimination. It is only energy being released, Hold a safe and sacred space for the client. If you as the practitioner get scared, pray for the client's well-being.

As suppressed pain is released, suggest the client breathe peace and safety into the space. Dr. Stone writes: "Comforting suggestions and inspirational uplift will help the mind to concentrate its scattered energy and to repolarize its various centers."[37] As resistance is released the energy fields of the body can come into harmony with the peace and guidance of the finer forces of higher intelligence of the subtle body. The client may have released emotional armoring during the session and may be more feeling, sensitive, vulnerable, and open. Suggest that the client avoid anything intense after the session, and to take time to integrate and make space to nurture him- or herself. Dr. Stone writes: "This is the type of therapy where there must be a reaction of soreness due to greater activity of the reestablishing life force and circulation in these tissues."[38]

Tamasic currents are the evolutionary return currents which move from without to within, from below to above. These motor currents rule structure in the body, moving from below to above. They have their foundations in the feet, heels, soleus and Achilles tendons, knees, psoas, sacrum, and occiput. Stimulating these return currents is a key to the facilitation of structural integration and grounding. Dr. Stone writes: "The feet are the final reflexes connecting man with the earth and gravity."[39]

Three Approaches to Tamasic Contacts

There are three approaches to stimulating the negative field. First, one may work the area rajasically and sattvically to heighten the level and quality of resonance as we do in foot reflexology. Second, deep, penetrating, focused energy may be directed into the resistance, to release spasms and tension. Third, one may use tamasic contacts which are painful for an instant and are meant to evoke the client's resistance and facilitate catharsis.

Painful tamasic contacts should only be held for a moment, one or two seconds, and should always end in a sattvic contact to ameliorate the pain and support a safe space for emotional release. Tamasic contacts are most effective in the areas of the body that resonate predominantly with the tamasic harmonics of force at the negative fields—toes, feet, and soleus/Achilles. Extreme care should be taken not to re-traumatize the client. Painful tamasic contacts should only be stimulated for a second or two. When emotions surface, support the client in experiencing the feelings in a safe space for a moment, then immediately move into perineal therapy to release trauma from the sympathetic nervous system and to foster a parasympathetic space of profound well-being.

In the skilled approach of contemporary somatics the emphasis is not so much on the catharsis of the release that is surfacing as it is on facilitating the experience that it is safe to be present, feeling, breathing, and alive in the body. Have the client breathe peace and safety into the opening, filling the new-found somatic space with positivity.

End each tamasic contact by holding sattvically. This facilitates the break-through into resonance with the finer forces of the source, a movement from the unconscious fixation of tamas guna to the superconscious integration of sattva guna.

Always ask the client for permission before doing painful tamasic contacts. Make it clear that the client is in control of the process, and that he or she need only say, "Please stop," to end the process. Explain that the evocative painful contact will only be held for two or three seconds. You may offer the client an "edge" scale, in which the number five represents a safe level of stimulation. If the client says "three" as you work the contact he or she is communicating to you to go deeper: An eight or nine indicates that you must lighten up as he or she is near the edge. A ten means STOP! If the practitioner or client does not feel comfortable with tamasic work he or she may do this session rajasically.

It is vital that you empower the client, communicate consistently, and respect his or her wishes in working with tamasic contacts. Hold a safe and sacred space.

The practitioner should only use tamasic contacts with clear intention to realign structure or release blocked emotion. Tamasic contacts resonate with resistance and usually bring blocked emotions into awareness to be experienced and cleared. Tamasic contacts release blocks in the structural return currents which move from below to above. Generally we only do one to four tamasic contacts in an energy balancing session.

The Tamasic Session

1. **Making Contact.** Make eye contact and ask about the client's physical and emotional well-being. Invite the client to set an intention for the session.

2. **Position Client.** Have the client lie on his or her back, head facing toward the north or northeast, with left hand on his or her heart.

   Define the Situation: Let the client know that the bodywork session is a safe and appropriate space for emotional clearing. Invite the client to be present for his or her feelings. Encourage the client to fully experience and express any feelings that surface. Let him or her know that this is a safe and sacred space for emotional release and clearing. Let the client know that he or she is in control and that at any time, if anything that is happening is too intense or inappropriate, to simply say "Please stop."

3. **Take Pulses.** Contact carotid artery on each side of the neck and contact the F.L.H. behind the ankle.

4. **Quiet Heart Cradle.** Right hand over left, hands crossed under neck at occiput.

5. **Quiet Mind Cradle.** Left hand above forehead, Fire and Water fingers contacting third eye, Earth and Air fingers over supraorbital ridge, right hand securely cradling head at occiput.

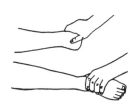

6. **Balance Fire Principle.** Left palm on forehead, right palm on solar plexus below navel, rocking sattvically.

7. **Stimulate Core Reflexes at F.L.H.** Stimulate for a minute, then hold. Continue this pattern until the client feels warmth or energy.

8. **Rotate and Pull Legs.** Check to see which leg is longest. Work long leg side, which is the most relaxed side, first. Work one foot at a time.[40]

9. **Flex Foot and Rotate Ankle.** Rotate rajasically; first in one direction, then hold sattvically, then reverse direction.

   The ankles are important diaphragm and kidney reflexes. Around the outside of the maleoli are reflexes to the ovaries, testes, hips, psoas magnas, ileocecal, sigmoid, iliacus and pelvis. The inside of the maleoli reflexes to the womb, prostate, sciatic nerve, and rectum.[41]

10. **Tweak and Pull Toes Tamasically.** Using two hands pull toes. Start with the Earth toe and work toward Ether. Use deep pressure on the base of

the toes. Tweak the end of each toe for a second, then rajasically stimulate each toe for a minute or so and then hold sattvically.

The toes are the most negative field of the long currents from the chakras. Rajasic and tamasic work at the toes releases resistance in the long currents and stimulates the chakras. Intense work at the toes realigns the electromagnetic fields with the potency of the core to facilitate a vibrant and tingly body. The Earth toe rules elimination, the rectum, and bladder. The Water toe is a reflex to the pelvis and reproductive organs. The Fire toe reflexes to the abdomen and rules digestion and assimilation. The Air toe resonates with circulation and the heart. The Ether toe reflexes to the brain, neck, and spine.[42]

> *Focus on the Out-Breath*
> *It is very important to have the client focus on breathing during release work, especially on the out-breath. Breath is the omniprsence of universal law. A fundamental organizing force, breath moves through the body much like waves move through the ocean. Breath rules the cycles of respiration, which are reflected in the peristalsis, oscillation, and vibration of every body cavity, organ, cell, molecule, and atom of the body. The out-breath resonates with elimination. A wave of centripetal, contracting, releasing, eliminating force moves through every atom, molecule, cell, organ, and cavity of the body with each out-breath. This wave releases tension and resistance. The radiant in-breath "respires" the vibrations of energy in centrifugal attunement to the core.*

11. *Optional:* **Iron Big Toes and Ball of Foot Tamasically.** Hold foot flexed to bring resistance to the surface. Use an oil or cream on the contact that allows you to move slowly with a great deal of presence and control. In ironing we use a slow deep contact that moves along the surface. End each movement by holding sattvically.[43]

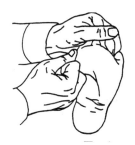

Tamasic contacts, Dr. Stone writes, are: "dispersive, scattering, eliminating, forceful, deep, and penetrating, eliciting a definite reaction through the resisting action of the negative pole, thus arousing it and turning it into a circuit."[44]

12. *Optional:* **Iron Down Achilles Tendon Tamasically.** Hold foot flexed. First outside then inside. The soleus and achilles tendon reflex to the lumbar vertebrae and sacrum. This is a powerful contact that influences the motor currents which move from below to above, supporting structure in the body.[45]

13. *Optional:* **Iron Top of Foot Between Tendons.** On top of foot separate tendon with thumb.

The top of the foot reflexes to the thoracic cavity and back of the body. The space between the Ether and Air toes reflexes to the brachial plexus, liver and diaphragm, heart, and stomach. Air and Fire reflexes to respiration and the solar plexus, diaphragm, heart, and lungs. The space between Fire and Water is a reflex to the kidneys, duodenum, ilium, and colon. The space between the Water and Earth toe relates to the prostate, perineum, uterus, and rectum.[46]

14. *Optional:* **Iron Bottom of Feet Down Tendons.** Use knuckles to iron bottom of foot.[47]

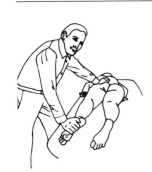

15. **Deep Stimulation of Spine Reflexes Along Arch of Foot.**

These are very powerful spine reflexes which represent the limiting factor in the spine's vibrancy and in the capacity to ground. Tamasic contacts on these reflexes and the acute spine reflexes on the thumb will often facilitate a spontaneous self-correction of the spine.[48]

Remember you always end each tamasic contact by holding sattvically so that your contact resonates with two octaves of vibration: tamas, which releases resistance and facilitates elimination, and sattva, which restores the resonance with the source.

16. **Hold Ten Fingers on Ten Toes Sattvically.**

Like wheels within wheels, the long currents radiate from the ultrasonic core as five fields of energy which define the body from top to bottom. The waves of these bipolar currents emanating from the chakras respond best to stimulation at the extremities.[49]

17. **Rajasic Contact Working Buttocks and Feet.**

Client turns over and lies on belly. Rajasically rock the right-hand contact for a minute while inhibiting the left hand contact, then inhibit both contacts for a few seconds, then rajasically rock left-hand contact and inhibit right-hand contact. Continue this pattern until the contact is complete.[50]

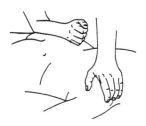

18. **Six-Pointed Star.** Client turns over and lies on abdomen.

a. Fingers of left hand consciously contact trapezius tension. Right forearm stimulates sacroiliac in a circular motion. Tenderly, with presence, stimulate trapezius, while client breathes out through the muscle telling it to let go.

b. Stimulating buttock with an energetic circular motion using the forearm, fingers of the left hand are present with the tension at the trapezius on opposite shoulder inviting client to breathe into the muscle, releasing tension with the breath.

c. Wide rocking from hip to trapezius, tenderly focusing on trapezius tension until trapezius tension is released.

Balances posterior structural fields and motor currents. The sacrum reflexes to the brain, buttocks to the scapula, and hips to the shoulders.[29] Trapezius to buttock contacts balance parasympathetic and sympathetic nervous systems.[51]

19. **Cradle of the Sages of Old.** Hands on face with Earth fingers contacting jaw, Water finger contacting lips, Fire finger contacting base of nose, Air finger at third eye, and thumbs aligned with the Etheric core.[52]

20. **Recheck Pulses.**

21. **Sit Client Up and Brush Off.** To break physical and psychic contact.

22. **Hug Client.**

23. **Wash Hands and Forearms in Cold Water to Ground.**

Summary of the Tamasic Session

1. Making Contact.
2. Position client.
   Define the situation: inform the client that this is a safe and appropriate space for emotional clearing.
3. Take pulses.
4. Quiet heart cradle.
5. Quiet mind cradle.
6. Balance Fire principle.
7. Stimulate core reflexes at F.L.H.
8. Rotate and pull legs.
9. Flex foot and rotate ankle.
10. Tweak and pull toes tamasically.
11. *Optional:* Iron big toes and ball of foot.
12. *Optional:* Iron down achilles tendon and heel.
13. *Optional:* Iron top of foot between tendons.
14. *Optional:* Iron bottom of feet down tendons.
15. Deep stimulation of spine reflexes along arch of foot.
16. Hold ten fingers on ten toes sattvically.
17. Rajasic contact working buttocks and feet.
18. Six-pointed star.
19. Cradle of the sages of old.
20. Recheck pulses.
21. Sit client up and brush off.
22. Hug client.
23. Wash hands and forearms in cold water to ground.

Foot Reflexology Session

Dr. Stone writes: "TRULY, THE *WHOLE BODY* CAN BE *REACHED AND BENEFITED THROUGH THE FEET!* The whole body is represented in the feet, the water energy, which goes deep and seeks out the foundation of things."[54] "The feet are the most negative pole of the entire body and because of this fact many negative energy blocks are found here ... crystallization and hardness spell old age and decay and death."[55] Chronic health conditions manifest as inertia, resistance, and crystallization in the more material negative energy fields. Crystallization in the negative fields inhibits the ability of the body to vibrate with the finer forces of nature and the cosmos. Dr. Stone writes, "All chronic and subnormal conditions are treated beginning at the extremities, in an inward and upward direction, to augment weakened return currents."[56] We die first at the feet. Sensitivity and pain at the feet are reflections of chronic conditions in corresponding areas of the body. Acute symptoms in the body may manifest as pain in the analogous areas of the hands.

Dr. Stone writes: "The FOOT TECHNIQUE according to POLARITY THERAPY PRINCIPLES is a new application of FOOT BALANCE IN RELATION TO THE VITAL CURRENTS. It works like a charm when done with a view to polarizing, *through each contact,* and not merely massaging to rub out the sore spots. To illustrate: When working on the foot, over the area marked 'colon, ascending, descending, transverse' ... grip the top of the outside and top of the foot firmly, to also balance the sensory and motor currents of flow up and down. This makes *foot therapy* a BI-POLAR APPLICATION."[57]

It is valuable in any condition of blocked energy to work the corresponding reflexes in the feet first. The feet are the most negative pole of the body. At each level of wholeness and organization, the body is the reflection of a universal pattern of wholeness. This universal pattern of wholeness is a pattern of attunement to the elemental resonances in nature and the cosmos. Resistance in the negative field blocks the cycle of polarity in its manifestation from the core. Releasing the tamas guna allows rajas guna to man-

The universal energy field is reflected in every level of wholeness in the body. Through this insightful illustration by Mark Allison R.P.P., we can understand the basis of foot reflexology in the holographic body.[53]

ifest and opens the energy fields to an attunement wjth sattva guna. Releasing resistance in the negative field restores healthy elimination.

The Foot Reflexology Session

1. **Making Contact.** Make eye contact and ask about the client's physical and emotional well-being. Invite the client to set an intention for the session.

2. **Position Client.** Have the client lie on his or her back, head facing toward the north or northeast, with left hand on his or her heart.

   Define the Situation: Let the client know that the bodywork session is a safe and appropriate space for emotional clearing. Invite the client to be present for his or her feelings. Encourage the client to fully experience and express any feelings that surface. Let him or her know that this is a safe and sacred space for emotional release and clearing. Let the client know that he or she is in control and that at any time, if anything that is happening is too intense or inappropriate, to simply say "Please stop."

3. **Take Pulses.** Contact carotid artery on each side of the neck and contact the F.L.H. behind the ankle.

4. **Quiet Heart Cradle.** Right hand above left, hands crossed under neck at occiput.

5. **Quiet Mind Cradle.** Left hand above forehead, Fire and Water fingers contacting third eye, Earth and Air fingers over supraorbital ridge, right hand securely cradling head at occiput.

6. **Balance Fire Principle.** Left palm on forehead, right palm on solar plexus below navel, rocking pelvis rajasically.

7. **Stimulate Core Reflexes at F.L.H.** Stimulate for a minute, then hold. Continue this pattern until the client feels warmth or energy.

8. **Rotate and Pull Legs.** Check to see which leg is longest. Work long leg side, which is the most relaxed side, first. Work one foot at a time.

9. **Flex Foot and Stretch Soleus and Achilles Tendon.** Stretch as the client releases the out-breath.

10. **Rotate Ankles.** Rotate rajasically, first in one direction, hold sattvically, then reverse direction. As you rotate the ankles, stimulate the reflexes around the inner and outer maleoli.

    The ankles are important diaphragm and kidney reflexes. Around the outside of the maleoli are reflexes to the ovaries, testes, hips, psoas magnas, ileocecal, sigmoid, iliacus, and

pelvis. The inside of the maleoli reflexes to the womb, prostate, sciatic nerve, and rectum.[58]

11. **Work Toes.** Start with the Earth toe and work toward Ether. Work toes rajasically stimulating each toe for a minute, expanding the mobility of the joints. End each contact with a sattvic focus of presence and stillness.

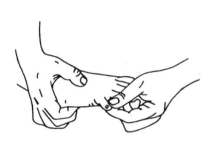

The toes are the most negative field of the long currents from the chakras. Rajasic and tamasic work at the toes releases resistance in the long currents and stimulates the chakras. Intense work at the toes realigns the molecular level of the electromagnetic fields, with the potency of the core to facilitate a vibrant and tingly body. The Earth toe rules elimination, the rectum, and bladder. The Water toe is a reflex to the pelvis and reproductive organs. The Fire toe reflexes to the abdomen and rules digestion and assimilation. The Air toe resonates with circulation and the heart. The Ether toe reflexes to the brain, head, neck, and spine.[59]

12. **Work Webbing Between Toes.** Reflexes to the shoulders and eyes.

13. **Work Top of Foot Between Tendons.**

The top of the foot reflexes to the thoracic cavity and back of the body. The space between the Ether and Air toes reflexes to the brachial plexus, liver, diaphragm, heart, and stomach. Air and Fire reflexes to respiration and the solar plexus, diaphragm, heart, and lungs. The space between Fire and Water is a reflex to the kidneys, duodenum, ilium, and colon. The space between the Water and Earth toe relates to the prostate, perineum, uterus, and rectum.[60]

14. **Work Reflexes Above Transverse Arch.**

Reflexes above the transverse arch on the bottom of the foot stimulate the thoracic cavity, diaphragm, heart, and lungs.

15. **Work Tendons on Bottom of Foot.** [61] Stimulates elemental long line currents from the chakras.

16. **Work Reflexes Below Transverse Arch of Foot.** Reflex to solar plexus and organs of assimilation: liver, gallbladder, stomach, spleen, and pancreas. Focus on transverse ascending and descending colon reflexes.

17. **Stimulation of Spine Reflexes Along Medial Arch of Foot.**

These are very powerful spine reflexes which represent the limiting factor in the spine's vibrancy and capacity to ground. Deep contacts on these reflexes and on the spine reflexes on the thumb balance energy to foster a spontaneous self-correction of the spine.[62]

18. **Hold Ten Fingers on Ten Toes Sattvically.** Like wheels within wheels, the long currents radiate from the ultrasonic core as five fields of energy which define the body from top to bottom. The waves of these bipolar currents emanating from the chakras respond best to stimulation at the extremities.[63]

19. **Chakra Balancing through Toning and Visualization.** Picture red/brown light radiating from the Earth chakra at the sacrococcygeal junction, as both client and practitioner tone using the seed mantra Lam. Picture crystal clear light, radiating from the lumbosacral junction while toning with the mantra Vam to balance the Water element. Picture a brilliant golden luster radiating from the Fire chakra in the spine behind the navel while toning with the seed mantra Ram. Picture green, the color of spring grass, emanating from the spine behind the heart as you tone Yam sweetly. Picture sapphire blue light radiating from the spine behind the throat as you tone Ham. Picture violet light radiating from the third eye as you tone Ksham. Picture diamond white light radiating from the crown as you tone Aum. Invoke/invite blessings from transcendental intelligence by honoring Mother Earth, the Angel of Water, Grandfather Fire, the Angel of Air, Father Sky, at each appropriate chakra.

20. **Rajasic Contact Working Buttocks and Feet.** Client on turns over and lies on belly. Rajasically rock the right-hand contact for a minute while inhibiting the left-hand contact, then inhibit both contacts for a few seconds, then rajasically rock left-hand contact and inhibit right-hand contact. Continue this pattern until the contact is complete.[64]

21. **Six-Pointed Star.** Client on stomach.
    a. Fingers of left hand consciously contact trapezius tension. Right forearm stimulates sacroiliac in a circular motion. Tenderly, with presence, stimulate trapezius, while client breathes out through the muscle telling it to let go.
    b. Stimulating buttock with an energetic circular motion using the forearm, fingers of the left hand are present with the tension at the trapezius on opposite shoulder inviting client to breathe into the muscle, releasing tension with the breath.
    c. Wide rocking from hip to trapezius, tenderly focusing on trapezius tension until trapezius tension is released.

    Balances posterior structural fields and motor currents. The sacrum reflexes to the brain, buttocks to the scapula, and hips to the shoulders. Trapezius to buttock contacts balance parasympathetic and sympathetic nervous system.

Rajasic contact with elbow stimulating scapula reflexes on buttock and shoulder reflexes on hip for excellent release of the trapezius. Reverse on other buttock and shoulder.

In the interlaced triangles of the six-pointed star, "the upper triangle conveys the fiery energy of the head downward as warmth and light for direction and motion . . . the base below (is the) energy reserve in the sacrum.[70]

22. **Cradle of the Sages of Old.** Hands on face with Earth fingers contacting jaw, Water finger contacting lips, Fire finger contacting base of nose, Air finger at third eye, and thumbs aligned with the Etheric core.[71]

23. **Recheck Pulses.**

24. **Sit Client Up and Brush Off.** To break physical and psychic contact.

25. **Hug Client.**

26. **Wash Hands and Forearms in Cold Water to Ground.**

Summary

1. Making contact.
2. Position client.
   Define the Situation.
3. Take pulses.
4. Quiet heart cradle.
5. Quiet mind cradle.
6. Balance Fire principle.
7. Stimulate core reflexes at F.L.H.
8. Rotate and pull legs.
9. Flex foot and stretch soleus and Achilles tendon.
10. Rotate ankles.
11. Work toes.
12. Work webbing between toes.
13. Work top of foot between tendons.
14. Work reflexes above transverse arch.
15. Work tendons on bottom of foot.
16. Work reflexes below transverse arch of foot.
17. Stimulation of spine reflexes along medial arch of foot.
18. Hold tell fingers on ten toes sattvically.
19. Chakra balancing through tolling and visualization.
20. Rajasic contact working buttocks and feet.
21. Six-pointed star.
22. Cradle of the sages of old.
23. Recheck pulses.
24. Sit client up and brush off.
25. Hug client.
26. Wash hands and forearms in cold water to ground.

Core Session

Introduction

A major focus in our approach to energy balancing is supporting the integrity of the ultrasonic core. Dr. Stone writes:

> The center line through the body is the location of the ultrasonic energy substance as the primary life current and the core of being.[67]
>
> It is the primary energy which builds and sustains all others ... This core is the center of attraction and emanation of all currents from the brain to the extremities.[68]
>
> The soul is the essence of being and life in the body, and functions through the brain and the center of the spinal chord, to the end of the coccyx ...The cerebrospinal fluid seems to act as a storage field and a conveyor for the ultrasonic and light energies. It bathes the spinal chord and is a reservoir in these finer essences ... Through this neuter essence, mind functions in and through matter as the light of intelligence ... Physical functions are stepped-down currents from the primary energy.[69]

The physical body is an emanation of this extrasensory spiritual radiance. This core harmonic of attunement is embodied in the sacrum, spine and cranium and resonates as the nucleus of every atom, molecule, and cell. When we release tension and resistance around this core, powerful healing dynamics are set in motion. Dr. Stone continues:

> The human body is a magnified cell, the expansion of the brain pattern of the airy neuter, all-present Life Energy Principle, which has its base in the cerebrospinal fluid in the brain and in the spinal chord and through the entire nervous system. The middle of human the body is the neuter pole—from the top of the head down through the spinal chord and its three coverings. The cerebrospinal fluid flows between the pia mater and the arachnoid sheets, and follows the nerves all through the body.[70]

The focus in Polarity Therapy is the relationship between Spirit and matter. Our work as Polarity Therapists is to release tension and resistance in the body to bring the physical into a deeper harmony with the spiritual. The spiritual is an inner force whose radiant potency sustains the fields of force found

The ultrasonic core is the nucleus of the body as a field of energy. Energy currents circulate around this potent fulcrum.[71]

in matter. While the source of Spirit is eternal and unchanging, its *Shakti* or ever-evolving potency is the force that sustains creation. We are always working with an evolutionary cycle of phases of harmonics, in which the spiritual potency steps down in vibratory rate to the resonant frequency of physical expression and then returns back up to the spiritual.

In our model every energy field is a cycle embodying creation. Cycles are made of an involutionary movement from within to without and from above to below, and then the evolutionary return from without to within and from below to above. At the center of every field of energy is a field of force that resonates with the neutral, sattvic, Etheric harmonic which is the vehicle for manifesting Spirit in matter. Like wheels within wheels, the potency of this field sustains the next coil of resonant frequency of the Air elemental harmonic. The combined radiance of Ether and Air sustain the ring of Fire. The potency of Ether, Air, and Fire sustain the resonant frequency of the field of Water. The potency of the four inner fields of force sustain the gross Earth elemental resonant frequency. At the Earth phase of resonance the vibrations are crystallizing into matter and lose their ability to resonate with the omnipotent spiral of vibration. At this phase, elimination takes place and resistance is released and the inner potency is drawn back inward to step back up to ether through the evolutionary phase.

The creative dynamic is always a cycle of involution into matter of a centrifugal, yang, electric, building radiance and then a centripetal, yin, releasing, eliminating, purifying phase of evolution. Energy fields in the body can best be understood as a spectrum of harmonics embodying the vibratory path between Spirit and matter. The fundamental process underlying structure on every level is based on the above dynamics. Every unit of wholeness, every atom, molecule, cell, and organ of the body is sustained by this underlying dynamic of Spirit moving into matter.

The *involutionary* cycle steps down energy into the physical body through the seven chakras. The chakras emanate *sensory fields* of force which define the body cavities. The body cavities resonate with seven phases of Spirit stepping down into matter. We begin at the pineal gland as the nucleus, which resonates with the spiritual source, The potency of this nucleus sustains the fields of resonance of the cranium and brain, personifying the functions of the mental plane. This field steps down into the neck and resonates with the Etheric. The thoracic cavity is the home of the Air element. The Fire field is centered in the solar plexus, Water in the pelvis, and Earth/Ether in the colon, bones, and marrow.

In the *evolutionary* cycle the body cavities are characterized by the motor currents which define the functional *oval fields* of the head ruled by Fire neck Ether, thoracic Air, abdomen Earth, and pelvis Water. (See Chapter 8 for further discussion.)

The fingers and toes embody the spectrum of harmonics of Spirit's emanation through the involutionary cycle. The big toe and thumb predominate in the Etheric resonant frequency of the core of the body. The index finger and second toe resonate with the Air element which predominates in the thoracic cavity. The middle finger and the third toe resonate with the Fire harmonics of the body which predominates in the organs of the solar plexus. The ring fingers and fourth toe predominate in the Water element and resonate with the cleansing and reproductive organs in the pelvis. The little finger and toe resonate with the Earth element and the processes of elimination. The seven chakras sustain this primary level of organization in the body. Energy moves from within to without and from above to below: Spirit, mental, Etheric, Air, Fire, Water, and Earth.

In the body the nucleus of every atom, molecule, cell, and the "core" of the body—the spine, cranium, sacrum, reproductive organs, and bone marrow—vibrate with the neutral harmonic of Ether. The thoracic cavity, lungs, and nervous system resonate with the Air element. Throughout the body, through sympathetic vibration, the resonant frequencies of Ether and Air form a fabric of resonance of the sattvic harmonic.

The Fire element predominates in the building and assimilating organs of the solar plexus and the red muscles throughout the body. Every atom of every muscle is a microcosm of creation able to manifest energy to sustain function in the body. These processes resonate with the rajasic harmonic of universal force. The resonant frequency of Water predominates in the cleansing organs, the reproductive organs of the pelvis, and in the white tissue and tendons.

The colon and the bones are the home of the Earth element which rules boundaries and elimination. Earth is the crystallization of Ether, and the colon and bones resonate with a phase dynamic that cycles between wave and form, Ether and Earth. The archetype of Ether is a *circle* of balanced energy where all forces are equal. The archetype of Earth is a *square* where all forces have crystallized into form.

Neutronic Force: Core Energy Balancing

The hands are the most physical expression of the powerful *neutronic force* in the body. It is the potency of their radiance of the sattvic field of force that we use to realign the energy fields of the body. The pulsations of the phase dynamic between the two hands has a powerful influence on the energy fields within the area of contact. This phase dynamic breaks through tension and resistance, reestablishing the harmony and alignment of the physical fields with their neutral source. Though the physical fields are created and sustained by the neutral potency, they are not always in harmony with the finer subtle harmonics of energy and the higher intelligence conveyed through these levels of attunement. It is very important to understand that creation is a fabric of attunement. Everything has its being through its attunement with the *tattwas*, the essences or universal principles of force which underlie all vibration.

Releasing the Core Radiance Through the Spine

Releasing tension at the occiput through the A/O (atlanto-occipital) release technique is valuable in restoring the integrity of the craniosacral rhythm. The occiput is the negative pole of the cranium. Excess and confused mental and emotional impulses create chronic tension and resistance in this area, which can be effectively released after perineal therapy through the A/O release. Releasing tension at the occiput, in conjunction with releasing tension and enhancing the subtlety and range of motion of the sacrum, is a profoundly effective way of working with the primary respiratory rhythm.

Therapeutic work which focuses on balancing the spine and releasing tension inhibiting the core radiance is very powerful and effective. Balancing the core energy field of the spine using the Polarity relationship above and below, between vertebrae, and with contacts between the feet and spine reflexes will often facilitate a spontaneous self-correction in the spine. Patterns of tension at the vertebrae are typically indications of blocked energy and an inability to integrate experiences and expression of the energies centered at the block. Issues of personal power, for example, will leave the vertebrae behind the umbilical center (between the second and third lumbar vertebrae) tense as well as the Polarity opposites in the throat chakra (between the third and fourth cervical vertebrae). The umbilical center is the center of vitality in the body, while the throat is the center of personal power. Releasing tension and balancing the energy between these two personal power centers facilitates deep levels of personal integration and enhanced well-being.[72]

Contacts such as the six-pointed star which stimulate the powerful helix of energy fields between hips (centripetal predominant) and shoulders (centrifugal predominant) can be done in a way that emphasizes releasing tension around the rim of the sacrum, which supports the integrity of the core energy. The sacrum and brain are positioned in a Polarity relationship and can be viewed as the nuclei of the physical and emotional bodies. Balancing these fields has profound implications for the body-mind. The thumb is neutral in polarity and resonates with the higher potency of spirit. Using the thumb as a focal point in energy-balancing contacts invokes this resonant frequency as an element of the contact. Energy steps down into physical manifestation through the bone marrow which is Ether-predominant, the joints which are Air-predominant, muscles which are Fire-predominant, tendons which are Water-predominant and bones which are Earth-predominant. Making contact at the joints is very powerful, as each of them is a nexus between Ether and Air and a source point for physical energy. We work with the powerful neutronic core energy when we choose a joint as one point of our polarity contact, especially when thumbs are used for the contact.

Releasing the Ultrasonic Core

The etheric core session offers an introduction to some of the most effective contacts in Polarity Therapy. This session balances the sympathetic nervous system, which is a step-down of the Fire principle and governs vitality and drive. It rules the autonomic functions of body building and maintenance and provides energy for action in the world. The ganglion of impar, located behind the coccyx, is the negative pole of the sympathetic nervous system. To balance the sympathetic nervous system, Dr. Stone recommends contacts between the sphenoid, occiput, and shoulders, as the positive field, sympathetic ganglionic chain and plexuses as the neutral field, and the sacrum, ganglion of impar, innominate, glutei, Achilles tendon and heel as the negative field.[73] The core session is excellent for any emotional imbalance as well as enhancing digestion and vitality. It releases sympathetic impulses from the musculature to foster physical well-being and spiritual attunement.

The Core Session

1. **Making Contact.** Squat or hold hands with client. Make eye contact and ask about the client's physical and emotional well-being. invite the client to set an intention for the session.

2. **Position Client.** Have the client lie on his or her back, head facing toward the north or northeast, with left hand on his or her heart.

3. **Quiet Heart Cradle.** Right hand over left, hands crossed under neck at occiput.

4. **Balance Fire Principle.** Left palm on forehead, right palm on solar plexus below navel, rocking sattvically.

5. **Check Legs.** See which leg is longer. Work long leg side first. Work one foot at a time.

6. **Rajasically Rotate Ankles.** Rotate clockwise for a minute then hold sattvically, rotate counterclockwise and hold sattvically. Work pelvic reflexes on inner and outer heels as you rotate ankles.

7. **Rajasically Stimulate All Toes.** Pay particular attention to the base of the toes and areas between toes. Work Ether and Air toes and ball of foot thoroughly.

8. **Work Spine Reflexes on Arch of Each Foot Deeply.** Contact reflexes to each vertebra tamasically with the out-breath, hold sattvically.

   Reflexes on the arch of the foot stimulate the spine, the corresponding abdominal organs and the brain areas which govern them.[74]

9. **Ten Fingers on Ten Toes.**

10. **Sympathetic Nervous System Release**
    a. Client on belly. Right thumb contact at buttock alongside coccyx moving towards ganglion of impar. Left elbow slides slowly and gently along lamina groove along side of spine. Ask client if any point is sensitive. Be gentle over sensitive areas. *Never put pressure directly on the spine.* Do slide six times, alternating sides.
    b. Hold sattvic contact from brow center to coccyx.
    Releases the sympathetic chain of ganglia alongside the spine to the negative pole reflex behind the coccyx at the ganglion of impar. Excellent for respiratory, liver, stomach, digestive, lumbago, sciatica, prostate, and uterus.[75]

11. **Six-Pointed Star.** Shoulder buttock rock. Hold trapezius for release. Balances parasympathetic nervous system which reflexes to shoulders and innominate rim and gluteals.[76] Balances structural currents which move from below to above and without to within.[77]

12. **A/O (Atlanto-Occipital) Release.** Client on back. Support hand at shoulder, line of force of thumb toward atlanto-occipital joint. Palm of other hand on crown with thumb on third eye. Tone (Ah) and hold for release. Repeat on other side.

    Release of tension in the perineum facilitates an opening at the occiput to release neck tension and parasympathetic nervous system.[78]

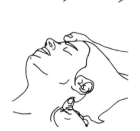

13. **Modified North Pole Stretch.** Right wrist vertical at occiput. Left palm

on forehead. Client takes deep breath and holds. With the in-breath the practitioner rotates wrist slightly and applies traction to occiput. A gentle stretch is used first to test. If sensitive, no traction should be used. (Note: most of the stretch comes from the client's breath stretching the vertebrae.)

Releases compression between the vertebrae.

14. **Cradle of the Sages of Old.** Hands on face with Earth fingers contacting jaw, Water finger contacting lips, Fire finger contacting base of nose, Air finger at third eye, and thumbs aligned with the Etheric core.

15. **Sit Client Up and Brush Off.**

16. **Hug Client.**

17. **Wash Hands and Forearms in Cold Water to Ground.**

Summary of the Core Session

1. Making contact.
2. Position client.
3. Quiet heart cradle.
4. Balance Fire principle.
5. Check leg length.
6. Rotate ankles.
7. Rajasically stimulate all toes.
8. Work core reflexes to spine on arch of foot.
9. Ten fingers on ten toes.
10. Sympathetic nervous system release.
11. Six-pointed star.
12. A/O (Atlanto-occipital) release.
13. North pole stretch.
14. Cradle of the sages of old.
15. Sit client up and brush off.
16. Hug client.
17. Wash hands and forearms in cold water to ground.

Spine Session

Dr. Stone writes: "This spinal balance is a most surprising and effective art … In your next *Kidney, Liver, Heart* or *Asthma* case, just use this vital balancing principle for about fifteen minutes in a few cases and you will be astonished at the remarkable results. Find the active sore area or vertebrae and balance them with the double contacts as shown on the chart."[79] Energy balancing using these contacts facilitates a spontaneous self-correction of the spine and is an effective remedy for functional problems throughout the body.

Dr. Stone writes: "The center line through the body is the location of the ultra-sonic energy substance as the primary life current and the core of being."[80] "It is the primary energy which builds and sustains all others … This core is the center of attraction and emanation of all currents from the brain to the extremities."[81] "The soul is the essence of being and life in the body, and functions through the brain and the center of the spinal chord, to the end of the coccyx … The cerebrospinal fluid seems to act as a storage field and a conveyor for the ultrasonic and light energies. It bathes the spinal chord and is a reservoir in these finer essences … Through this neuter essence, mind functions in and through matter as the light of intelligence … Physical functions are stepped-down currents from the primary energy …"[82]

Polarity of the Spine		
(+)	(0)	(−)
Sphenoid	T 5/6	Coccyx
Occiput	"	Sacrum
C1	T3	L5
C2	T4	L4
C3	T5	L3
C4	T6	L2
C5	T7	L1
C6	T8	T12
C7	T9	T11
T1	T2	T10

The Spine Session

1. **Making Contact.** Squat or hold hands with client. Make eye contact and ask about the client's physical and emotional well-being. Invite the client to set a healing intention for the session.

2. **Position Client.** Have the client lie on his or her back, head facing toward the north or northeast, with left hand on his or her heart.

3. **Quiet Heart Cradle.** Right hand over left, hands crossed under neck at occiput.

4. **Balance Fire Principle.** Left palm on forehead, right palm at navel, rocking sattvically.

5. **Check Legs.** See which leg is longer. Work long leg side first. Work one foot at a time.

6. **Rajasically Rotate Ankles.** Rotate clockwise for a minute, then hold sattvically, rotate counterclockwise and hold sattvically. Work pelvic reflexes on inner and outer heels as you rotate ankles.

7. **Rajasically Stimulate All Toes.** Pay particular attention to the base of the toes and areas between toes. Work Ether and Air toes and ball of foot thoroughly.

8. **Work Spine Reflexes on Arch of Each Foot.** Curl thumb up, behind spine reflexes on medial longitudinal arch of foot. Use other thumb to apply pressure, with the out-breath in a quick tamasic stimulation to the reflex and a long sattvic hold, work down arch of foot. Reflexes on the arch of the foot stimulate the vertebrae, the corresponding abdominal organs, and the brain areas which govern them.[83]

9. **Hold Ten Fingers on Ten Toes.** Toning.

10. **Sympathetic Nervous System Release.**

    a. Client on belly. Thumb contact at buttock alongside coccyx, line of force is towards ganglion of impar under coccyx. Elbow slides slowly and gently in the lamina groove alongside spine. *Never put pressure directly on the spine.* Ask client if any point is sensitive. Be gentle over sensitive areas. "Contacts must be held long enough to make a change in the molecular energy circuits and effect the release of spastic muscles attached to vertebrae."[84] Rock pelvis at the end of each slide with elbow at sacral base, then hold sattvically with long out-breath. Do slide six times, alternating sides.

    b. Hold sattvic contact from brow center to coccyx.

      Releases the sympathetic nervous system through contacts at the sympathetic ganglion along the spine, to the negative pole sympathetic reflex behind the coccyx at the ganglion of impar. Excellent for any functional imbalance including: respiratory, liver, stomach, digestive, kidney, lumbago, sciatica, prostate, and uterus.[85]

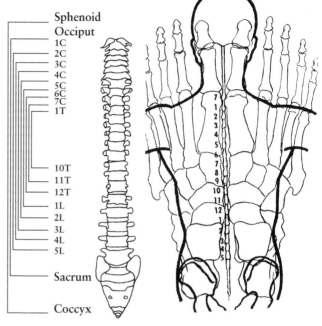

Balancing the polarity of the spine facilitates a spontaneous self correction of the vertebrae. (Stone, Book V, charts 2, 19)

11. **Perineal Therapy.** Client on left side in fetal position. Contacts from perineum to occiput, duodenum, two finger widths below umbilicus, knee, and sacrum. Releases imbalances in the pelvis and sacrum the base of the spine. The perineum is the negative field of the parasympathetic system.[86]

12. **Spinal Balance.**

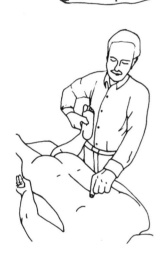

    a. Assessment: Left hand contacts occiput with Ether and Air fingers, and right hand contacts fifth lumbar vertebra at sacral base with Ether and Air fingers. Both hands move toward center symmetrically, two vertebrae at a time (one superior, one inferior), feeling for tension or sensitivity to identify the first triad of vertebrae superior/positive, inferior/negative, and neutral that are sensitive or hard.[87]

    Used with the client sitting up as "Country Side" technique these contacts facilitate gas release and are excellent for acute symptoms of indigestion.[88]

Contacts:

b. Hold contact from positive to negative vertebrae on spine. Stimulate rajasically, hold sattvically, and tone.

c. Contact from neutral to negative vertebrae on spine. Stimulate rajasically, hold sattvically, and tone.

d. Identify corresponding positive, neutral, and negative reflexes on the spine reflexes on the arch of the feet. Look for stress line (wrinkle).[89]

e. Hold contact from negative vertebra on spine to negative spine reflex on the arch of the foot. Rajasic, sattvic, tone.

f. Hold contact from positive vertebra on spine to positive spine reflex on the arch of the foot.[90] Rajasic, sattvic, tone.

g. Hold contact from neutral vertebra on the spine to neutral spine reflex on the arch of the foot. Rajasic, sattvic, tone.

h. i. j. Hold negative, positive, and neutral contacts from spine to other foot. Rajasic, sattvic, tone.

13. **Six-Pointed Star.** Shoulder buttock rock. Hold trapezius for release with wide rocking from ilium.

    Releases parasympathetic nervous system which reflexes to shoulders and innominate rim and glutials.[91] Balances motor and structural currents which move from below to above and without to within.[92]

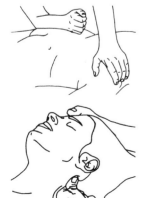

14. *Optional:* **A/O (Atlanto-occipital) Release.** Client on back. Support hand at shoulder, line of force of thumb toward atlanto-occipital joint. Palm of other hand on crown with thumb on third eye. Tone (Ah) and hold for release. Repeat on other side.

Release of tension in the spine, sacrum and perineum facilitates an opening at the occiput to release neck tension and parasympathetic nervous system.[93]

15. **Modified North Pole Stretch.** Right wrist vertically at occiput. Left palm on forehead. Client takes deep breath and holds. With the in-breath the practitioner rotates wrist slightly and applies traction to occiput. A gentle stretch is used first to test. If painful or sensitive, no traction should be used. (Note: most of the stretch comes from the breath stretching the vertebrae.) Releases compression between the vertebrae.

16. **Cradle of the Sages of Old.** Hands on face with Earth fingers contacting jaw, Water finger contacting lips, Fire finger contacting base of nose, Air finger at third eye, and thumbs aligned with the Etheric core.

17. **Sit Client Up and Brush Off.** To break physical and psychic contact.

18. **Hug Client.**

19. **Wash Hands and Forearms in Cold Water to Ground.**

Summary of the Spine Session

1. Making contact.
2. Position client.
3. Quiet heart cradle.
4. Balance Fire principle.
5. Check legs.
6. Rajasically rotate ankles.
7. Rajasically stimulate all toes.
8. Work core reflexes to spine on arch of foot.
9. Ten fingers on ten toes.
10. Sympathetic nervous system release.
11. Perineal therapy.
12. Spinal balance.
13. Six-pointed star.
14. *Optional:* A/O (Atlantooccipital) release.
15. Modified north pole stretch.
16. Cradle of the sages of old.
17. Sit client up and brush off.
18. Hug client.
19. Wash hands and forearms in cold water to ground.

The Parasympathetic, Sympathetic, and Cerebrospinal Nervous Systems

The principles Air, Fire, and Water step-down to function in the body through the three nervous systems: parasympathetic, sympathetic, and cerebrospinal or central nervous system."The Parasympathetic system is the neutral conduit that channels thoughts from their pre-physical field of origin into the physical realm. The sympathetic system, in turn, becomes the positive channel to shuttle thoughts into action through the ... Cerebrospinal nervous system, [which is] the negative pole of the nervous system."[94] The parasympathetic nervous system is a step-down of the Air principle and relates to the east-west currents which flow from the periphery of the body to the core, serving to integrate and harmonize the body as a whole. The transverse currents emanate from the core energy system functioning in communication and coordination, integrating the periphery of the body with the core and carrying sensory input from without to within. It is thus associated with the parasympathetic nervous system, which performs these functions in the body.

In Yogic theory, ida and the left nostril are associated with the parasympathetic chain of ganglia in the autonomic nervous system which controls the relaxation response, while pingala and the right nostril are associated with the sympathetic ganglia, which control the fight-or-flight response.

The sympathetic nervous system is a step-down of the Fire principle and governs vitality and drive. It rules the autonomic functions of body building and maintenance and provides energy for action in the world. "The sympathetic nervous system ... repairs the body and keeps it in tune with the natural forces."[95] The ganglion of impar, located behind the coccyx, is the negative pole of the sympathetic nervous system. To balance the sympathetic nervous system, Dr. Stone recommends contacts at the sphenoid sutures or occiput as the positive field, sacrum and sympathetic ganglion as the neutral field, and ganglion of impar, Achilles tendon and heel as the negative field.[96]

Water is the receptacle of life and the medium of form. The Water principle predominates in the cerebrospinal system which steps down through the chakras, whose pulsating energies move from within to without as "wheels within wheels" emanating the elemental long currents. These currents pulsate from each of the five chakras to sustain five zones of force radiating out from the core of the body. It is at this "step-down" into matter that energy assumes substantial solidity. The long or bipolar currents define the body from top to bottom. This system can be balanced through spine, chakra, and craniosacral therapies. Dr. Stone writes:

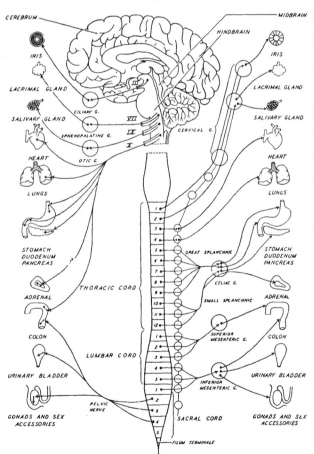

*The Autonomic
Nervous System*[102]

Parasympathetic impulses must flow into the work in conjunction with the Sympathetic, in order to work at all; for it is the parasympathetic system that expresses balance and conscious mind impulses, and conveys them to the sympathetic. What are mental frustrations? Parasympathetic or cranial impulses that could find no expression or response in the Sympathetic system ... Emotional frustrations are sympathetic and heart center impulses which are suppressed by the conscious mind impulses of the Parasympathetic system.[97] Sympathetic and parasympathetic nerve reflexes ... respond from below upward.[98]

Perineal therapy releases the parasympathetic nervous system, which is a key to balancing the three nervous systems. It releases the sympathetic nervous system which rules the autonomic functions of body maintenance and building. Dr. Stone continues: "parasympathetic emotional outlets can only be expressed when Nature, the sympathetic nervous system responds."[99] "Parasympathetic and sympathetic releases allow the cerebrospinal system to function harmoniously."[100] In Polarity Therapy the perineum is a negative pole of the parasympathetic system, the sacrum is neutral, and the shoulders near the neck are positive.[101]

Perineal Session

Introduction

One of the foundations of Polarity Therapy is perineal therapy. The pelvis and perineum rule elimination. The muscles of the perineum are the most negative field of the torso and are the negative pole of the diaphragm. The negative field resonates with the centripetal, releasing, purifying harmonic, and facilitates elimination. When the negative field is tense, elimination is blocked and the whole evolutionary cycle of Spirit's movement into matter is blocked. The individual becomes tense, anxious, depressed, sensitive, defensive, or numb to life, and a likely host to illness and injury. When elimination is blocked, the individual expresses negativity, and when threatened, may be run by fear and unconscious forces.

Dr. Stone writes: "The perineum is the major negative pole of the body and, as such, holds the key to all negative and irrational impulses and perversions of the currents in the energy field."[103] "The perineal contact is the best treatment for nervousness and hysteria . . ."[104] Releasing tension in the perineum facilitates elimination, and perineal therapy is often a turning point in reestablishing healthy elimination. Tension and resistance are released, and emotions can be processed-experienced, expressed, and eliminated.

When the integrity of the energy cycle is restored, the faithful reflection of Spirit manifests in an energetic, joyous, inspired life in tune with higher Intelligence. When tension and resistance have been released from the body, we are attuned to spirit, attuned to Creative Intelligence, inner peace, and an intuitive sense of wholeness, meaning, and purpose expressed in a joyful and fulfilling life.

Perineal therapy "resets" the nervous system and is profoundly effective for any emotional condition. The perineum is the negative field of the parasympathetic nervous system. Perineal therapy releases the sympathetic nervous system and so releases chronic anxiety as an expression of unresolved trauma. The coccyx and ganglion of impar constitute the negative pole of the sympathetic nervous system. Deep perineal contacts release tension in the glutei, the negative field of the cerebrospinal nervous system.

All the muscles between the pubic arch and the coccyx are considered, for purposes of Polarity work, as perineal muscles. These figure-eight shaped muscles have been called the "infinity muscles" in Chinese medicine. The muscles of the perineum can be understood as a diaphragm that moves in harmony with the diaphragms at the solar plexus, the thoracic inlet, and the third eye. The health and resilience of the perineal muscles are a key factor in the body's ability to reflect a finer attunement with higher mind. When these muscles are tense and resistant, the full elimination and radiance of aliveness in the breath and energy cycles cannot be achieved.

The muscles of the perineum may be pictured as a hammock attached to the sacrum. The craniosacral rhythm ripples through this diaphragm. Tension and resistance in the perineal floor inhibit the subtlety and range of motion of the sacrum, thereby preventing the body from resonating with the finer forces emanating from the sacrum and embodied in the potency of the craniosacral system. A key to healing with life-force is restoring the integrity of the perineal energy field.

Energy-balancing contacts between the perineum and the occiput as well as contacts between the perineum and reflex points at the diaphragm, umbilical center, knee, inner ankle, and other joints are a key to effective Polarity Therapy. Perineal therapy is profound for any emotional condition.

The sacrum is the cornerstone of the body structure; perineal therapy, which

Polarity contacts to Release and Balance
The parasympathetic/Sympathetic/Cerebrospinal Nervous Systems
One of the keys to releasing trauma is working with the polarity of the energy fields of the nervous system.

	+	o	-
Parasympathetic	X Cranial nerve Parasympathetic nerves at neck and shoulders also III, VII, IX, XI cranial nerves, top of head, occiput.	S 2, 3, 4	Perineum
Sympathetic	Occiput, shoulders	Sacrum, innominate	Heels, Achilles Tendon, glutei
	C-1 to T-1 Sphenoid, nasal mucosa	T-2 to T-9 Sympathetic ganglion	T-10 to L-5 Coccyx, ganglion chain and ganglion of impar
Cerebrospinal	Head-neck, back	Shoulders, musculture	Glutei, hips, feet

contributes to the harmony and range of motion of the sacrum, is often the key to dealing with sciatica, spine, and chronic back pain. Deep perineal contacts under the coccyx release feelings of inadequacy, fear, and resentment as well as relieving constipation and hemorrhoids. Perineal therapy call be excellent for throat and neck problems. Contacts toward the pubic arch alongside the reproductive organs release sexual trauma and fear associated with sexuality.

Perineal Therapy is one of the most effective healing techniques in bodywork. This session is excellent for any emotional imbalance and fosters deep relaxation, physical well-being, and receptivity to essence. Perineal therapy releases the parasympathetic nervous system, which facilitates deeply relaxed, meditative, and ecstatic states and is a key to balancing the three nervous systems. It releases the sympathetic nervous system, which rules the fight-or-flight mechanism, as well as autonomic functions of body maintenance and building. Dr. Stone writes: "Parasympathetic and sympathetic releases allow the cerebrospinal system to function harmoniously."[115]

The perineum is the most negative reflex to the parasympathetic nervous system, which carries emotional impulses to the musculature (sympathetic) for expression. When it is not safe for the muscles to express our feelings in the world, both systems get blocked Dr. Stone writes: Parasympathetic emotional outlets can only be expressed when Nature, the sympathetic nervous system responds. This is the key to mental and emotional and functional treatment for frustration."[106] "The perineal contact is the best treatment for nervousness and hysteria ..."[107] "Sympathetic and parasympathetic nerve reflexes ... respond from below upward."[108] Dr. Stone continues: "Deep perineal contacts to balance the three nervous systems from below upward is the answer to most ... pains in the deep muscle fibers of the pelvis."[109] "In treating lumbago, sciatica and low back pains ... perineal technique works like a miracle."[115] "In pregnancy, Perineal Technique is invaluable in relieving pains and tensions due to pressure in the pelvis. It gives almost miraculous relief in all leg symptoms and pains due to pregnancy ..."[110] Perineal therapy relieves menstrual pain. "When the perineum is relaxed the neck muscles ... let go."[111] The perineum receives the final impulse of the caduceus which steps down energy from Spirit into the body. Dr. Stone believed that he had identified a key to the healing process in perineal therapy.

The Perineal Session

1. **Making Contact.** Squat or hold hands with client. Make eye contact and ask about the client's physical and emotional well-being.

2. **Position Client.** Have the client lie on his or her back, head facing toward the north or northeast, with left hand on his or her heart.

3. **Quiet Heart Cradle.** Right hand over left, hands crossed under neck at occiput.

4. **Balance Fire Principle.** Left palm on forehead, thumb at brow center, right palm on solar plexus below navel, rocking rajasically with energy rippling through the body.

5. **Rotate and Pull Legs.** See which leg is longer. Work long leg side first. Work one foot at a time.

6. **Rajasically Rotate Ankles.** Rotate clockwise for a minute, then hold sattvically, rotate counterclockwise and hold sattvically. Work pelvic reflexes on inner and outer heels as you rotate ankles.

7. **Rajasically Stimulate All Toes.** Pay particular attention to the base of the toes and areas between toes. Work Water and Earth toes thoroughly.

8. **Ten Fingers on Ten Toes.** Chakra toning.

9. **Deep Perineal Contact.**

   a. Have client lie on left side in a fetal position. With Ether and Air finger of left hand make contact on spine at occiput. With Fire finger of right hand contact the soft tissue of the perineum between ischial tuberosities. The contact should be on the muscles alongside the reproductive organs toward the thighs. Always use a glove or tissue over the Fire finger and wash hands carefully after each session. For hygienic reasons it is imperative that no fecal matter contact the practitioner.

   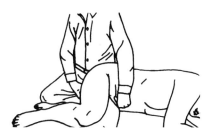

   The perineum is the negative field of the parasympathetic nervous system which carries emotional impulses to the musculature. Excellent for emotional blocks and frustration.[113]

   b. Move left hand to duodenum.

   The duodenum is the emotional seat of the Fire principle and the key to opening the left side of the body.

   c. Move left hand to Fire center; stimulate rajasically, then hold sattvically until release on perineum.

   The perineum represents the final impulse of the parasympathetic nervous system.

   d. Move left hand to knee.

   To balance the Earth triad—an astrological harmonic of Capricorn (knee, neutral), Virgo (bowels, negative), and Taurus (neck, positive).[114]

CHART NO. 30. PERINEAL CONTACTS IN RELATION TO THE NECK, SHOULDER, ELBOW AND HIP.

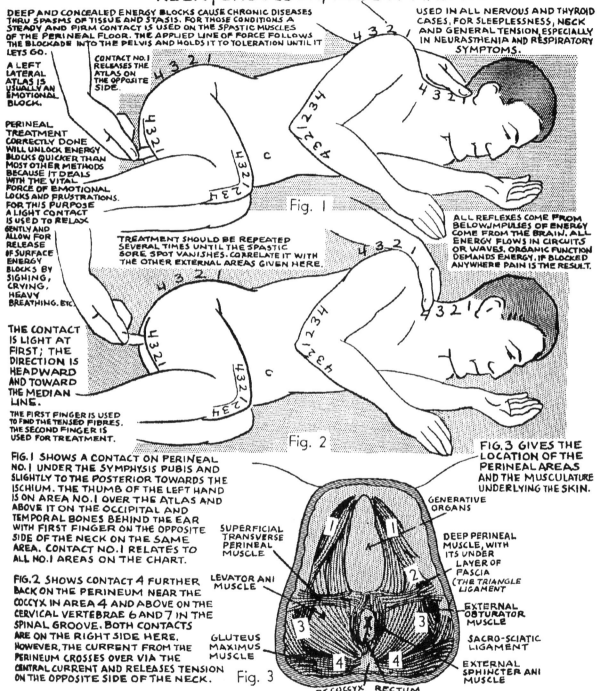

DEEP AND CONCEALED ENERGY BLOCKS CAUSE CHRONIC DISEASES THRU SPASMS OF TISSUE AND STASIS. FOR THOSE CONDITIONS A STEADY AND FIRM CONTACT IS USED ON THE SPASTIC MUSCLES OF THE PERINEAL FLOOR. THE APPLIED LINE OF FORCE FOLLOWS THE BLOCKADE INTO THE PELVIS AND HOLDS IT TO TOLERATION UNTIL IT LETS GO.

A LEFT LATERAL ATLAS IS USUALLY AN EMOTIONAL BLOCK.

PERINEAL TREATMENT CORRECTLY DONE WILL UNLOCK ENERGY BLOCKS QUICKER THAN MOST OTHER METHODS BECAUSE IT DEALS WITH THE VITAL FORCE OF EMOTIONAL LOCKS AND FRUSTRATIONS. FOR THIS PURPOSE A LIGHT CONTACT IS USED TO RELAX GENTLY AND ALLOW FOR RELEASE OF SURFACE ENERGY BLOCKS BY SIGHING, CRYING, HEAVY BREATHING. ETC.

CONTACT NO. I RELEASES THE ATLAS ON THE OPPOSITE SIDE.

USED IN ALL NERVOUS AND THYROID CASES, FOR SLEEPLESSNESS, NECK AND GENERAL TENSION, ESPECIALLY IN NEURASTHENIA AND RESPIRATORY SYMPTOMS.

Fig. I

TREATMENT SHOULD BE REPEATED SEVERAL TIMES UNTIL THE SPASTIC SORE SPOT VANISHES. CORRELATE IT WITH THE OTHER EXTERNAL AREAS GIVEN HERE.

ALL REFLEXES COME FROM BELOW. IMPULSES OF ENERGY COME FROM THE BRAIN. ALL ENERGY FLOWS IN CIRCUITS OR WAVES. ORGANIC FUNCTION DEMANDS ENERGY. IF BLOCKED ANYWHERE PAIN IS THE RESULT.

THE CONTACT IS LIGHT AT FIRST; THE DIRECTION IS HEADWARD AND TOWARD THE MEDIAN LINE.

THE FIRST FINGER IS USED TO FIND THE TENSED FIBRES. THE SECOND FINGER IS USED FOR TREATMENT.

Fig. 2

FIG. 3 GIVES THE LOCATION OF THE PERINEAL AREAS AND THE MUSCULATURE UNDERLYING THE SKIN.

FIG. I SHOWS A CONTACT ON PERINEAL NO. I UNDER THE SYMPHYSIS PUBIS AND SLIGHTLY TO THE POSTERIOR TOWARDS THE ISCHIUM. THE THUMB OF THE LEFT HAND IS ON AREA NO. I OVER THE ATLAS AND ABOVE IT ON THE OCCIPITAL AND TEMPORAL BONES BEHIND THE EAR WITH FIRST FINGER ON THE OPPOSITE SIDE OF THE NECK ON THE SAME AREA. CONTACT NO. I RELATES TO ALL NO. I AREAS ON THE CHART.

FIG. 2 SHOWS CONTACT 4 FURTHER BACK ON THE PERINEUM NEAR THE COCCYX IN AREA 4 AND ABOVE ON THE CERVICAL VERTEBRAE 6 AND 7 IN THE SPINAL GROOVE. BOTH CONTACTS ARE ON THE RIGHT SIDE HERE. HOWEVER, THE CURRENT FROM THE PERINEUM CROSSES OVER VIA THE CENTRAL CURRENT AND RELEASES TENSION ON THE OPPOSITE SIDE OF THE NECK.

SUPERFICIAL TRANSVERSE PERINEAL MUSCLE

LEVATOR ANI MUSCLE

GLUTEUS MAXIMUS MUSCLE

Fig. 3

GENERATIVE ORGANS

DEEP PERINEAL MUSCLE, WITH ITS UNDER LAYER OF FASCIA (THE TRIANGLE LIGAMENT

EXTERNAL OBTURATOR MUSCLE

SACRO-SCIATIC LIGAMENT

EXTERNAL SPHINCTER ANI MUSCLE

OS COCCYX   RECTUM

CHART FOR PERINEAL TREATMENT DESCRIBED ON PAGES 50, 51, 52 IN THE NEW ENERGY CONCEPT OF THE HEALING ART. PAGE 49 GIVES A DIAGRAM OF PERINEAL FLOOR AND CONTACT POINTS.

(Stone, *Wireless Anatomy*)

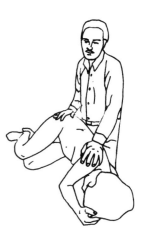

e. Move left hand to inner ankle.

Negative field reflexes to the core, perineum, and sympathetic nervous system.

f. Shoulder hip rock. Contacts from the neck and shoulders to the hips and gluteal muscles release the sympathetic nervous system and stimulate the parasympathetic system.[115]

10. **Posterior Perineal Contact.**

a. Work seventh cervical vertebra to release trapezius with Air and Fire finger. With Fire finger of right hand make contact under coccyx. Hold for release under coccyx.

Reflexes functionally to the Earth triad of the abdomen and knees. Useful for digestive disturbances, constipation, hemorrhoids, pregnancy, leg problems. Coccyx area location of the Earth chakra, releases boundary, survival issues, resentment, fear, negativity.[116]

b. Move left hand to ankle. Hold for release at coccyx. Releases sympathetic nervous system.[117]

11. **Anterior Perineal Contact.**

a. Make contact with Ether and Air fingers at occiput with left hand. Contact perineum with Fire finger of right hand directly behind pubic arch, alongside reproductive organs. Tone into contact (Vam) and hold for release.

b. Move down spine, toning at each vertebra. Halfway down spine move to other side of perineum and continue toning.

Releases erector spinea muscles and sympathetic nervous system.[118]

c. Move Air finger of left hand to heel and hold sattvically for release. The heel is a reflex for emotional release and leg problems especially in pregnancy.[119]

12. **Sympathetic Nervous System Release.** Client on belly.

a. Thumb contact at buttock alongside coccyx. Elbow slides slowly and gently in lamina groove along side spine. Ask client if any point is sensitive. Be gentle over sensitive areas. Never put pressure directly on spine. Do slide three times on each side, alternating sides.

b. Hold sattvic contact from brow center to coccyx. Releases the sympathetic nervous system which has its most negative pole behind the coccyx at the ganglion of impar. Excellent for respiratory, liver, stomach, digestive, lumbago, sciatica, prostate, and uterus.

13. **Six-Pointed Star.** Shoulder buttock rock. Hold trapezius for release. Releases sympathetic nervous system, which reflexes to shoulders, innominate rim, and glutials.[120]

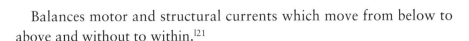

Balances motor and structural currents which move from below to above and without to within.[121]

14. **A/O (Atlanto-occipital) Release.** Client on back. Support hand at shoulder, thumb at sternoclidomastoid muscles at occiput. Palm of other hand on crown with thumb on third eye. Tone and hold for release. Repeat on other side.

Release of tension in the perineum facilitates an opening at the occiput to release neck tension and the parasympathetic nervous system.[122]

15. **Modified North Pole Stretch.** Right wrist vertical at occiput. Left palm on forehead. Client takes deep breath and holds. With the in-breath the practitioner rotates wrist slightly and applies traction to occiput. A gentle stretch is used first to test. If too sensitive, No traction should be used. (Note: most of the stretch comes from the client's breath stretching the vertebrae from within.)

Releases compression between the vertebrae.

16. **Cradle of the Sages of Old.** Hands on face with Earth fingers contacting jaw, Water finger contacting lips, Fire finger contacting base of nose, Air finger at third eye, and thumbs aligned with the Etheric core.

17. **Sit Client Up and Brush Off.** To break physical and psychic contact.

18. **Hug Client.**

19. **Wash Hands and Forearms in Cold Water to Ground.**

Summary of the Perineal Session

1. Making contact.
2. Position client.
3. Quiet heart cradle.
4. Balance Fire principle.
5. Rotate and pull legs.
6. Rajasically rotate ankles.
7. Rajasically stimulate all toes.
8. Ten fingers on ten toes. Toning.
9. Deep perineal contact.
10. Posterior perineal contact.
11. Anterior perineal contact.
12. Sympathetic nervous system release.
13. Six-pointed star.
14. A/O (Atlanto-occipital) Release.
15. Modified north pole stretch.
16. Cradle of the sages of old.
17. Sit client up and brush off.
18. Hug client.
19. Wash hands and forearms in cold water to ground.

Pelvic Session

Elimination in the body is centered in the centripetal field of the pelvis. Releasing energy blocks in the pelvis releases fixation and trauma to restore healthy elimination. This session is appropriate for any problem with the Water predominant cleansing organs of the pelvis. The session releases the perineum and the psoas and iliacus muscles which embody the negative field of the five pointed star, releasing "tension and emotional frustration" with profound implications for the functioning of the vital organs.[123] This release realigns an anterior or anterior lateral rotation of the sacrum.

The alignment of the sacrum is the key to structural integration in the body. Dr. Stone writes: "In the study of the entire bony structure of the body, the sacrum is the most vital and the most neglected bone."[124] "The sacrum is the key wedge between the innominate bones, like the keystone in an arch, between the two pillars on which this arch rests"[125] The sacrum is the keystone and foundation ... of the spinal column and of structure in the entire body." Structure supports from below upward. The inverted triangle of the sacral base is "the level weight bearing line of the entire spinal column and its structure. The sacrum is the opposite pole to the brain expressing action and structural balance."[126] "Vital structural and motor forces are locked up in the cerebrospinal fluid in the sacrum ... The sacrum is the vital balancer of the individual energy in relation to gravity. It is the balance wheel of the body's own center of POLARITY which *governs the position of the spinal column ... The sacrum is the key triangle of the pelvis,* registering all the impulses, reactions and disturbances of the water element in the body ... Correction of the anterior sacral base is the simple answer to the complicated problems of the spine."[127] "Anterior muscle pull of the psoas is the *real culprit* ... The deep front muscles like the psoas, pull the body out of line and the back muscles try to straighten the body. It is this ... attempt to right the body ... which causes constant back pain."[128]

The session facilitates a release of resistance, tension, and trauma in the pelvis, the negative pole of the torso. Through opening the centripetal field of the pelvis the session reestablishes healthy elimination and a receptivity to the finer forces of higher intelligence for clarity and well-being. This session often marks a turning point in the client's life where the past is released and healthy elimination established.[129]

The Pelvic Session

1. **Making Contact.** Hold hands or squat with client. Make eye contact and ask about client's physical and emotional well-being.

2. **Body Reading.** Dr. Stone writes: "Anterior rotation usually accompanies laterality. In most cases, the anterior *side of the body* is also the *anterior sacral base side*."[130] When the sacrum is distorted laterally there is usually a crease in the tissue on the high base side and a compensatory spinal curve on the opposite side. The inferior anterior sacral base is usually the contracted side with the low shoulder and the short leg. In chronic cases the center line between the buttocks is distorted as well as the inferior level of the buttocks.[131]

3. **Position Client.** Have the client lie on his or her back, head facing toward the north or northeast, with left hand on his or her heart.

4. **Quiet Heart Cradle.** Right hand over left, hands crossed under neck at occiput.

5. **Measure Legs.** Grasp both ankles, pull and rotate legs. Bring ankles together to assess short leg. Begin foot work on longer, more relaxed leg.[132]

6. **Ankle Rotation.** Rotate ankle vigorously in each direction. Hold sattvically between rotations. Stimulate pelvic reflexes around ankle as you rotate ankle.

    The outside of the heel near the ankle reflexes to the psoas, hip, kidneys, ileocecal, ovaries, and testes. The inside reflexes to the bladder, womb, prostate, sigmoid, and rectum.[133]

7. **Work Toes Rajasically.** Especially concentrate on the Water toes and pelvic reflexes in the base of the toes. Feel for crystals and work them out.

8. **Stretch and Iron Soleus and Achilles Tendon.** While holding foot in flexion stretch, iron down tendon with thumb, tamasically-focusing on juncture of soleus and Achilles tendon and on psoas and pelvic reflexes behind and under ankle. Hold sattvically at the end of each stroke in the contact.

    Excellent contact for lumbago and lower back pain. The soleus and Achilles reflex to the fifth lumbar vertebra and the sacrum.[134]

9. **Perineal Therapy.**

   a. **Sympathetic Nervous System Release:** Right thumb at ganglion of impar below coccyx. Using oil, left elbow slowly slides in groove between erector spinae muscles and spinus process. At end of each slide rock pelvis with elbow at sacrosciatic notch. Hold sattvically. Three slides on each side of the spine.

   b. **Parasympathetic Nervous System Release:**

      i. Have client lie on left side in a fetal position. With Ether and Air fingers of left hand make contact on spine at occiput. With Fire finger of right hand contact perineum between ischial tuberosities. Hold for release in perineum.

      ii. Move left hand to sternum, contact with thumb.

      iii. Move left hand to Fire center, stimulate rajasically. Hold sattvically until perineum releases.

      iv. Move left hand to knee.

      v. Move left hand to ankle. Hold for release.

      vi. Shoulder hip rock.

   c. *Optional:* **Deep Perineal Therapy.**

      i. Make contact with Ether and Air fingers at occiput with left hand. Contact perineum with Fire finger of right hand directly behind pubic arch alongside reproductive organs. Tone into contact and hold for release. Move down spine, toning at each vertebra. Halfway down spine move to other side of perineum and continue toning.

      ii. Move Air finger of left hand to heel and hold sattvically for release.

        "Perineal treatment . . . will unlock energy blocks quicker than most other methods because it deals with the vital force of emotional locks and frustrations."[135]

   d. *Optional:* A six-pointed star deepens the parasympathetic release.

10. **On Short Leg Side: Rock Rajasically Shoulder to Hip.** (Or six-pointed a star.) "Spastic muscles over the anterior superior spine of the ilium cause the anteriority of the innominate and the sacrum. The positive pole is the tension over the shoulder:"[136]

   Dr. Stone writes: "The upright pillar is the spine, and it must be the center line . . . [for] Sacroiliac Balance . . . Anterior muscle pull of the psoas is the *real culprit* . . . It is the first pull or lead that causes any distortion. The deep front muscles like the psoas pull the body out of line . . ."[137]

11. **Psoas Release.**

    a. Hold left hand and right hand together with fingers focused toward psoas area.

    b. Have client raise leg toward head against the resistance of the practitioner's hand on the femur. You call then feel the psoas move under your finger tips.

    c. Focus finger tips toward the psoas, between the colon and the ASIS (anterior superior iliac spine)

    d. Hold contact steady on shallow in-breath and follow down on deep out-breath. Tone. Slowly, over as many breaths as necessary, gently push finger tips toward psoas. At the end of each tone. Bring the client's awareness into the contact by asking: "What sensations are you experiencing? or What are you in touch with in your body?" A goal of therapy is to facilitate the experience that it is safe to be present, safe to be feeling, safe to be breathing and alive in the pelvis.

    e. In a few breaths you will feel the psoas under your finger tips. Hold contact until you feel heat released or psoas soften.

    f. Contact from psoas to psoas reflex behind heel on the outside.

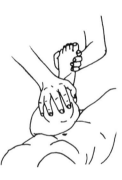

12. **Stretch of Femur in Acetabulum.**

    a. Upper hand on knee, lower on foot, lifting bent leg toward upper torso for a gentle stretch. Take care not to hyperextend the knee. Use the femur as a gentle lever to increase range of motion at acetabulum. On the out-breath the leg is circled into a wide rotation stretch, first in one direction, then the other.

    b. Using the femur as a gentle lever to increase the range of motion of the femur in the acetabulum, stretch first diagonally toward opposite shoulder, then to same shoulder.

13. **Innominate Stretch.** Place outside hand on knee, place inside hand on foot, bend knee, and position foot on opposite knee. On out-breath apply downward pressure on knee as a gentle stretch widening the pelvis. Use activation, if appropriate, to release tension. Have client breath in through pursed lips while lifting leg with twenty-five percent of strength against practitioner's resistance. Muscles let go on out-breath through pursed lips.

14. **Pubic Pop.** Realigns pubic arch. Have client bend knees and hold them together with one hand on each knee, have client pull legs together as the practitioner makes one quick effort to pull them apart.[138]

15. **Leg Pull to Align Pelvis.** Check for short leg. Work on short leg side. Grasp foot with thumbs under transverse arch and Air fingers pressing

against anterior tibia. On in-breath push leg toward torso. With out-breath stretch leg gently, pulling out "slack " in the ankle, knee, and acetabulum. Then tug to align pelvis. Check length of legs again.

16. **T.M.J. Release.** Air fingers contact T.M.J. as client yawns widely, then clenches jaw. Hold for a minute. Palpation of muscles of T.M.J. as a pelvic reflex.

17. **Balance Pelvis With Jaw.** Contact immediately behind pubic arch with Fire finger to Air finger at center of chin.

18. **Cradle of the Sages of Old.**

19. **Sit Client Up and Brush Off.** To break physical and psychic contact.

20. **Hug Client.**

21. **Wash Hands and Forearms in Cold Water to Ground.**

Summary of the Pelvic Session

1. Making contact.
2. Body reading.
3. Position client.
4. Quiet heart cradle.
5. Measure legs.
6. Ankle rotation.
7. Work toes rajasically.
8. Stretch and iron soleus and Achilles tendon.
9. Perineal therapy.
10. On short leg side: Rock rajasically shoulder to hip.
11. Psoas release.
12. Stretch of femur in acetabulum.
13. Innominate stretch.
14. Pubic pop.
15. Leg pull to align pelvis.
16. T.M.J. release.
17. Balance pelvis with jaw.
18. Cradle of the sages of old.
19. Sit client up and brush off.
20. Hug client.
21. Wash hands and forearms in cold water to ground.

Sacral Base Indicators[139]
Anterior/Inferior (AI) and Posterior/Superior (PS)

Sacral Base	Lateral (SB)	Lateral skin fold curve Gluteal lumbar base
AI		— on posterior side points to superior side
PS	on PS side	— —
	Atlas	**Occiput tension Mastoid pain with gluteal tension**
AI	lateral and posterior	— —
PS	—	on PS side on PS side
	Anterior side of body	**Shoulder**
AI	on anterior side	low (inferior SB)
PS	—	high (superior SB)
	Buttock with good tone	**Innominate**
AI	low (inferior SB)	PI (high iliac crest, low ischial tuberosity)
PS	high (superior SB)	AS
	Leg length	
AI	short (not a definitive indicator)	
PS	long (not a definitive indicator)	

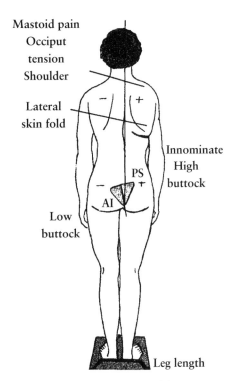

Mastoid pain
Occiput tension
Shoulder
Lateral skin fold
Innominate
High buttock
PS
AI
Low buttock
Leg length

Lateral distortion of the sacrum.[140]

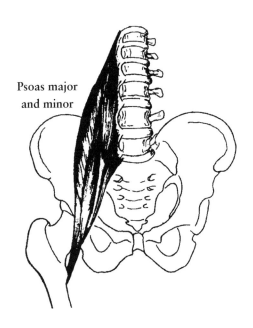

Psoas major and minor

Sacrum Session

The sacrum session balances energy to realign an anterior or posterior sacral base or a rotation of the pelvis. This session works to release tension in the perineum, glutei, innominate rim, and musculature of the back and shoulders, which affect the sacrum. The sacrum is the foundation of the spinal column and of structure in the entire body. Dr. Stone writes: "In the study of the entire bony structure of the body, the sacrum is the most vital and the most neglected bone."[141] "The sacrum is the key wedge between the innominate bones, like the keystone in an arch, between the two pillars on which this arch rests."[142] Structure supports from below upward. The inverted triangle of the sacral base is "the level weight bearing line of the entire spinal column and its structure"[143] "The sacrum is the opposite pole to the brain expressing action and structural balance."[144]

The sacrum session works with the alignment of the musculature of the back and the Earth's gravity lines to release the currents that flow from below to above and without to within, which support structure. The glutei are the negative field of the cerebrospinal nervous system. "DEEP perineal contacts also release the glutei tension. Place one hand on the perineum and with the other hand work out the soreness found all over the gluteal region. This polarizes the internal and external muscles, and balances the sensory and motor systems."[145]

The Sacrum Session

1. **Making Contact.** Hold hands or squat with client. Make eye contact and ask about client's physical and emotional well-being. Invite the client to set an intention for the session.

2. **Body Reading.** Dr. Stone writes: "Anterior rotation usually accompanies laterality. In most cases, the anterior side of the body is also the anterior sacral base side."[146] When the sacrum is distorted laterally there is usually a crease in the tissue on the high base side and a compensatory spinal curve on the opposite side. The inferior anterior sacral base is usually the contracted side with the low shoulder and the short leg. In chronic cases the center line between the buttocks is distorted as well as the inferior level of the buttocks.[147]

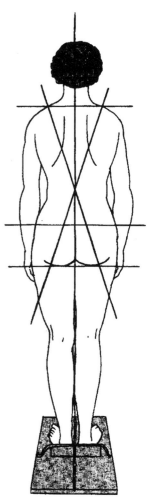

Perfect body Polarity and gravity lines on the test board.[138]

3. **Position Client.** Have the client lie on his or her back, head facing toward the north or northeast, with left hand on his or her heart.

4. **Quiet Heart Cradle.** Right hand above left, hands crossed under neck at occiput.

5. **Measure Legs.** Grasp both ankles, pull and rotate legs. Bring ankles together to ascertain short leg. Begin foot work focusing on longer, more relaxed leg.[149]

6. **Ankle Rotation.** Support ankle with outside hand. Grasp toes with inside hand. Rotate ankle vigorously in each direction. Hold sattvically between rotations. Stimulate pelvic reflexes around ankle as you rotate ankle. The outside of the heel near the ankle reflexes to: psoas, hip, kidneys, ileocecal, ovaries, and testes. The inside reflexes to the bladder, womb, prostate, and sigmoid rectum.[150]

7. **Work Toes Rajasically.** Especially concentrate on the Water toes and pelvic reflexes in the base of the toes. Feel for crystals and work them out.

8. **Stretch and Iron Soleus and Achilles Tendon.** While holding foot in flexion stretch, iron down tendon with thumb tamasically, focusing on juncture of soleus and Achilles tendon or on psoas or pelvic reflexes behind or under ankle. Hold sattvically at the end of the contact. Excellent contact for lumbago and lower back pain.

   The soleus and Achilles tendon reflex to the lumbar vertebrae and the sacrum.[151]

9. **Deep Perineal Therapy.**
   a. Have client lie on left side in a fetal position. With Ether and Air fingers of left hand make contact on spine at occiput. With Fire finger of right hand contact perineum between ischial tuberosities. Hold for release in perineum.
   b. Move left hand to sternum and make contact with the thumb.
   c. Move left hand to Fire center, stimulate rajasically. Hold sattvically until perineum releases.
   d, e. Move left hand to knee, inner ankle.
   f. Shoulder hip rock.

      "The foundation of all imposed strains is accumulated at the negative pole and must be released there, in the perineal floor."[152] "Perineal treatment . . . will unlock energy blocks quicker than most other methods because it deals with the vital force of emotional locks and frustrations."[153]

10. **Mobilize Sacrum.** Client on stomach. Support leg at thigh and apply traction, make at least three contacts along lateral margin of sacrum, releasing tension at sacroiliac.

"The sacrum is the opposite pole to the brain expressing action and structural balance. "[154]

"In all cases of *anteriority of the lumbar vertebrae and sacral base,* pillows or a bolster should be placed under the clients abdomen ..."[155]

11. **Buttock Clearing.** Client on stomach. Perineal contact with Fire finger of right hand under coccyx. Left thumb on glutei, coccygeal, and pyriformis muscles stimulating points along sacrum, hip, and iliac crest in a spiral fashion.

"Spastic muscles over the anterior superior spine of the ilium cause the anteriority of the innominate and the sacrum.... The positive pole is the tension over the shoulder ... [on the same side] *DEEP perineal contacts* also release the glutei tension. Place one hand on the perineum and with the other hand work out the soreness found all over the gluteal region. This polarizes the internal and external muscles, and balances the sensory and motor systems."[156]

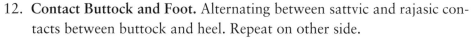

12. **Contact Buttock and Foot.** Alternating between sattvic and rajasic contacts between buttock and heel. Repeat on other side.

"The gluteal muscles must also be balanced with the feet, especially the sore spots on the soles and the tension in the toes, as well as the fixation of the transverse arch ... The feet are the final reflexes and very important ones connecting man with the Earth and gravity."[157]

13. **Six-Pointed Star.** Work each side from buttock to scapula and hip to trapezius. Focus contact on vertebral border and along the spine of the scapula.

"The posterior part of the shoulder is covered by the trapezius muscle which is supplied by the spinal accessory nerve and is one of the cranial parasympathetic reflexes." The upper portion of the glutei is also supplied by the three sacral parasympathetic nerves. These must be balanced with the upper parasympathetics ... of the neck region and the occipital area ... The sympathetic current must be coordinated with this, as the parasympathetic, emotional outlets can only be expressed when nature, the sympathetic nervous system, responds. This is the key to mental-emotional and functional treatment for frustration.[158]

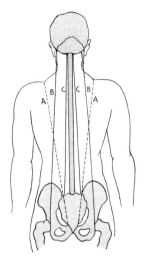

(Stone, Mysterious Sacrum)

14. **Sacrum Alignment Points.** Hold sattvic contacts with Ether and Air fingers of both hands.

"The sacrum is the opposite pole to the brain expressing action and structural balance."[159]

From C to C1 balances "the ocipito-atlas articulation with the sacral articulation in a horizontal relationship across the base."[160]

From A to Al balances the trapezius with the sacral base on each side and the sacral apex.

From B to B1 Balances the T.M.J. with the hip joint and innominate bone.[161]

15. **A/O (Atlanto-occipital) Release.**

16. **North Pole Stretch.**

17. **T.M.J. Release.** Air fingers contact T.M.J. as client yawns widely. Hold for a minute. Palpation of muscles of T.M.J. to stimulate jaw as a pelvic reflex.

18. **Quiet Heart Cradle.**

19. **Cradle of the Sages of Old.**

20. **Sit Client Up and Brush Off.**

21. **Wash Hands and Forearms in Cold Water to Ground.**

Summary of the Sacrum Session

1. Make contact.
2. Body reading.
3. Position client.
4. Quiet heart cradle.
5. Measure legs.
6. Ankle rotation.
7. Work toes rajasically.
8. Stretch and iron soleus and Achilles tendon.
9. Deep perineal therapy.
10. Mobilize sacrum.
11. Buttock clearing.
12. Contact buttock and foot.
13. Six-pointed star.
14. Sacrum alignment points.
15. A/O (Atlanto-occipital) release.
16. North pole stretch.
17. T.M.J. release.
18. Quiet heart cradle.
19. Cradle of the sages of old.
20. Sit client up and brush off.
21. Wash hands and forearms in cold water to ground.

Diaphragm and Heart Session

Introduction

A diaphragm block is a form of character armoring that reflects the ego's inability to integrate the feelings centered in the thoracic cavity and their expression through the solar plexus, pelvis, and thighs. Tension held in the diaphragm embodies our resistance to experiencing or expressing our feelings and our resistance to mobilizing energy in the solar plexus in order to act upon our feelings. Thus, this pattern of tension blocks the flow of energy to the solar plexus, cutting off vitality and wholeness. We tense the diaphragm because we are afraid to breath fully, afraid to be fully alive and feeling. The diaphragm session releases tension and improves circulation to the heart and solar plexus and is effective in releasing chronic anxiety and fear. Dr. Stone writes: "When the diaphragm is free, the heart is free to act without fear or apprehension."[162]

> The breath contains the life energy and goes to every cell. These impulses travel over the life waves and go very deep. Deep breathing with a concentrated idea distributes the impulse of the Pranic mental waves. A balancing of the three poles of the diaphragm can be accomplished by this therapy; the mental, emotional tension release in the upper areas; the head, neck, over the shoulders; the respiratory, emotional energy under the shoulders and at the diaphragm; and the vital emotional energy in the pelvis, sacrum and perineum.[163]

The diaphragm is the neutral field of the torso. Again, from Dr. Stone:

> The diaphragm is the main respiratory muscle doing the most important work in life. Every cell needs the life energy contained in the breath; without which they cannot survive. Its polarity function and minute distribution of fine energy waves is a most important factor to reach cellular tissue ... The diaphragm is the functioning neuter pole of life ... truly the diaphragm is the firmament which divides the energy (waters) above and below, it is the bridge where mind and life cross into the emotional vital field. Its rhythmic motion truly supports the energies above, lifts and activates the contents below.[164]
>
> A rigid diaphragm is the first and most vital block in nature's economy and flow of the four elements and their natural vital forces. Any treatment that releases the diaphragm is of the greatest value; however the

PERINEUM, AS THE NEGATIVE POLE [of the diaphragm], IS A LOCK THAT ALSO NEEDS TO BE *RELEASED AND POLARIZED AT THE SAME TIME.*[165]

The buttock lines posteriorly are extensions of the perineal line of the negative pole of the diaphragm.[166]

The quality of the breath is a foundation of our psychological well-being. Richard C. Miller, Ph.D., writing in *Yoga Journal* explains:

In abdominal breathing, during inhalation the abdomen extends outward and the diaphragm internally descends. These two actions create a drop in internal pressure that draws air into the lungs. During exaltation, the abdominal muscles contract and pull in slightly and the diaphragm ascends, forcing air out of the lungs. The overall effect is a depth and regularity of breath that allows for almost complete ventilation of the lungs. On the other hand, thoracic breathing—in which only the upper chest and the intercostal rib muscles participate—tends to be shallow rapid and irregular, with minimal exchange of air ... On the psychological level, diaphragmatic breathing increases ego strength, emotional stability, confidence, alertness, and perceived control over one's environment: thoracic breathing decreases them. Diaphragmatic breathing reduces anxiety, phobic behavior, depression and psychosomatic problems; thoracic breathing increases them.[167]

Breathing patterns are intimately tied to the body's emotional chemistry. A blocked diaphragm may contribute to anxiety, fear, and depression.

The Diaphragm and Heart Session

1. **Making Contact.** Squat with client. Make eye contact and ask about the client's well-being. Invite the client to set a healing intention for the session.

2. **Position Client.** Have the client lie on his or her back, head facing toward the north or northeast, with left hand over his or her heart.

3. **Quiet Heart Cradle.** Right hand over left, hands crossed under neck at occiput.

4. **Stretch Soleus and Achilles Tendon and Rotate Ankles Rajasically.** Stimulate diaphragm reflexes at anterior tibia at ankle.

    The resistance and range of motion of the ankles is a good measure of diaphragm tension. This is a powerful reflex to the duodenum and gallbladder, the emotional seats of the Fire principle.

5. **Work All the Toes Rajasically.** Pay attention to the webbing between the Ether and Air Toes which reflexes to the heart, brachial plexus, and diaphragm.[168]

6. **Work the Air and Fire Toes Tamasically and Rajasically.** Pay particular attention to the Air and Fire toes on the left side, which reflex to the heart. The first flange of the Air toe is a heart reflex.[169]

7. **Work Diaphragm Reflexes at Ankles.** Rajasically and/or tamasically.

8. **a. Iron Anterior Tibia.** Reflexes to the duodenum, gallbladder, and diaphragm. Dr. Stone writes: "The duodenum is the 'key' to opening the left side of the body. "[170]

   b. Iron Interspaces Between Tendons on Top of Foot. Excellent heart reflex, opens upper back and thoracic cavity.[171]

9. **a. Iron Transverse Arch and Ball of Foot.** The ball of the foot reflexes to the brachial plexus and diaphragm.

   b. Toning Ten Fingers on Ten Toes. Use the tone *Yam* which resonates with the heart chakra.

10. **Perineal Therapy.** The perineum is the negative pole of the diaphragm.[172]

11. **Scapula Release with Client on Left Side.** Support shoulder with inside hand. Work fingers of outside hand under scapula. On out-breath, lift scapula.

    Valuable in respiratory, heart, and nervous conditions.[173]

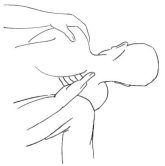

12. **Six-Pointed Star.**

    The positive field of the diaphragm is the shoulders. The neutral center is T-10, and the negative is the perineum, thighs, and the line of the perineum at the buttocks. Look for tender spots on the buttocks and thighs.[174]

13. **Scapula Release with Client on Right Side.** Support shoulder with inside hand. Work fingers of outside hand under scapula. On out-breath, lift scapula.

14. **Diaphragm Rock.**

    a. Thumb of right hand on anterior tibia, Air and Fire finger at ball of foot. Left hand makes contact at tense areas of diaphragm under rib cage. Alternate rajasic and sattvic between upper and lower contacts as client breathes into and releases tension at diaphragm. Use toning and activations to facilitate release.

b. Left hand makes contact at tense areas of diaphragm under rib cage, right hand works out tension in diaphragm reflexes at thighs near knees.

c. Left hand at clavicle right hand under diaphragm. Fingers radiating towards each other. Work contact rajasically and hold sattvically."[175]

d. Move to other side, switch hands and repeat a., b., c.

e. *Optional:* "The thumb of the right hand pushes under the sternum, at the tip of the xiphoid process, in an upward direction toward the left shoulder. Then it follows all along the floating ribs, moving little by little, until all spastic spots are released under the stomach."[176]

   This contact promotes the three vital functions of respiration, circulation and digestion.

15. **Diaphragm Release.** Fingers of right hand contact three inches below xiphoid process. Have client take deep breath, hold breath, and sit up (with your assistance through support with left hand at seventh cervical vertebra). The client sits up, collapsing forward, and releases breath. Practitioner's left hand moves behind T-10 and right hand follows the out-breath, moving deeply into the center of the diaphragm. Client breathes and tones into release of tension.

16. **Shoulder Rock.**

a. Place hands on shoulders, thumbs on trapezius, rock rajasically and hold sattvically.[177]

b. Hold one hand on shoulder/trapezius contact and move other downward diagonally to opposite hip, rock rajasically and hold sattvically.

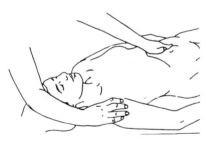

17.   **North Pole Stretch.**

18.   **Cradle of the Sages of Old.**

19.   **Sit Client Up and Brush Off.**

20.   **Hug Client.**

21.   **Wash Hands and Forearms in Cold Water to Ground.**

Summary of the Diaphragm and Heart Session

1. Making contact.
2. Position client.
3. Quiet heart cradle.
4. Stretch soleus and Achilles' tendon and rotate ankles rajasically.
5. Work all the toes rajasically.
6. Work the Air and Fire toes tamasically and rajasically.
7. Work diaphragm reflexes in ankles.
8. a. Iron anterior tibia
   b. Iron between tendons on top of foot.
9. a. Iron transverse arch and ball of foot.
   b. Toning ten fingers on ten toes (Yam)
10. Perineal therapy.
11. Scapula release with client on left side.
12. Six-pointed star.
13. Scapula release with client on right side.
14. Diaphragm rock.
15. Diaphragm release.
16. Shoulder rock.
17. North pole stretch.
18. Cradle of the sages of old.
19. Sit client up and brush off.
20. Hug client.
21. Wash hands and forearms in cold water to ground.

Joint Session

Dr. Stone writes: "Every joint is a polarized cross over point for energy waves." Centripetal sensory currents move from without to within, joint to joint, from the circumference back to the center. Every joint is a neuter crossover point of these polarized positive and negative currents.[178] "Each joint has a relationship to every other joint in the unity of energy flow of graceful motion."[179]

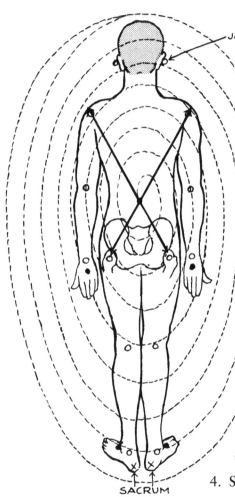

Dr. Stone points out that reactions are from below upward, with the inferior supporting the superior through return energy flow and gravity pull. This session balances sensory and motor currents in relationship to the Fire center at the umbilicus. Recommended for arthritis, hot flashes, and conditions where there is an excess of energy or Fire in the joints.[185]

(Stone, Evoutionary Energy Charts)

The Joint Session

1. **Making contact.** Squat or hold hands with client. Make eye contact and ask about the client's well-being. Invite the client to set an intention.

2. **Position Client.** Have the client lie on his or her back, head facing toward the north or northeast, left hand over his or her heart.

3. **Quiet Heart Cradle.** Right hand over left, hands crossed under neck at occiput.

4. **Stimulate Core Reflexes at F.L.H.**

5. **Work All the Toes Rajasically.** Pay attention to the webbing at the base of the toes.

6. **Joint Contact from Sole of Foot.** Thumb of right hand contacts the center of the sole of the foot. The positive middle finger of left hand is placed above it on a triangle line. Hold firmly for a half-minute or until energy flow is established. Thumb then moves upward to the place where the

CHART NO.59.   PRINCIPLES OF LOCAL WIRELESS CURRENT FLOW IN THE BODY

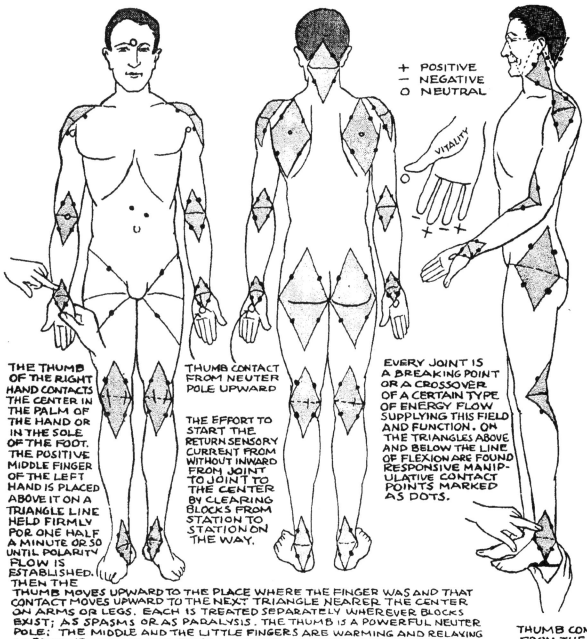

+ POSITIVE
− NEGATIVE
O NEUTRAL

VITALITY

THE THUMB OF THE RIGHT HAND CONTACTS THE CENTER IN THE PALM OF THE HAND OR IN THE SOLE OF THE FOOT. THE POSITIVE MIDDLE FINGER OF THE LEFT HAND IS PLACED ABOVE IT ON A TRIANGLE LINE HELD FIRMLY FOR ONE HALF A MINUTE OR SO UNTIL POLARITY FLOW IS ESTABLISHED. THEN THE

THUMB CONTACT FROM NEUTER POLE UPWARD

THE EFFORT TO START THE RETURN SENSORY CURRENT FROM WITHOUT INWARD FROM JOINT TO JOINT TO THE CENTER BY CLEARING BLOCKS FROM STATION TO STATION ON THE WAY.

EVERY JOINT IS A BREAKING POINT OR A CROSSOVER OF A CERTAIN TYPE OF ENERGY FLOW SUPPLYING THIS FIELD AND FUNCTION. ON THE TRIANGLES ABOVE AND BELOW THE LINE OF FLEXION ARE FOUND RESPONSIVE MANIP-ULATIVE CONTACT POINTS MARKED AS DOTS.

THUMB MOVES UPWARD TO THE PLACE WHERE THE FINGER WAS AND THAT CONTACT MOVES UPWARD TO THE NEXT TRIANGLE NEARER THE CENTER ON ARMS OR LEGS. EACH IS TREATED SEPARATELY WHEREVER BLOCKS EXIST; AS SPASMS OR AS PARALYSIS. THE THUMB IS A POWERFUL NEUTER POLE; THE MIDDLE AND THE LITTLE FINGERS ARE WARMING AND RELAXING AS POSITIVE ACTIONS. THE FIRST AND THE RING FINGER ARE COOLING AND TONIC FOR FLACID TISSUE AS NEGATIVE POLES.

THUMB CONTACT FROM THE NEGATIVE POLE UPWARD.

(Stone, *Wireless Anatomy*)

finger was and the Fire finger moves upward across the ankle, then knee, and hips. (See chart 59, Principles of Local Wireless Current Flow in the Body for pattern of movement.)

Dr. Stone writes: "The effort (functions) to start the return sensory current from without inward from joint to the center by clearing blocks from station to station on the way.[181]

7. **Joint Contact from Palm.** Thumb of right hand contacts the center of the palm. The positive middle finger of left hand is placed above it on a triangle line. Hold firmly for a half-minute or until energy flow is established. Thumb then moves upward to the place where the finger was, and the Fire finger moves upward toward the elbow, shoulder, T.M.J., and fontanel.

8. **Joint Contact from Sole of Foot.** Client on belly. Thumb of right hand contacts the center of the sole of the foot. The positive middle finger of left hand is placed above it on a triangle line across the ankle. Hold firmly for a half-minute or until energy flow is established. Thumb then moves upward to the place where the finger was, and the Fire finger moves upward toward the knee and then hip.

9. **Joint Contact from Palm.** Thumb of right hand contacts palm. The positive middle finger of left hand is placed above it on a triangle line. Hold firmly for a half-minute or until energy flow is established. Thumb then moves upward to the place where the finger was, and the Fire finger moves upward to the next triangle toward the elbow, shoulder, T.M.J., and fontanel.

10. **Chakra Balancing through Toning and Visualization.** Hold ten fingers on ten toes. Picture red/brown light radiating from the Earth chakra at the sacrococcygeal junction, as both client and practitioner tone using the seed mantra Lam. Picture crystal clear light, the color of the water in a bubbling brook, radiating from the lumbosacral junction while toning with the mantra Vam to balance the Water element. Picture a brilliant golden luster radiating from the Fire chakra in the spine behind the navel, while toning with the seed mantra Ram. Picture green, the color of new grass, emanating from the spine behind the heart as you tone Yam sweetly. Picture dark blue light radiating from the spine behind the throat as you tone Ham. Picture violet light radiating from the third eye as you tone Ksham. Picture diamond white light radiating from the crown as you tone Aum. Invoke/invite blessings from transcendental intelligence by honoring Mother Earth, the Angel of Water, Grandfather Fire, the Angel of Air, Father Sky, at each appropriate chakra.

11. **North Pole Stretch.**

12. **Cradle of the Sages of Old.**

13. **Sit Client Up and Brush Off.**

14. **Hug Client.**

15. **Wash Hands and Forearms in Cold Water to Ground.**

Summary of the Joint Session

1. Make contact.

2. Position client.

3. Quiet heart cradle.

4. Stimulate core reflexes at F.L.H.

5. Work all the toes rajasically.

6. Joint contact from sole of foot:

7. Joint Contact from palm.

8. Joint contact from sole of foot. Client on belly.

9. Joint contact from palm. Client on belly.

10. Chakra balancing through toning and visualization.

11. North pole stretch.

12. Cradle of the sages of old.

13. Sit client up and brush off.

14. Hug client.

15. Wash hands in cold water.

PART II

ESSAYS ON

ESOTERIC ANATOMY

CHAPTER FIVE

Polarity Therapy: A Transpersonal Approach to Natural Healing

The Soul is a unit drop of the ocean of Eternal
Spirit which is the dweller in the body as the
knower, seer, doer; it experiences all sensation and
action. It alone is the power in the body which
reacts to any mode of application of therapy or
action. Consciousness and intelligence reside
within the Soul.
—Dr. Randolph Stone[1]

Polarity Therapy is a comprehensive health-care system based on principles of natural healing and esoteric philosophy. Esoteric philosophy relates to higher knowledge which is hidden from ordinary understanding and revealed only to those who have evolved their consciousness through an earnest quest for deeper insight into the great mystery of life and sentience. Polarity Therapy is unique as a Western healing art in that it offers an illuminating transpersonal approach to healing, health building, and somatic psychology. Transpersonal is defined as transcending the personal or individual. Transpersonal refers to ancient esoteric schools of philosophy, psychology, and self-development which view individuality—in a certain profound sense—as not real.[2]

In recent years many traditions of transpersonal psychology have been brought to our attention in the West. Yoga, union with the higher self, has become a household word. Westerners are practicing a myriad of techniques of yogic Self-cultivation. Yoga psychology has been systematically formulated and practiced for millennia. Its efficacy for healing and for the acceleration of personal evolution has been replicated over thousands of years. In this essay, our goal is to make clear a model of esoteric psychology which underlies yogic practices. Our intention is to illuminate an ancient model for the practice of energy medicine and to underscore the appropriateness of Polarity Therapy for the practice of somatic psychology.

The Ancient Wisdom

The most lucid of the esoteric traditions which inspire the Polarity paradigm is the Sanatana Dharma, the timeless philosophy of ancient India. The Sanskrit term *Sanatana*, which is often translated as "eternal," means "prior to time." *Dharma*, which is often translated as "righteousness, religion, or law," comes from Sanskrit roots which mean "to uphold." It relates to a way of living that unites one with the Transcendental Self.

The Sanatana Dharma represents the largest, most coherent corpus of the ancient wisdom presently available. These teachings are said to have come out of a golden age, from a time before the great civilizations of Egypt, Babylon, Athens, and Rome. The Sanatana Dharma is what remains of an ancient system of higher knowledge, which in another age was the inheritance of the whole of humanity. Historian Will Durant points out that "India was the motherland of our race, and Sanskrit the mother of Europe's languages; she was the mother of our philosophy, mother, through the Arabs, of much of our mathematics, mother ... of the ideals embodied in Christianity, mother, through the village community, of self-government and democracy. Mother India is in many ways the mother of us all."[3]

The Sanatana Dharma insists that it is not a religion, but is: "the holder of a tradition common to all men, encompassing all that revelation and men's effort have produced in terms of knowledge."[4] Sages of the East worship the Sanatana Dharma as the culmination of human knowledge. In many quarters the Sanatana Dharma is venerated as ultimate truth.[5]

The Sanatana Dharma has been preserved in the *Vedas* (from Sanskrit *Vid*, "to know") which are the oldest known Indo-European religious and philosophical writings.[6] While Western scholars date these teachings to a time thirty to fifty centuries past, tradition emphasizes their revealed and eternal character. What is most remarkable about these primordial teachings is that they have not been lost in the rubble of history but have been preserved with profound integrity as a contemporary esoteric tradition. While: "the Pyra-mids have been eroded by desert wind ... broken by earthquakes ... stolen by robbers ... the *Vedas* remain, recited daily by an unbroken chain of gen-erations, traveling like a great wave through the living substance of the mind."[7]

The final teachings of the *Vedas* is called *Vedanta. Veda* means "wisdom" and *anta,* "the culmination or essence of wisdom." The German philosopher Schopenhauer attests: "In the whole world there is no religion or philosophy so sublime and elevating as Vedanta. This Vedanta has been the solace of my life, and it will be the solace of my death."[8]

Aldous Huxley, a seminal thinker in the contemporary consciousness movement, asserts that there is a timeless truth revealed crossculturally and transhistorically through mystical contemplation. He refers to this wisdom as the Perennial Philosophy. Huxley writes: "In Vedanta and Hebrew prophecy, in the Tao Teh King and the Platonic dialogues, in the Gospel according to St. John and Mahayana theology, in Plotinus and the Areopagite, among the Persian Sufis and the Christian mystics of the Middle Ages and the Renais-sance ... Hindu, Buddhist, Hebrew, Taoist, Christian, or Mohammedan, were attempting to describe the same essentially indescribable Fact." Huxley goes on to introduce *The Bhagavad Gita* (The Song Celestial), an ancient Gospel cherished by many as the quintessence of the Sanatana Dharma. As Huxley continues, *The Bhagavad Gita* "is one of the clearest and most comprehen-sive summaries of the Perennial Philosophy ever to have been made."[9]

The Bhagavad Gita offers an understanding of The Great Mystery which has inspired many of the giants of the Western intellectual tradition. The tran-scendentalist poets Emerson, Thoreau, and Whitman were students of *The Bhagavad Gita.* Thoreau wrote: "In the morning I bathe my intellect in the stupendous cosmological philosophy of *The Bhagavad Gita,* in comparison with which our modern world and its literature seem puny and trivial." Emer-son, referring to the *Gita,* wrote: "It was the first of books. It was as if an empire spoke to us, nothing small or unworthy, but large, serene, consistent, the voice of an old intelligence which in another age ... had pondered and thus disposed of the same questions which exercise us."[10]

Advaita: a Vision of Oneness

The wisdom of the Sanatana Dharma has its foundation in the principle of *advaita.* This Sanskrit term means "not two." The doctrine of advaita asserts an identity of all-consciousness, a vision of the universe as one living con-scious being.

Advaita reveals that beyond the endlessly changing mind, emotions, intellect, and body is the unchanging awareness of subjective consciousness. Advaita points out that our sense of self refers to this immutable witness. This presence of living consciousness, which is our essential nature, is called the Self. "What then is the reality behind all our experiences? There is only one thing that never leaves us—deep consciousness. This alone is the constant feature of all experience. And this consciousness is the real, absolute Self."[11]

Advaita asserts that the consciousness of each individual is a portion of one universal consciousness, the *Atman* (from the roots *an*, "to breathe," and *manas*, "to think"), which is a portion of the Divine manifesting as individual awareness. "The *Atman* or Self is all-pervading, immense, the ground and agent of knowledge. It is thus the seat of consciousness."[12] Advaita postulates that the subjective awareness, the Self, the cognizance, the intelligence in each

being is a portion of a single, transpersonal, timeless, immutable, omniscient, universal consciousness called *Brahman*. The identity of Brahman, the Supreme Being, and Atman, the portion of the Ultimate supporting individual life, is the basic truth of the Sanatana Dharma. The revered sage Shankara (A.D. 686) explains: "The wise men of true discrimination understand that the essence of both Brahman and Atman is Pure Consciousness, and thus realize their absolute identity. The identity of Brahman and Atman is declared in hundreds of holy texts."[13] The ancient esoteric texts, *The Upanishads*, declare: "Of everything, God is the innermost Self."[14]

Shankara continues:

> Just as a clay jar or vessel is understood to be nothing but clay, so this whole universe, born of Brahman, essentially Brahman, is Brahman only—for there is nothing else but Brahman, nothing beyond That. That is the only Reality. That is our Atman. Therefore, "That art Thou"—pure, blissful, supreme Brahman, the one without a second.[15]
>
> There is no higher knowledge than to know the Atman is one, and everywhere.[16]
>
> The wave, the foam, the eddy and the bubble are all essentially water. Similarly, the body and the ego are really nothing but pure consciousness. Everything is essentially consciousness ... This entire universe of which we speak and think is nothing but Brahman.[17]

Advaita asserts an identity between the Supreme Being, *Paramatman*, and the individual Self, Atman. R. C. Zaehner, Professor of Eastern Religions at Oxford University, writes: "When the mystic says 'I am Brahman,' he is

The revered sage Shankara (AD 686). At the time of Shankara's birth, India was under the influence of Buddhism. In a short life of thirty-two years, Shankara traversed the Indian subcontinent, engaged in debate with Buddhist scholars. His influence moved scholars from the nontheism of Buddhism to the worship of the Supreme Being.

speaking the literal and only truth; the source and ground of the whole universe of appearances is identical with the innermost essence of man. There is nothing else."[18] The ancient wisdom invites us to become a mystic, to know the Self, for without this knowledge everything else is ignorance. The foundation of the ancient wisdom is Self-realization, the knowledge of the Transcendental Self. The sage Shankara declares: "There is no higher knowledge than to know that the Atman is one and everywhere."[19] The ancient sages harmonized their bodies and quieted their minds to realize the Transpersonal Self. "Brahman is all—this universe and every creature."[20]

In the scripture, *The Bhagavad Gita,* the Avatar (divine incarnation) Sri Krishna reveals: "My Prakriti [natural phenomena] is of eightfold composition; earth, water, fire, air, ether, mind, intellect and ego. You must understand that behind this, and distinct from it, is That which is the principle of consciousness in all beings, and the source of life in all. It sustains the universe."[21]

The venerated sage Sri Ramana Maharshi explains: "Since, however, the physical body cannot subsist [with life] apart from Consciousness, bodily awareness has to be sustained by pure Consciousness. The former, by its nature, is limited and can never be co-extensive with the latter, which is infinite and eternal. Bodily-consciousness is merely a monad-like, miniature reflection of pure Consciousness ... body-consciousness is only a reflected ray, as it were, of the self-effulgent infinite Consciousness."[22]

Advaita reveals a profound vision of the sacredness and unity of all life. The universe is understood to be a unitary field of consciousness. God did not merely create the world, God became the world. God, as ultimate consciousness, manifests this world as a tiny portion of Being.[23] The universe, solar system, nature, and our bodies are a single conscious living being. Atman, the living consciousness of the individual, is a drop of the ocean of consciousness of Brahman, the Supreme Being. This Supreme Being manifests the body and the world as a vehicle to express its creative potential. In this ancient paradigm, all of nature is a living Being, ever seeking greater consciousness, creative unfoldment, and freedom. Each being is sacred as nothing less than a leaf on The Tree of Eternal Life. The living intelligence within each of us, the sentience, the presence, the cognizance, the awareness, the sense of "I am," is nothing less than the personification of the Great Mystery. Advaita offers a profound monism, a vision of one God ... who is the one life and one consciousness of all beings.[24]

There was never a time when I did not exist,
nor you ...
Nor is there any future in which we shall cease
to be.
—Sri Krishna[25]

Spirit and Matter

The Sanatana Dharma reveals that we exist in two dimensions—a dimension of Spirit and a dimension of matter. Our five senses are exclusively attuned to the dimension of matter. Spirit, which transcends matter, is the source of our consciousness and aliveness. The body is an emanation of Spirit. It is the body's attunement with Spirit that sustains life.

Derived from the Latin *spirari,* meaning "to breathe," Spirit is the breath of life. The body is lifeless and unconscious without its presence. In Sanskrit the word Atman, the portion of the Divine which supports individual existence, is also from the verb root *at,* meaning "to breathe."[26] When the body falls out of harmony with Spirit, we are left with a corpse which begins to decompose immediately. Without Atman, which in the West we call the Soul (the animating force), the body is seen as something unclean and decaying, to be disposed of as soon as possible. It is only the presence of Spirit that gives this mortal coil value.

In the Sanatana Dharma, Spirit, the animating force in the body, is transpersonal and exists in a timeless dimension beyond the body and outside the cycle of birth, death, and decay.[27] The spiritual dimension is an all-pervasive field of living, breathing, cosmic consciousness, beyond description and beyond change. Spirit is immutable. Dr. Stone referred to Spirit as The One Life.

In *The Ultimate Medicine,* the revered twentieth-century sage Sri Nisargadatta (Mr. Natural State) reveals: "Treat the life force as God itself, there cannot be consciousness without life force. So consciousness and life force are two components, inextricably woven together, of one principle ... Once you consider that the life force is God itself, and that no other God exists, then you raise the life force to a status enabling it, together working with consciousness, to give you an understanding of the working of the whole."[28]

Sri Nisargadatta, author
of The Ultimate Medicine

The Body as an Epiphenomenon of Consciousness[29]

By the single sun,
This whole world is illumined:
By its one Knower,
The field (of creation) is illumined.[30]

The Sanatana Dharma offers us a transpersonal paradigm. The universe is one living sentient being. In *The Bhagavad Gita,* Sri Krishna explains: "Know this body to be the 'field' and know me to be 'the Knower' of the field. Know also that I am the sole knower of all fields, of all bodies."

A contemporary student of *The Bhagavad Gita,* Mathaji Vanamali, explains:

> The whole of physical existence must be considered as a field of the Spirit's constructions. The "field" is the object upon which consciousness operates. The "field" in the individual is the body, mind, intellect and senses; and universally, it is the whole of cosmic manifestation. The "knower of this field," both individually as well as in the universe is the Lord Himself, in each embodied creature, there is only one and the same "Knower" ... one and the same consciousness. But its knowledge of the world differs because it is deflected by the consciousness passing through the limited body. In the human being this seemingly small consciousness, which is really an all consciousness, can enlarge itself to its universal size. Physically all of us are the microcosm in the macrocosm, and the macrocosm, also, is a body or "field" inhabited by the same spiritual "Knower." The totality of Nature, as we behold it, is the field of action, of the consciousness or Spirit.[31]

In the cosmology of the ancients, all creation is a play of the consciousness of One Being. Creation is a moment-to-moment appearance within this One Transpersonal Being. Dr. Deepak Chopra, referring to Ayurveda, the ancient healing science based on the Sanatana Dharma, explains: "Ayurveda's great contribution was to see the physical body as an epiphenomenon of consciousness. Consciousness shapes, controls, and even creates matter."[32] "The body, the vital energy, the sense organs, the mind, the intellect and the ego—these are the coverings of the Atman."[33] All the material energy fields of the body are emanations from this innermost yet all-pervasive center whose presence is the sentience and awareness in the body.

At the nucleus of every atom, molecule, and cell of the body is an energy field of a higher potential which is in tune with the spiritual realm. This is the animating principal of the body, the Soul.[34] The physical field is a radiation into time and space from the nucleus of the Soul. Matter and energy are an emanation of Spirit.[35] Creation is a field of consciousness. Matter is a crystallization of consciousness. The ancient wisdom offers us a sacred science. The laws that rule the crystallization of spirit into matter are universal.

The Tantric scholar Arthur Avalon (Sir John Woodroffe) reveals: "The ultimate or irreducible reality is `Spirit' in the sense of Pure Consciousness ... from out of which as and by its Power ... Mind and Matter proceed. Spirit is one. There are no degrees or differences in Spirit. The Spirit which is in man is the one Spirit which is in everything ... Mind and Matter are many and of many degrees and qualities."[36]

Know me, eternal seed
Of everything that grows ...
I am the essence of waters,
The shining of the sun and moon:
ॐ *(Aum) in all the Vedas,*
The word that is God.
It is I who resound in Ether
And am potent in man.
I am the sacred smell of the earth,
The light of the fire,
Life of all lives
—*Sri Krishna*[37]

God Becomes the World

S. Radhakrishnan, distinguished philosopher and former Prime Minister of India, explains: "God does not create the world but becomes it. Creation is expression. It is not so much making as becoming. It is the self-projection of the Supreme."[38]

The universe and nature are aspects of the Being of God. "God" as consciousness, *Satchitananda* (existence/knowledge/bliss), is wholly present as every level of creation. According to the Sanatana Dharma, God is the very stuff, the actual essence, of each and every stage/level. God, or *Brahman*, is not the highest level, nor a different level itself, but the reality of all levels.[39] "Brahman is Existence, Knowledge and Bliss; but these are not attributes. Brahman cannot be said to exist. Brahman is existence itself. Brahman is absolute and infinite—the eternal, immutable, illimitable, omniscient, formless Great Mystery: Being absolutely present, Brahman is within all creatures and objects."[40] Brahman is the Being in human beings. In the embodied creature, this Brahman is known as Atman.

> *Know this Atman, Unborn, undying,*
> *Never ceasing, Never beginning,*
> *Deathless, birthless, Unchanging for ever.*
> *… Knowing It birthless, Knowing it deathless,*
> *Knowing it Endless, Forever unchanging*
> *… Innermost element, Everywhere, always,*
> *Being of Beings, Changeless, Eternal,*
> *For Ever and Ever.*[41]

Although Brahman, the Absolute itself, is indescribable and beyond qualities, its creative potency is called *Brahma*, the Evolver, Creator, or Emanator. Brahma, derived from the Sanskrit *brih*, means "to grow," "to burst forth." It is the ever-expanding breath whose rhythm is the energy of all creation.[42] The nature of Cosmic Intelligence is to universally unfold its creative potential as growth/evolution. All the systems and subsystems of the cosmos, nature, and life processes are the ever-expanding breath of Brahma. This Cosmic Breath is the primal utterance, the divine Logos, the creative word, whose voice is perpetually creating the universe. All vibration on every level is a part of the unified field of the breath of Brahma. Our own breath is a microcosm of the evolutionary force of the macrocosm. As the life-breath or *prana* animates the body, the breath of Brahma is the energy which animates a conscious living cosmos. [43]

In the Sanatana Dharma, Brahma (the universe) is an ever-changing field of creative evolution. Brahman (the Absolute) is the unchanging omnipresent Knower of the field. In each embodied creature, there is only one and the same knower, one and the same consciousness. Brahman is the one awareness, the Self of everyone: "That which is the principle of consciousness in all beings, and the source of life in all."[44]

Brahma, the personification of the Principle of Creation, sits on his divine vehicle Hamsa, the highest flying swan of ham (out-breath) and sa (in-breath).

Healing Is Spiritual

The body is sustained by its attunement with the Soul. The Soul is that which is actually present, cognizant, conscious in each of us as the Self. Healing is a process of attunement to the Self. We experience this attunement to higher consciousness as lucidity, positivity, inspiration, presence, and joy. All health practices promote this attunement to the Life Field, to the healing presence of the Soul. Dr. Stone explains: "True health is the harmony of life within us, consisting of peace of mind, happiness and well-being. It is not merely a question of physical fitness, but is rather a result of the Soul finding free expression through the mind and body of that individual. Such a person radiates peace and happiness, and everyone in his presence automatically feels happy and contented."[45]

Meditation: The Healing Process of Self Realization

In every energy system there must be a fulcrum, a center of movement, or there is chaos. The Self is the natural center of our Being. It exists within each of us, as the source of our life and our consciousness. The Self is intelligence: a sentient, cognizant, self-organizing force. The body, mind, and senses emanate from this innermost Intelligence.

We suffer from a chaos of identification with every object of the mind and senses, from a lack of connection to our Source. It is profoundly important to cultivate a relationship with the Self on a daily basis to clear the chaos of the mind. Relating to the fulcrum of Being offers us the opportunity to find a stable center for healing our lives.

The Self is pure Being. It exists in a higher realm than the mind and five senses. It exists outside of time and space. Thus it cannot be grasped directly by the mind and senses. Yet it is always present as the ground state of our consciousness. The Self is the actual cognizer and intelligence which is present from moment to moment as our awareness. But our minds are too disturbed by chaos and obsession for us to experience this truth. Soul communion purifies the mind. The potency of the Unitive Field harmonizes the lower vibrational fields of the body and mind. We can commune with the Self through the creative power of our intention. Through a passionate desire to know the Self, we can build a vehicle within our lives that is receptive to our innermost essence. Daily practices such as contemplation and worship exalt our consciousness and foster the integration of the Self into our lives. The clarity and inner peace that one experiences through meditation and yoga offer this higher knowledge as does devotion and the rapture of worship. Gratefulness, service to life, and love of the Mystery all facilitate Soul communion. A personal experience of Divine Love can become the center of our lives and the basis of a genuine sense of Self-love and Self-esteem.

Contemplation of the Higher Self

Let go of any expectation of having a phenomenal experience. Union with the Self is supersensory—beyond the mind and senses. The process is one of evolving your consciousness through a day-by-day, organic process of Self-realization. We make the Self real in our lives by creating the space to be with the Self. We actualize our Higher Self by investing our lives in it. The sages tell us: What your mind contemplates, you become! Every thought is a prayer. You build your world through your moment-to-moment thoughts, feelings, and desires. Passionately desire to know the Self. Take the energy you have invested in obsession with your pain and suffering and refocus it into a passion to be free. Move the energy which has been invested in the distractions and trivialities of life to an earnest quest for freedom.

The eyes are the windows to the Soul. We can catch a glimpse of the Light of Life, the Presence, through the eyes. Many seekers have made a leap in consciousness through the simple practice of mirror work. It is valuable to take a few minutes each day, sitting in front of a mirror at eye level, to practice the classic "Who am I?" contemplation, given to us by the great twentieth-century sage Sri Ramana Maharshi.

Soul Communion: The Ultimate Medicine[46]
Who Am I?
I have a body, but I am not my body.
I can see and feel my body,
and what can be seen and felt is not the true Seer.
My body may be tired or excited,
sick or healthy, heavy or light,
but that has nothing to do with my inward I.
I have a body, but I am not my body.

I have desires, but am not my desires.
I can know my desires,
and what can be known is not the true Knower.
Desires come and go, floating through my awareness,
but they do not affect my inward I.
I have desires, but I am not desires.

I have emotions, but I am not my emotions.
I can feel and sense my emotions,
and what can be felt and sensed is not the true Feeler.
Emotions pass through me,
but they do not affect my inward I.
I have emotions, but I am not emotions.

I have thoughts, but I am not my thoughts.
I can know and intuit my thoughts,
and what can be known is not the true Knower.
Thoughts come to me and thoughts leave me,
but they do not affect my inward I.
I have thoughts, but I am not my thoughts.

I am what remains, a pure center of awareness,
an unmoved witness of all these
thoughts, emotions, feelings, and desires.

CHAPTER SIX

The Theory of Theomorphic Resonance

Energy is the real substance behind the
appearance of matter and form.
—Dr. Stone[1]

Polarity Therapy seeks to understand health and healing from the point of view of energy. By *energy* we mean the animating force in nature, the source of life, growth, and healing. It is valuable to appreciate that we could easily substitute the word "Soul" or "God" for the word energy. Energy is a great mystery. Energy is everywhere. Energy is the medium of the universe, but we do not "know" where it comes from. Energy cannot be created or destroyed, it can only transform itself. Our knowledge of energy is merely a description of this process of transformation. In the following we will be unfolding a paradigm for research in understanding energy in the healing arts based on the transpersonal cosmology of the Sanatana Dharma.

In Chapter 5, the first essay, we outlined the transpersonal perspective of the Perennial Philosophy. We underscored the axiom of the identity of all consciousness, sharing the realization of the ancient rishis that individual cognizance is transpersonal and that all life and consciousness are one. The ancient esoteric texts, the *Upanishads,* profess that Brahman the Godhead is the innermost Self of everything. In the Sanatana Dharma, creation is a moment-to-moment emanation from the Self, the unified field of Cosmic Intelligence. All creation is the living body of this one Spirit.

In the Sanatana Dharma God becomes creation. All of nature emanates from the one unified field of Ultimate Intelligence. In the ancient wisdom this intelligence *is* the energy of creation. The creative potency/evolution/growth expansion of this consciousness *is* the universe. The ancients understood the universe to be one living conscious being. All creation emanates from the One Life. All life and all consciousness are one. In the ancient wisdom the universe is an ever-expanding Creative Intelligence.

The theomorphic principle is central to The Sanatana Dharma and a foundation of our understanding of energy. Theomorphic (*theos*, meaning, in Latin and Greek, "God," and the Greek *morphe,* meaning "form") translates as the form of God. A theomorphic perspective avers that we are made in the image of creation. In the theomorphic paradigm every aspect of creation is a reflection of the process of creation. In a theomorphic perspective we assert that in nature all energy moves in dynamics which embody the process of creation.

The Ved

In the Sanatana Dharma, God as Ultimate Intelligence becomes creation. The substance of the universe is this Ultimate Intelligence. Every atom of creation is made of the One Divine Living Consciousness. The ancients expressed the theomorphic principle in their formulation of *Ved,* the holographic knowledge which manifests the Universe.[2] The Ved is the seed of Creation, just as DNA contains the "knowledge" to manifest every aspect of a life form. The Ved contains the "Omniscience/Knowledge" to manifest universes. In the involutionary phase of creation Ved unfolds creation. At dissolution, creation spirals back into the Ved. Like the DNA that is present in all life, the Ved is omnipresent in all creation, and every level of creation is a microcosm of Ved. Indeed what contemporary scientists call "DNA" on the microbiological level may be understood to be a universal phenomenon paralleling the Ved by future researchers. The creative potency of this holomovement is a quality of the Ultimate Intelligence, which is the medium of creation.[3]

The Sanskrit word *ved* comes from the Sanskrit root *vid,* "to know." In this case, knowledge is nothing less than the omniscience of an omnipresent God whose holographic consciousness is the medium of creation. All life and consciousness emanate from the unified field of Ultimate Intelligence, Brahman.

Robert Keith Wallace, Ph.D., a leading quantum physicist and student of Ayurveda, offers a synthesis of Vedic science and quantum physics in his book *The Physiology of Consciousness.* He explains the teachings of the contemporary sage, Maharishi Mahesh Yogi, about the ancient understanding of the *Ved,* the knowledge of pure consciousness. Dr. Wallace writes:

Maharishi's Vedic Science describes the self-interacting dynamics of the unified field as the eternal dynamics of consciousness knowing itself. From this self-interaction within the field of pure consciousness, the laws of nature begin to unfold.[4]

Transformations [of pure consciousness] unfold in a precise, perfectly orderly sequence. It is this sequential unfoldment, Maharishi explains—the mechanics of transformation of unity into diversity—that forms the fundamental processes of creation. These fundamental rules of transformation are repeatedly seen at every level of creation. Through these mechanics are created all the laws of nature that structure and govern the diverse forms and phenomena in the entire universe. It is this fundamental structure of natural law unfolding within pure consciousness, and creating the universe, that is the Ved.[5]

In the Sanatana Dharma the Ultimate Ground of creation is pure consciousness as a unified field of Ultimate Intelligence. Dr. Wallace concludes: "In the Vedic paradigm, the unified field is therefore not only the source of matter, but also—because it is pure consciousness—the source of mind. It is the common source of mind and body, of subjective experience and material creation."[6] In the ancient wisdom the body/mind of life emanates from the unified field of omniscience and omnipresence of the creative potency of Universal Consciousness.

In the ancient wisdom life emanates from the omniscience and omnipresence of the creative potency of Atman, the Self. The sage Shankara declares the divinity and identity of all consciousness as nothing less than Brahman, the Godhead! Transpersonal psychologist Ken Wilber refers to this ultimate ground of Cosmic Consciousness as the "Unitive Field," the field that unites.

Dr. Stone points out that "consciousness and intelligence reside within the soul, the animating force of the body. Stone refers to the Supreme Being as the "One Life."[7] Remembering the profound insight of Sri Nisargadatta Maharaj quoted in the first essay, that consciousness and life are inextricably bound, we will refer to this ultimate ground of the oneness of sacred living consciousness from which creation arises as the Life Field.

The omnipresent holomovement of the Ved is the foundation of the harmony of natural law. All energy in the universe and nature holographically contains the creative potency of the Supreme Intelligence expressing its nature as the Ved. All of the energy of nature shares this innermost core of living intelligence. All of nature emanates from the Unitive Field of Ultimate Intelligence. This is the center of the theomorphic basis of all creation.

The Word: The First Born

In the beginning was the Word,
The word was with God . . .
The word was God . . .
             —John 1:1

In the Sanatana Dharma the energy of the universe is the Word of God. All energy in creation moves as the breath of Brahma. Brahma, whose nature is ever-evolving creativity, expresses that potential through the "word." The Creative Intelligence and creative will of Brahma is potent in every spiral of manifestation, just as your intelligence and will permeate your words. The intelligence and will of God impregnates the "Logos." Like the Greeks, the Hindus understand that divine reason/wisdom/intelligence is the controlling principle of the universe.

The Sanskrit word for universe is *jagat*, which can be translated as "continually moving." Though the universe as Ultimate Consciousness is unmoving, its creation is a paradoxical opposite of unceasing change. In this ancient paradigm all movement is the breath of Brahma. The rhythm of this cosmic breath underlies the vibratory and cyclical nature of all creation. Contemporary philosopher Alan Watts explains: "the Hindu world-view is that the whole universe of multiplicity is the *lila* or play, of a single energy known as *Paramatman*, the Supreme Self. The coming and going of all worlds, all beings, and all things is described as the eternal out-breathing and in-breathing of this One Life . . . "[8]

The intelligent will of the Cosmic Being is expressed in this breath as the Sacred Word of creation. In this ancient cosmology, creation is understood as being fundamentally composed of sacred sound/vibration.[9] The *Anahad Shabd*, called in the West, the "Lost Chord," the "Word of God," or "Logos," is the nonvibratory sound of the Unitive Field of Brahma.[10]

The Life Breath of the Creator is the Holy Word. Mystics have understood the "breath of Brahma" as the Logos or Divine Word whose song is the all-pervasive field of sacred sound, the mother of all vibration in the universe.[11] Dr. Stone writes:

> Sound, as supersonic and ultrasonic energy, is the first principle in nature. 'God spoke' Sound, not light, is the first principle of whirling, fiery action, as wheels of energy in space, creating rivers of whirls, entities, bubbles

and individualizations from the All-ness or the One, in its centrifugal primal outpouring. Sound, as ultrasonic energy, is the creative principle and is the cause of all creations. This Sound Energy or The Word . . ." was in the beginning with God, and was God, and by it all things were made that were made.[12]

All energy is an echo of the primordial (*Aum*) an emanation from Brahma. In the Sanatana Dharma,[13] creation is *The Bhagavad Gita,* The Song of God. The Divine Being "lives" through all that moves. All vibration is His becoming. In the paradigm of the ancients, all creation as vibration is a sacred mystery of God's creative will.

Prana, the life force in the body, steps down from *Sabda-Brahman* (the Divine as primal sound power) through the *pranava,* the unified field of sound vibration, which is the medium of the life of *a living universe.* Brahma is the inner potency of vibration, the omnipresent center of all manifestation, the resonance of Creative Intelligence at the nucleus of every atom of creation. The phenomenal world is an emanation from this primary creative ground. All vibration in the material world aligns itself with the vibration of the Unitive Field within creation. The resonance of material phenomena with this unified field of breath, pulsation, oscillation, and vibration is the reflective identity between above and below, macrocosm and microcosm. The harmony of this holomovement is universal law.

Brahma is the principle of growth, of conscious becoming. Brahma is the power of evolution reflected in all natural processes. Brahma is an ever-expanding cycle of conscious breath whose force encompasses all energy in creation. It is a universal system of energy. All the energy systems in the cosmos and in nature are subsystems within the universal system and align themselves with the holomovement of this unified field of vibration.

Brahma is the potency of Creative Intelligence[14] which is the ground of all being. The Sanatana Dharma offers us an evolutionary model in which all creation is an expression of an ever-evolving universe. In nature this evolutionary force manifests as the ubiquitous processes of *growth*. All energy in the cosmos is the creative potency of evolution. In the Sanatana Dharma, the Creator is not in some heavenly realm in the sky but is an all-pervasive, ever-evolving Creative Intelligence.

The Dance of Shiva

In profound states of yoga (union with the Self) the ancient rishis (seer-sages) directly experienced the mysteries of the universe. They used the techniques of yoga to attune the body and concentrate the mind to facilitate a direct experience of the higher states of consciousness of the Self. In ecstatic communion they witnessed the "dance of Shiva."

In the Sanatana Dharma the breath of the universe is expressed as two universal forces: Shiva and Shakti, in-breath and out-breath. All energy in creation is a spiraling vortex of Shiva and Shakti, centripetal and centrifugal forces. Universally, energy spirals as a whirling vortex which resonates as a womb to manifest Divine creative will. The in-breath and out-breath of Shiva and Shakti, centripetal and centrifugal, are the electromagnetic fields undulating through every spiraling atom of creation.[15]

On the atomic level all creation is the passion of the Gods. Shiva (centripetal) and Shakti (centrifugal) undulate as proton and electron around the nucleus of Brahman, just as our words spiral out in the creative vortex of the breath and are an expression of our intelligence and will. All creation is pregnant with Sat-chit-ananda, the Being-Intelligence-Will of God.

At the center of every undulating cycle of vibration is a stillpoint where Shiva and Shakti unite. Centripetal and centrifugal become one! This point of equilibrium—neutrality, emptiness—is the bindu, the womb of creation. This bindu gives birth to the next octave of evolution. From the whirling galaxies above to the spinning atoms below, all creation is a spiral helix, the "Dance of Shiva." All vibration is theomorphic and embodies the process of creation.

As above, so below ... humans, following the theomorphic principle, personify the process of creation. Our upper lip is Shakti (centrifugal), and the lower lip is Shiva (centripetal). As we speak, breath whirls between our lips as a centrifugal/centripetal vortex of energy. This vortex of breath is the womb of creation. Our words are pregnant with our intelligence and will. Our breath carries the vibrations expressing our intention into the creative process. Our living is a microcosm of the creative process of the Godhead. We are this Ultimate Intelligence expressing itself in personhood. *Our Word* is theomorphic—the personification in time of the mystery of creation!

Energy waves, spiraling as the currents of energy fields, sustain patterns of resonance which universally step energy down from consciousness to matter. All energy resonates in a universal gestalt embodying the process of creation. All of the systems of the universe, and the subsystems of nature and the human body, are attuned to this universal cosmic process. As above, so below— all of the spiraling energy fields of creation are a reflection of this ubiquitous holomovement. Every rhythm of creation is theomorphic and embodies the process of creation.[16]

Espansion +
Shakti
Centrifugal

Bindu ∅

Centripetal
Shiva
Contraction –

Energy fields are a balance of Shiva and Shakti, centripetal and centrifugal, electromagnetic forces cycling around a neutral source.

In a theomorphic perspective every rhythm of vibration embodies the process of creation. Vibration is cyclical. All creation, from a vibrational perspective, is made up of energy fields. In nature, energy pulsates centrifugally from a center though a field and returns centripetally to its source. This cycle universally sustains a spiraling helix of polarity. This spiraling helix resonates as the vortex of all creation. In the theory of theomorphic resonance, everything in the universe is understood as being made up of energy fields organized around cycles of harmonics of resonance, embodying the process of creation. Every level of microcosm and macrocosm embodies this process of creation. The theomorphic principle is that vibration spirals as a vortex of resonance which is a microcosm of the creative process. All structure and process throughout creation are based on this principle. This is the fundamental dynamic which underlies "form" in nature. This is the fundamental feature of universal law.

Consciousness precipitates into manifestation through five quantum steps. These five harmonics of resonance are the fundamental archetypes (types-of-arcs) that weave the fabric of matter. (Purce, The Mystic Spiral)

The Unity of All Being

The theory of theomorphic resonance offers an evolutionary perspective in which all of creation is a single process of evolution. Every galaxy, every solar system, every planet, and every system of nature is contained within the unified evolutionary force field of Brahma, the Creator. The energy fields of every atom, molecule, cell, and organism take place within the body of Brahma and are emanations from the Unitive Field of Cosmic Consciousness, Brahman.

The immutable nature of this universal field of living consciousness is expressed in the *Brihad Aranyaka Upanishad* (5:1):

> That which lies beyond is Plenum [full and undiminished]. That which appears as this here is Plenum [the universal] is also plenum, equally full and undiminished. Out from Plenum, Plenum arises. Plenum having been taken away out of Plenum, what remains is still the same [undiminished].[17]

As the Self of both the microcosm of the individual and the macrocosm of the universe, Cosmic Consciousness is unchanging.

All the energy of creation moves as one universal field of Creative Intelligence on every level of manifestation. The creative potency is everywhere manifesting as a theomorphic dynamic underlying all vibration and all structure in nature and the cosmos. All creation is the evolutionary expression of the imminent consciousness. All personal *karma* (the creative force that brings beings into existence[18]) is likewise a movement of universal karma. Each person is the embodiment of the creative process in its eternal play of consciousness.

Each Being is sacred, the leading edge of an ever-evolving universe. Each is a theomorphic personification of the Creator.

In yoga the One is experienced as *Sat-chit-ananda* (Ultimate Being, Ultimate Consciousness, and Ultimate Love).[19] We assume that Ultimate Being, in its nature as Ultimate Consciousness and Ultimate Love, chose to give its creation, its child, the greatest of gifts—its own Self.[20]

When That which was beyond creation
sought to manifest Itself
through the act of creation,
In Its Ultimate Wisdom,
expressing Itself as Ultimate Love,
It made Its creation nothing less than Itself

Everything is theomorphic, a microcosm of the creative process. All structure and process throughout creation are based on this principle. This holomovement is the fundamental feature of universal law, and the fundamental assumption of a theomorphic perspective.

As Above, So Below

As is the atom, so is the universe,
As is the human body, so is the cosmic body,
As is the human mind, so is the cosmic mind,
As is the microcosm, so is the macrocosm.
—*The Sanatana Dharma* [21]

It follows from this theomorphic model that we can study the universe by viewing its expression on any level of creation. The microcosm embodies the totality of the dynamics of the universe, so that by studying the microcosm we can comprehend the macrocosm.[22] Dr. Stone writes: "The infant has a complete zodiac in its own make-up, an exact duplicate of the cosmos in which it lives."[23] As the Tantrics tell us, "What is here is elsewhere. What is not here is nowhere."[24] Again, from Dr. Stone:

In most of the ancient writings, man was the key to the universe and the grand symbol of all the mysteries. Every effort was made to explain man to himself and to make him aware of all the forces and energies locked up in this being that he calls himself. The building of the pyramids and temples everywhere expressed this urge in man ... Man stands at the top of the ladder of all Creation. He alone is endowed with *all* the faculties necessary to understand his Source, his Being, and his relationship to Nature.[25]

Every principle or energy field within us, our little world, has a universal plane of being or field of operation.[26]

When the ancients tell us to "know thyself," they are telling us not only what to study but how to study it. Each of us is the embodiment and personification of the universe. Everything about us is an expression of all of the aspects of universal law, and is a key to understanding the mystery of creation. Paracelsus, the renowned mystic, alchemist, and physician, revealed that "Gods are immortal men, and men mortal gods."[27]

Our fundamental proposition is that the Universe is one living being. God becomes the world. God, as illimitable consciousness, is the stuff of creation. Creation is a field of consciousness, and matter, a crystallization of consciousness. The Sanatana Dharma asserts that the laws that rule the crystallization of Spirit into matter are universal. This process, the evolution of consciousness, the ever-expanding breath of Brahma, is fully present on every level of creation.

Swami Muktananda, an Indian sage who taught in the U.S. for many years, proclaimed, "God lives within you, as you." We are atoms in the body of God. Each facet of creation is theomorphic, the embodiment of Ultimate Intelligence—in our case—the personification of the creative process.

The Implicate Order of the Universe Creates a Synergistic System

In Indra's Heaven there is a set of pearls so contained that if you pick any pearl you will see all the other pearls. This pearl is all the other pearls.[28]

In the following discussion we will be blending concepts from the ancient wisdom and "new physics." While this may appear to do an injustice to both paradigms, it may be valuable to understand that Niels Bohr and David Bohm, fathers of the new physics, were exposed to the Sanatana Dharma. Bohm was a student of the mystic J. Krishnamurti, and Bohr was influenced by a series

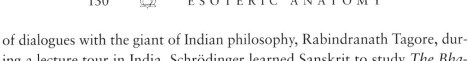

of dialogues with the giant of Indian philosophy, Rabindranath Tagore, during a lecture tour in India. Schrödinger learned Sanskrit to study *The Bhagavad Gita*. Transpersonal psychologist Ken Wilber writes: "Men like Einstein, and Schrödinger, and Hisenberg were spiritual mystics of one sort or another. Louis de Broglie, Max Planck, Niels Bohr, Wolfgang Pauli, Sir Arthur Eddington, and Sir James Jeans all shared a profoundly spiritual or mystical worldview, that in the deepest part of your own being, in the very center of your own pure awareness, you are fundamentally one with Spirit, one with Godhead, one with All, in a timeless, eternal, and unchanging fashion."[29]

Leading-edge theoretical physicists like David Bohm have been drawn to the conclusion that there is a pattern of wholeness enfolded within all being. Bohm calls this the "implicate order," a universal pattern in which "everything is enfolded into everything." Thus "what is" can be thought of as an intrinsic hologram, and everything is to be explained in terms of forms derived from this holomovement.[30]

The ancients also based their science on an understanding of the holomovement of the implicate order. The concept of the Ved, the omniscience of Ultimate Intelligence, was a foundation of the cosmology of the Sanatana Dharma. The understanding, "as above, so below," in which the macrocosm is enfolded within every level of the microcosm, is the fundamental axiom of hermetic research.[31] In our model, it is the implicate order which fosters the unity and integration of nature and the cosmos. One universal energy field is implicate in every part of being.

In the Sanatana Dharma that which keeps the universe together is a universal field of living consciousness. In this transpersonal paradigm all creation emanates from the field of the One Life. This holographic field of Ultimate Intelligence is eminent as the nucleus of every atom of nature. This is the Soul of creation. All creation emanates from this Unitive Field of omniscience.

Intelligence is the ability to learn.[32] The evolutionary cycle of polarity embodies the essence of intelligence. All vibration cycles from the neutral source of Cosmic Intelligence into the positive field of intelligent expression, and then the intelligence is drawn back to the source in the negative intelligent feedback phase. This input influences the next cycle of expression, in a process of "learning." All movement in nature is cyclical, expressing this play of intelligence.

Appreciate that this cycle of intelligence is happening, from subatomic to galactic, through every cycle of cosmic evolution. Intelligence is the medium of creation, the process and the fruit of cosmic evolution. The ubiquitous cycle of awareness, at the speed of light, may be the basis of what we call intelligence. In the ancient wisdom all movement expresses this essence of intelligence.

As the ancient esoteric texts the *Upanishads* declare, intelligence is the quintessence of being. In the Sanatana Dharma the cosmos is intelligence. A profound harmony emanates from this self-organizing field of Cosmic Intelligence.

A Synergistic System

The universe is a synergistic system in which every level of being integrates with and supports every other level of being. For the purposes of our model, *"synergy"* is defined as a dynamic in which every level and every part maximally supports the whole, and every level of the whole integrates with and maximally supports the parts. The flow of Cosmic Intelligence underlies all movement of energy. In the Sanatana Dharma the evolutionary cycle of Cosmic Intelligence weaves the fabric of creation. The basis of the profound synergy of nature is that the substance of the universe is Cosmic Intelligence. The cybernetic dynamic of Cosmic Intelligence is a universal principle.

Our bodies, nature, and the cosmos are a synergistic system—self-organizing fields of ultimate integration between the parts and the whole. The microcosm is an undistorted reflection of the macrocosm. In the ancient wisdom, humanity personifies the cosmos. To "know thyself" is to know creation.

The Unitive Field

Recall, then, that in the Sanatana Dharma, God,"absolute being, consciousness, bliss," is known as the Self. In the Timeless Way, God is fully present as "the innermost Self of all that is." God is the consciousness that dwells within you, as you. The tiniest drop of water is a microcosm of the ocean. The tiniest drop of consciousness is a microcosm of the ocean of consciousness. Describing the ancient paradigm offered in Sankhya-Vedanta in his classic work, *The Positive Sciences of The Ancient Hindus,* Dr. Brajendranath Seal explains: "In fact the original Energy is one and ubiquitous, and everything therefore exists in everything else."[33] In the ancient wisdom it is the identity of all consciousness through the omnipresence of the One Consciousness (chit) that is the basis of the unity and harmony of creation. All creation is a moment-to-moment emanation from the One Life. Transpersonal psychologist Ken Wilber refers to this mystery as the "Unitive Field."

In this paradigm, we view all phenomena as energy. Energy manifests as self-organizing fields of Cosmic Intelligence. These energy fields express the universal order that sustains the harmony of all creation. When we explain the dynamics of any phenomenon as energy fields, we will be illustrating the

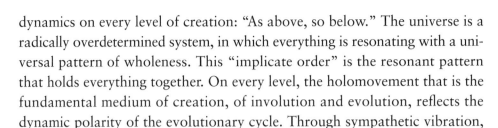

dynamics on every level of creation: "As above, so below." The universe is a radically overdetermined system, in which everything is resonating with a universal pattern of wholeness. This "implicate order" is the resonant pattern that holds everything together. On every level, the holomovement that is the fundamental medium of creation, of involution and evolution, reflects the dynamic polarity of the evolutionary cycle. Through sympathetic vibration, every system and subsystem of macrocosm and microcosm resonates with this pattern of wholeness.

Consider the biological evidence that "ontogeny recapitulates phylogeny," that each creature, in the course of its embryological development, experiences the entire developmental history of its species, nature, and the universe. In the course of development of each individual organism, the whole of the history of the evolution of life and creation is reenacted. This is a profound example of wholeness embodied in every living part, and an illustration of universal law embodied in all of nature's being. Every level of macrocosm and microcosm is a whirling vortex of energy reflecting an underlying theomorphic-resonant level of organization. Integration on this level sustains the profound levels of synergy found throughout nature and the cosmos. How is this synergy possible? From the esoteric perspective of the Sanatana Dharma, what are the characteristics of energy fields in nature? These are the questions that we will be asking as we unfold a theomorphic model of creation.

CHAPTER SEVEN

The Evolutionary Cycle

This entire universe is pervaded by me, in that
eternal form of mine which is not manifest to the
senses. Although I am not within any creature, all
creatures exist within me. I do not mean that they
exist within me physically. That is my divine
mystery. You must try to understand its nature.
My being sustains all creatures and brings them to
birth, but has no physical contact with them.
The Bhagavad Gita[1]

Life wiggles! Everything in creation is vibrating. Energy is always moving. All life is a breath, a pulsation, an oscillation, a wave of phases resonating in attunement with the resonance of the universe underlying all forces of nature and the cosmos. In the Sanatana Dharma all motion is the expression of spiraling vibration. Brajendranath Seal in his classic work, *The Positive Sciences of the Ancient Hindus,* describes the standpoint of Vedanta and its concept of *Parispanda* which resolves all physical action into motion. He explains: "Parispanda ... is whirling or rotary motion, a circling motion, but it may also include simple harmonic motion [for example, vibration]. All action, operation, work, is ultimately traced to this form of subtle motion."[2]

Energy is lawful. The fundamental feature of vibration is that it is always expressed in cycles. All of nature is an expression of cycles—cycles of birth and death, summer and winter, night and day, in-breath and out-breath. In these cycles, energy radiates outward to create a field of force. Upon reaching the limits of the field's potency, the energy returns to the source once again. Dr. Stone writes: "Polarity is the law of opposites in their finer attraction from center to center ... Creation brings forth opposites by its centrifugal force, like a fountain spray of manifestation flowing out to the limits of the cosmos and of each pattern unit."[3] All energy moves in these cycles of involution and evolution, from subtle inner to gross material levels of vibration. All movement in creation is cyclical—from above to below and from within to without; then from without to within and from below to above. In the Sanatana Dharma the universe is composed of ubiquitous cycles of centrifugal and centripetal forces known respectively as involution (*pravritti*, from *pra*, "forth"; and the verb root *vrit*, "to turn"; hence "to spiral centrifugally"); and evolution (*nivritti*, from *ni*, "back"; hence "to spiral centripetally"). This ubiquitous helix of yang and yin, whose undulating spirals sustain all manifestation, has been called "the two hands of God."[4]

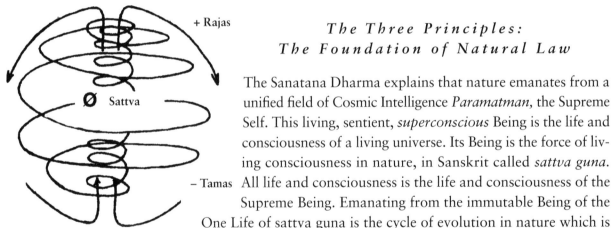

The Sanatana Dharma explains that three forces underlie nature. Sattva guna is the unchanging living intelligence which is the source of creation. Rajas guna is the ascending spiral of manifestation, and tamas guna is the descending spiral of dissolution. All energy in nature moves though this "evolutionary cycle." (Purce, The Mystic Spiral)

The Three Principles: The Foundation of Natural Law

The Sanatana Dharma explains that nature emanates from a unified field of Cosmic Intelligence *Paramatman,* the Supreme Self. This living, sentient, *superconscious* Being is the life and consciousness of a living universe. Its Being is the force of living consciousness in nature, in Sanskrit called *sattva guna.* All life and consciousness is the life and consciousness of the Supreme Being. Emanating from the immutable Being of the One Life of sattva guna is the cycle of evolution in nature which is made up of two forces called *rajas guna* and *tamas guna.* Rajas guna is the centrifugal force of creation. Its ascending spiral sustains fields of force emanating from a center as a fundamental organizing principal throughout nature. Rajas guna sustains the *conscious* mind. Tamas guna is the centripetal force of dissolution in evolution. Its descending spiral is the crystallizing force which sustains matter. Tamas guna rules the *unconscious* mind. In the paradigm of the ancients all of nature can be understood as resonating with these three *archetypes of consciousness.*

The Trinity: Life, Light, and Love

A trinity of forces is the foundation of this ancient paradigm of natural law. First is sattva guna, the force of God's Being in nature, which we experience as life. All energy in nature is understood as living subsystems within the evolution of the One Life.

> Hidden in all energy is the working of a conscious will. Energy is will in action in the outer world. Behind will is sentience or consciousness as the power of determination. Life is being, the conscious principle. It manifests in nature as the working intelligence behind the movement of energy. This natural or organic intelligence is conscious … intuitively and spontaneously a movement of pure harmony and beauty … All of nature is a living being, ever-seeking greater awareness, freedom and creative unfoldment.[5]

All creation is One Living Being, the Cosmic Person (*Purusha*), *Paramatman*, the Supreme Self. The Unitive Field of divine living Being which is the ground of all becoming is sattva guna.

The light of Cosmic Intelligence illuminates all creation. Rajas guna, the force of light, of radiance, is the second phase in the cycle of manifestation. "Energy is light. Energy as it moves, undergoes transformation and emits light and heat. There is a natural warmth to all life. And there is a natural light to all energy. There is in all life a principle of reflection, a transparency that manifests as intelligence and consciousness."[6] The glow of light, of passion, and of excitement, the hunger for experience, and the will to create, emanating from all life, is the fire of rajas guna.

Tamas guna, the principle of cohesion, unity, and the will to love, is the third of these forces. All manifestation is an interlinking of forces of cycles in a single rhythm. All creation is a vast spiral harmony which is the reflection of a conscious intent to sustain unity and cohesion. This force of cohesion, which underlies the crystallization of consciousness in form and the beauty of nature, is Love. "Love is the real force that holds things together."[7] This third phase of the cycle, the binding principle, which veils consciousness as form, is tamas guna.

Sattva, rajas, and tamas are levels of consciousness. Sattva—superconscious-

The Three Principles			
Guna	Consciousness	Element	Ayurvedic Dosas
Sattva	Omniscience	Ether, Air	Vatta
Rajas	Conscious Mind	Fire	Pitta
Tamas	Unconscious Mind	Water, Earth	Kapha

ness—is the omniscient ground of creation. Rajas is a step-down into the limited world of creation and ordinary consciousness. Tamas is the most veiled expression of the cosmic creative medium of intelligence as form, which is unconscious. Crossculturally and transhistorically this trinity is personified as: Vishnu, Brahma, Shiva; Mother, Maiden, Crone; Father, Son, Holy Ghost; Yang, Yin, Tao; and even the materialistic dialectic of thesis, antithesis, synthesis.

The supreme ultimate principle

In nature there is always a dynamic between male and female, positive and negative, building and cleansing, active and receptive, electric and magnetic, pushing and pulling. The creative cycle consists of the involution into matter of a centrifugal, yang, electric, building radiance, followed by a centripetal, yin, magnetic, releasing, eliminating, purifying phase of evolution back into Spirit. The interaction of these yang and yin forces in wholeness is known as the *Tao*.[8] In the cosmology of traditional China this fundamental axiom of universal law was called "the supreme ultimate principle" and was illustrated by the yin/yang symbol. The cycle of evolution, the cycle of polarity, the cycle of Cosmic Intelligence was recognized as the ubiquitous basis of universal law.

In the Sanatana Dharma all creation takes on its fundamental qualities through its attunement with the centrifugal force of the rajas guna, the centripetal force of the tamas guna, or the quiescent force of sattva guna. The word "guna" is derived from the Indo-European base *gere,* "twirl, wind."[9] In this cosmology in which everything is understood as "the breath of Brahma," divinely conscious vibration, the gunas describe the root archetypes of the cosmic breath.

Dr. Randolph Stone, the founder of Polarity Therapy, used the term *Polarity* for this dynamic underlying all creation. "Polarity" describes a pulsating movement pushing outward from a neutral source, through an expansive, radiant, electric, positive phase; then pulling back to the source in a releasing, magnetic, negative, contracting phase. All energy in nature is an expression of the electromagnetic dynamic of polarity.

The Two Hands of God
A centrifugal spiral, a centripetal spiral, and a sunflower. (Doczi, The Power of Limits)

Dr. Stone writes, "Whirling energy forms centers of concentration and flows outward as a positive or a negative charge, in relation to the other centers of action. This is 'polarity' or attraction and repulsion."[10] G. Zukav, in *The Dancing Wu Li Masters,* writes, "This dance of attraction and repulsion between charged particles is called the electromagnetic force. It enables atoms to join, to gather to form molecules and it keeps negatively charged electrons in orbit around positively charged nuclei. At the atomic and molecular level it is the fundamental glue of the universe."[11] According to Dr. Stone:

The old Ayurvedic Medicine and Acupuncture ascribed three modes of motion to energy in the three dimensions of space, and called them the three "Gunas"—"Satwa" the neuter, "Rajas" the positive, and "Tamas" the negative mode of motion. The Chinese called them the "Qi" (neuter), the "Yang" (positive) and the "Yin" (negative) poles of action. These became the Air, Fire, and Water elements of the Hermetic philosophers, who obtained this information from the orient ... This is the foundation of the mystery of the link of Consciousness with Matter.[12]

The Evolutionary Cycle

The mechanics of the process in which the primordial energy unfolds, the foundation of all creation, we call the *evolutionary cycle.* "Cosmic evolution proceeds from the unconscious, unmoving, unknowable and unmanifest macrocosm to the conscious, moving, knowable and manifest microcosm. Human evolution is a return journey from the gross physical plane of the microcosm back to the unmanifest macrocosm. In one case the force is centrifugal; in the other, centripetal."[13]

The principle of Polarity is found on every level of organization in nature and is the basis of the theomorphic holomovement which underlies manifestation. The involutionary/evolutionary cycle is fundamental to all vibration in creation. In Polarity Therapy we work with this trinity of forces, the Polarity Cycle, as the fundamental patterns of resonance which define the body. We translate sattva guna as the Air principle, rajas guna as the Fire principle, and tamas guna as the Water principle. These are the archetypal forces, the principles in the one mind which underlie all manifestation. The balance and

The Two Hands of God

Centrifugal	Centripetal
rajas guna	tamas guna
pravritti	nivritti
positive pole	negative pole
Sun	Moon
heating	cooling
flying upward	sinking downward
to the right	to the left
repelling	attracting
masculine	feminine
outward from center	surface inward
outgoing	ingoing
energy	form
dispersing	centralizing
red fiery nature	blue watery nature
stimulating	quieting
expression	space
electrical	magnetic
repulsion	attraction
motor current	sensory current
back	front

Three Primary Principles of Motion / Three Forces

	Polarity	Quality of Mind	Element	Motion	Body System
Sattva	0	Superconscious Mind	Air	Equilibrium	Nervous System
Rajas	+	Conscious Mind	Fire	Expansion	Muscles
Tamas	—	Unconscious Mind	Water	Contraction	Elimination

Rulership of the Principles

Guna	Principle	Element	Essence
Sattva	Air principle	Ether/Air	Life
Rajas	Fire principle	Fire	Light
Tamas	Water principle	Water/Earth	Love

harmony of this cycle is the basis of our attunement with what Dr. Stone called the "finer forces" of universal Creative Intelligence, which facilitate health and happiness. See Chapter 9 for further development of these ideas.

Sattva guna, the Air principle, involves two elements, Ether and Air, and rules the step-down of vibration from the core into manifestation. Ether, referred to as "wood," in the Chinese model (which is evolved from the Sanatana Dharma), moves from heaven to Earth, above to below, within to without. This is the nature of wood, whose rings move from within to without, growing *vertically*, like our spines, in the space between heaven and Earth. Energy always moves in cycles; Ether is the mother of Air, and Air is the child of Ether. Air is receptive to Ether, and we can influence the suprasensory Ether by balancing its material child Air. Sattva guna, which rules the Air element, facilitates a receptivity to the Ether element and its sattvic potency of healing equilibrium.

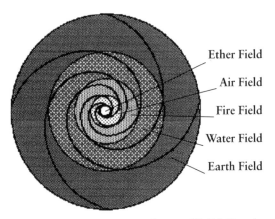

Ether Field
Air Field
Fire Field
Water Field
Earth Field

Energy Fields radiate from a center and step-down in quantum fashion from higher to lower rates of vibration from the core to the periphery.

Air is unlike Fire and Water, which rise and set like the Sun and the Moon. Air, called "metal," in the Chinese cosmology, fills space with the wholeness of sattva guna, the force of immutable equilibrium in nature, out of which the rhythms of Sun and Moon, rajas and tamas, emanate. Tamas guna, as well, rules two phases of the cycle of Polarity, Water and Earth. Water is always flowing downward, receptive, sinking into the entropy of Earth to release Spirit back to the centripetal vortex. The Earth phase of resonance is crystallized and slow to transform. Child Earth, though slow to respond, can be influenced through her mother Water. Thus the dynamic phases of the Moon rule the Water element and the deep, inner processes of the growth of form that are ruled by this centripetal force. The more stable Earth element rules the impenetrability of boundaries, the crystallization of Spirit into matter.

All energy fields are emanations from a center. In nature, wherever the two hands of God—the spiraling centrifugal and centripetal forces—intersect, a bindu, or seed, of sattvic (equilibrium) resonance is sustained, out of which emanates the next octave of the step-down process. The stillpoint of the bindu is nothing less than the omnipresent center of the universe. This bindu provides the mechanism for linking larger and smaller cycles, macrocosm and microcosm of the one coil of creation. The following discussion will explore how the elements of creation emanate from this omnipresent creative center, resonating with the all-pervasive ground of Creative Intelligence.

All material vibration is an emanation of spiritual vibration.[14] By *spiritual* we mean the transpersonal, subtle, inner, higher vibrational force of the Unitive Field of sattva guna out of which the more material forces emanate, "the finer forces" of higher intelligence. All physical manifestation is an emanation from the bindu (seed),[15] the all-pervasive center of the unified field of Creative Intelligence (*Sristi-kalpana*), whose holographic center is everywhere. The physical forces of vibration are sustained through their harmony and alignment with the all-pervasive spiritual forces that are the ground of nature and the cosmos.

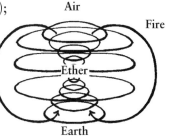

In the sunflower the seeds grow at the bindu, or stillpoint, at the intersection of the centrifugal and centripetal radii. (Doczi, The Power of Limits)

Everything in creation is woven of vibrations radiating in spiraling fields of energy. Physical energy fields are material emanations of the more subtle spiritual force field at the core. "Energy travels in a circuit, from center to circumference and back to its source or neuter center,"[16] and "all things proceed from a higher vibratory cause or center to the circumference, or lower center."[17] Physical vibrations are thus seen as steps down from the finer inner forces of spiritual energy into the slower, denser vibratory rates of material phenomena.

Energy fields in nature can best be understood as a spectrum of harmonics embodying this vibratory path between Spirit and matter. All vibration is theomorphic, embodying the step-down from Spirit to matter and back again—the primordial pattern underlying all vibration. All energy fields are likewise theomorphic, embodying this same evolutionary cycle. All structures in nature have as their fundamental organizing principle these patterns of theomorphic resonance in which energy moves from within to without, from Spirit to matter.

The step-down from Spirit to matter is a quantum process, through the keynotes of the elements. All vibration in creation is entrained to these archetypal harmonics of resonance.

In the evolutionary cycle, intelligence moves from higher to more material vibration—above to below (Ether/Air) and within to without (Fire); then returns via without to within (Water) and below to above (Earth). Energy projects from the Unitive Field into the "whirld" to experience its creation.

(Purce, The Mystic Spiral)

The Innermost Self of All That Is

In the Sanatana Dharma the universe, nature, and humanity are understood as emanations into materiality of the all-pervasive Creative Intelligence of the Absolute. There is no difference in substance between Brahman, the Absolute, and Atman, the Godhead in the individual. Nature, *prakriti*, is an emanation

The Polarity Principle			
	Rajas Guna	Tamas Guna	Sattva Guna
Polarity	+	—	0
Element	Fire	Water	Air
Pulsation	Centrifugal	Centripetal	Center
Force	Kinetic	Static	Equilibrium
Direction	Out	In	Neutral
Current	Pushing	Pulling	Stillness
Cycle	Involution	Evolution	Absolute
Manvantara	*Pravritti*	*Nivritti*	*Pralaya*
Ayurveda	*Pingala*	*Ida*	*Sushumna*
Ruling Deity	Brahma	Shiva	Vishnu
Light	Rainbow	Shadow	Light
Gender	Male	Female	Androgynous
Oriental	Yang	Yin	Tao
Embryo	Mesoderm	Ectoderm	Endoderm
Breath	Inhalation	Exhalation	Pause
Consciousness	Conscious	Unconscious	Superconscious
Astrology	Cardinal	Fixed	Mutable
Planet	Mars	Moon	Sun/Mercury
Seasons	Spring, Summer	Fall, Winter	Equinoxes
Temperature	Warming	Cooling	Unchanging
Fire	Gold flame	Blue flame	Violet flame
Color	Red-making	Blue-making	Green-making
Plane	Mind	Body	Spirit
Geometry	Triangle	Hexagon	Circle
Body	Right Side	Left side	Core
Music	First Interval	Third interval	Fifth interval

of the will, of the intelligent Being of *Purusha,* the all-pervasive Spirit, the Cosmic Person (see further discussion in Chapter 9).[18] The omnipresence of this intelligence is the key to understanding the mystery of life in nature.

The Ancients Viewed the Body as a Crystallization of Consciousness

Purusha, human consciousness, is the consciousness of an omnipresent God who lives within you, as you. The body is an emanation of this consciousness. The human physical body is made up of five Purushas (spirits, soul-witnesses, conscious-spiritual-persons) and their five *kosas,* or vessels, which are emanations from the *Jivatman* (the portion of the divine life that supports individual life in nature).

The word "kosa" is from the Sanskrit root *kush,* "to enfold." Each of these vessels, or sheaths, is a resonant frequency of the harmonics of the resonant anatomy of the body. Each of the kosas is a phase in the step-down of Spirit to matter, a more material expression of consciousness. The sheaths are fields of resonance within fields of resonance, whose matrices underlie the physical fields of force. While the kosas are analytically separable, in physical reality they are inseparable and function as a gestalt, as manifestations of the five elements of creation.

The sheaths are emanations from our spiritual core. The most gross of these energy sheaths is the physical body. Underlying the apparent reality of the physical body (*annamayakosa*) is the energy sheath (*pranamayakosa*), which sustains the physical body. The lower or sensory mind (*manomayakosa*) pervades the body to collect, organize, and interpret sensory data. This is conditioned mind, which is ruled by habit and language. The discriminating mind (*vijnanamayakosa*) is still higher in purity and vibration and capable of the levels of consciousness of intuition and yogic perception to develop spiritual insight and wisdom. The finest and most subtle aspect of mind is the balanced, blissful mind (*anandamayakosa*), which resonates with unitive consciousness.[19]

> The atman, which is in an absolute state of *Sachidananda* (infinite divine awareness), envelops itself in the body of infinite beatitude, the *Anandamayakosa*, and becomes the *Anandamayapurusha*, "the blissful witness." The atman embodies itself in the *Vijnamayakosa*, or "sheath built of discrimination." The atman emanates the mental part, the *Manomayakosa*, "sheath built of mind" (from *manas* or *mano*, meaning "mind") and becomes the mental consciousness of the person, the *Manomayapurusha*. The atman radiates the vital part of our being, the *Pranamayakosa*, "sheath built of life-force" (from *prana*, "breath," and *maya*, "formed") and becomes the vital and nervous consciousness, the *Pranamayapurusha*. Atman, the divine self, sustains the physical body or *Annamayakosa*, the "sheath built of food," and becomes the *Annamayapurusha*, the material consciousness of the person.[20]

The *anandamayakosa*, or bliss-sheath, is also known as the *karana sarira*, or "causal body." The vijnanamayakosa, manomayakosa, and pranamayakosa are often given the general name of *sukshma sarira*, or *linga sarira*, or "astral body." The annamayakosa, or "physical body," is called the *sthula sarira*, or "gross body."[21]

The individual proceeds from the universal condition of spirit (purusha) and matter (prakriti), manifesting as an individual soul or *jiva*. So it is that the individual consists of a subtle aspect and a gross aspect. All things must have a vehicle in

The Kosas Their Power and Purpose[22]

Body			
	Annamayakosa	Tool to enhance awareness	Health/ wellness
Energy			
	Pranamayakosa	Control mechanism of body/mind	Breath, Movement/balance Enhance concentration
Sensory Mind			
	Manomayakosa	Perceptual Emotions organization	Habits, language Creativity, instinct, imagination
Discriminating Mind			
	Vijnanamayakosa	Discrimination	The ability to discern, Intuition, decisiveness, critical thinking, cause/effect relations
Balanced Mind			
	Anandamayakosa		Experience of inner harmony and peace absolute self-confidence, dynamic will, equanimity

which to manifest themselves; these spiritual forces have their vehicle in the linga sarira. *Linga* means that invariable mark which proves the existence of anything, and *sarira* means "body." Together they denote the subtle body which accompanies the individual spirit or soul (jiva) and survives the destruction of the physical body. This subtle body, the linga sarira, is the invisible vehicle of the soul; it is constant and does not change throughout the cycles of life and death. However, the elements of which it is composed are not eternal. The linga sarira consists of eighteen elements: intelligence (*buddhi*), ego (*ahamkara*), mind (*manas*), five knowing-senses (*jnanendriyas*), five working senses (*karmendriyas*), and five subtle elements (*tanmatras*). The gross aspect, the *sthula sarira* or "gross body," is the material or perishable body.[23]

It is important to appreciate that all of the kosas, or bodies, are states of consciousness. In the Sanatana Dharma the sheaths that underlie form are essentially forces of consciousness. However, the five sheaths, with all their diverse faculties, are only forms and functions within consciousness (pure, unqualified being) and are not in themselves conscious entities.

The Intelligence that is expressed in the dharma of natural law is the omnipresence of Purusha—Spirit, the omnipresent Cosmic Person. Sri Krishna reveals in *The Bhagavad Gita:*

> All the objects of the universe depend upon Me, and are sustained by Me, even as precious gems depend upon the thread that passeth through them holding all together sustaining them. Moisture in the water am I ... light of the sun and moon am I ... the sound waves in the air am I ... the virility in men; ... the perfume of the earth ... the glowing flame in the fire. Yea even the very life of all living things, am I ... Know, that I am the eternal seed of all nature [24]

Mathaji Vanamali explains:

> There is finally, only one Purusha, whose heads is all heads, whose eyes are all eyes and whose ears are all ears. All thoughts are his thoughts. All deeds are his deeds. No one can think or exist except He. As the Veda proclaims: "What ever was, what ever is, what ever shall be, what ever can be anywhere, under any circumstances, is that Purusha alone." Into it all other Purushas melt as ice blocks into the ocean. And there is neither the individual nor the world of matter, neither the subject nor the object. There is the one indivisible oceanic experience of all comprehensive existence.[25]

CHAPTER 8

The Golden Spiral: A Key to Understanding Energy in Nature

All creation is the interweaving of cycles. From galactic manifestation to sub-atomic waves, the universe is a vast spectrum of cycles. The cycles of birth and death, summer and winter, day and night, in-breath and out-breath weave the fabric of life. The ancient rishis experienced the underlying unity of all cycles as the breath of Brahma and the periodicity of the universe as the rhythms of a single harmonious Being.

There is a profound body of scientific evidence that points to the fact that the universe is a single harmonious system. A key to understanding this unity is found in an aspect of natural law known variously as the "Divine Propor-tion," "Golden Section," "Golden Ratio," or "Golden Spiral."

Bruce Rawles explains:

> The Divine Proportion was closely studied by the Greek sculptor, Phidias, hence, it was given the name Phi. Also known as the Golden Mean, the Magic Ratio, the Fibonacci Series, etc., Phi can be found throughout the universe; from the spirals of galaxies to the spiral of a Nautilus seashell; from the harmony of music to the beauty in art. A botanist will find it in the growth patterns of flowers and plants, while the zoologist sees it in the breeding of rabbits. The entomologist views it in the genealogy of a bee, and the physicist observes it in the behavior of light and atoms. A Wall Street analyst finds it in the rising and falling patterns of a market, the mathematician in the examination of the pentagram ... The ancient Egyptians used it in the construction of the great pyramids and in the design of hieroglyphs found on tomb walls ... Plato in his Timaeus con-

"Looking down from above, from the perspective of the nucleus of our energy system the Sun, we see that the distribution of leaves marking the growth of plants follows the spiraling Fibonacci progression: 3 leaves in 5 turns, 5 leaves in 8 turns." (Lawlor, Sacred Geometry)

Top: Line spiraling out from a center following the Golden Ratio. Bottom: Chamber nautilus. (Doczi, The Power of Limits)

sidered it the most binding of all mathematical relations and makes it the key to the physics of the cosmos.[1]

The Golden Spiral can be described mathematically through a principle known in the West as the Fibonacci progression which is named after an Italian mathematician Leonardo Fibonacci da Pisa. Fibonacci, the father of Western mathematics, learned this principle from traders of the Aryan civilizations of Asia. The Fibonacci progression is a mathematical sequence that is produced by starting with 1 and adding the last two numbers in the progression to arrive at the next. The Fibonacci sequence begins: 1, 1, 2, 3, 5, 8, 13, 21, 34, 55, 89, 144, and so on. Each number is the sum of the previous two numbers.

The Fibonacci numbers, and especially ratios of two successive Fibonacci numbers, show up extensively in nature. For example, the petals of a Monterey pine cone are arranged in spirals crossing in both directions: eight spirals in one direction and thirteen in the other. Similar patterns arise in the seeds of sunflowers and in other plants whose leaves grow in a spiral around a central stem; each successive leaf may be on the opposite side (½ way around) or may be ⅔ of the way around, or ⅗, etc.[2]

The ratio of any two successive Fibonacci numbers from three on is about 1:1.618. This ratio occurs ubiquitously throughout nature, in logarithmic spirals that underlie the process of growth.

The Golden Spiral describes the radiation of energy from a center. In all natural processes energy radiates from a center in the logarithmic proportions of the Golden Spiral. One can witness the Golden Spiral in the curves of an elephant's tusk, a wild sheep's horns, and a canary's claw; and in the spirals in a pineapple or daisy. The spiral of your fingerprint or curl in your eye lash follow the Divine Proportion. The planets of our solar system radiate from our Sun, and galaxies manifest following the rhythms of the Golden Spiral. The Fibonacci progression describes the law that underlies the radiation of energy in nature. "It governs, for example, the laws involved with the multiple reflections of light through mirrors, as well as the rhythmic laws of gains and losses in the radiation of energy."[3]

The breath of Brahma sustains the harmony of the Golden Spiral. The source of all vibration is the word of God. The song of God animates the cosmic dance of Shiva and Shakti, centripetal and centrifugal currents that undulate through every spiraling atom of creation. The vibration of these spiraling yang and yin currents weaves the fabric of creation. David Tame in his valuable book, *The Secret Power of Music,* in a section titled "The vocal range of the one Singer," points out the unity of the electromagnetic spectrum:

Not only "solid" matter but all forms of energy, are composed of waves: which is to say, vibrations. All of the different kinds of electromagnetic energy—including radio waves, heat, X rays, cosmic rays, visible light, infra-red, and ultraviolet—are composed of wave like or vibratory activity, these vibrations traveling through the universe at 186,000 miles per second. The only difference between each of these phenomena is their frequency of vibration or wavelength. Each merges into the other at a certain wavelength: which obviously means, when one gets down to it, *they are each one and the same thing.*[4]

Like wheels within wheels, fields undulate within fields in the universal electromagnetic process of centrifugal and centripetal forces.

All vibration is aligned with the harmony of the electromagnetic spectrum. The mathematics that describe the radiation of energy throughout the vast spectrum of vibration of the universe is the Golden Spiral. Every level of macrocosm and microcosm of vibration is attuned to the harmony of the one singer.

The ancient rishis (seers) purified their bodies and concentrated their minds. They communed with the intelligence of the Cosmic Being through prayer and devotion. They actualized microclairvoyant powers (siddhi) in which they directly experienced the fundamental forces of creation. They witnessed "spirals, within spirals, within spirals …" as they described the vibratory nature of energy. They described the gunas, the fundamental modes of nature, which they likened to a rope spun of three threads. Centrifugal and centripetal spiraling around an unchanging core. Spiraling through all vibration are a balance of centrifugal and centripetal currents fractally emanating from an infinite core. These forces radiate from the core, universally, in the proportional harmonics defined by the Golden Spiral.[5]

All energy in creation vibrates as a Logarithmic Spiral.[6] (Picture this spiral three-dimensionally as a spiraling vortex, undulating in both directions, centrifugally and centripetally, simultaneously.)[7]

Understanding this undulating "dance of Shiva" is a key to the vibratory basis of creation. The infinity archetype (∞) embodies the key to this universal process in which balanced opposites sustain a unity. In this case the opposites are the balance of centrifugal and centripetal forces whose logarithmic undulations fractally weave the fabric of creation. Following the harmonic proportions defined by the Golden Spiral, the opposite undulating forces weave a pattern of seven intervals within eight phases. Esoterically this is called the Law of Seven.

Energy fields in nature are organized following the proportional harmonics of the Golden Spiral. (Doczi, The Power of Limits)

The Cosmic Octave: The Law of Seven

On every level of macrocosm and microcosm, vibration spirals as the cosmic octave. An octave describes a process with eight steps and seven intervals. Seven is a fundamental level of organization in the universe. The seven whole notes of the tones of the musical scale are a natural measure of the harmony of universal law. The musical octave is so "natural" that it defines the proportional distances between the planets in the solar system! The seven colors of the rainbow define the spectrum of light.

Musical Notes	Colors
G	Ultraviolet
F, F#	Violet
E	Indiagi Blue
D#	Cyanogen Blue
D	Greenish Blue
C#	Green
C	Yellow
A#	Orange-red
G#, A	Red
G	Infrared

The physicist Hermann von Helmholtz devised a correspondence between the visible spectrum and the musical scale.[8]

Tame reveals:

> Nature herself also indicates the close link between sound and light, the solar spectrum of colors displaying a number of the properties of tones. The resemblance is just as though the one phenomenon—light—were a higher state of the other. Just as audible tone organizes itself naturally into the seven notes of the diatonic scale, so too does the visible solar spectrum form the seven colors of the rainbow.[9]

In Sanskrit the letters of the alphabet that reflect the spectrum of vibration underlying creation are called *varnas*, or "colors." Sanskrit *varnas* coordinate with universal basic principles that build and unfold creation.

All growth in nature is ruled by the lunar cycle. "From one cell to two there is a cycle of change in eight phases with seven intervals, analogous to the musical octave, or the spectrum of light ... The lunar month, a perfect example of graduated phases within a continuous process, is dominated by seven and its multiples."[10] The twenty-eight days of the lunar cycle, of the two fortnights of the waxing and waning Moon, map a cycle of four seven-day cycles and twenty-eight graduated steps that underlie all biological processes. The cycle spiraling through the lunar month of four-times-seven is fundamental to all growth. Cells divide in an octavian process.

The planets spiral out from the nucleus of the solar system following what in astronomy is called Bode's law, in which the distance between the

Undulating through every spiral of periodicity in creation is the profound balance of the universal breath of ascending and descending currents of yang and yin, expansion and contraction, Fire and Water. (Doczi, The Power of Limits)

planets follows the Fibonacci progression. The planets of the solar system follow the proportional harmonics of the Golden Spiral and parallel our earthly musical scale. The seven days of the week is a "natural process" of the rhythms of the cosmos. Sun-day, Moon-day, Mars-day, Mercur-day, Thors-day (Jupiter), Fri-day (Venus), and Saturn-day reflect the same *cosmic octave.*

Seven is a fundamental number in nature. There are seven rows of stable elements in the periodic table of the elements, which is understood to be the scientific basis of matter. Indeed in Leadbetter's *Occult Chemistry,* the periodic table is portrayed as a Golden Spiral. There are seven types of crystal systems—cubic, rhombohedral, hexagonal, triclinic, monoclinic, trigonal, and orthorhombic—the crystallization within the mineral kingdom of the seven elemental harmonics of creation. There are seven major hormonal glands in the human body and seven ventricles or cavities of the skull.[11]

In *Sacred Geometry,* Robert Lawler asserts: "Numerous studies document the ubiquity of the Golden Ratio and the Fibonacci Series, which describe the rhythmic laws of the gains and losses in the radiation of energy."[13] For example, Professor Amstutz of the Mineralogical Institute at the University of Heidelberg maintains: "Matter's latticed waves are spaced at intervals corresponding to the frets on a harp or guitar with analogous sequences of overtones arising from each fundamental. The science of musical harmony is in these terms practically identical with the science of crystals."[14] Wilfried Krüger, in his book, *Das Universium Singt* (*The Universe Sings*), combines musical theory and atomic physics. He demonstrates that the structure of atoms contain ratios and numbers that parallel the harmonious principles of music. Through a wealth of highly detailed and painstakingly prepared notes and diagrams he demonstrates that it is impossible for chance alone to account for the consistently musical patterns of the song of the universe.[15]

The ancient cultures understood the profound harmony of the universe. Their science was based on this fundamental aspect of universal law. Universally, energy emanates from a center in a "step-down" process that follows the Golden Spiral. Seven fields of force radiate from a center, in a quantum step-down of graduated phases, within a continuous process. The first two steps are the inner harmonics of the causal and mental plane; the next five steps are the elemental harmonics of resonance that underlie matter: Ether, Air, Fire, Water, and Earth. The Law of Seven is a fundamental principle in the theomorphic perspective.

The Golden Ratio and the logarithmic succession of the Fibonacci pro-

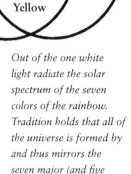

Out of the one white light radiate the solar spectrum of the seven colors of the rainbow. Tradition holds that all of the universe is formed by and thus mirrors the seven major (and five minor) cosmic Tones.

The candelabra of seven lights, the menorah, held as a sacred symbol by the Jews, offers a model of the universe. The Sun is represented by the center light; Mars and Venus, Mercury and Earth, Saturn and Jupiter are represented in the polarity of candles on each side.

gression are fundamental expressions of universal law. The Golden Spiral is a key to looking beyond the physical to the unified field that is the substratum underlying manifestation. All physical phenomena exist in alignment with this underlying field. The fundamental features, structure, and function of all organisms are reflections of the fact that they are subsystems within the larger energy systems of nature and the cosmos from which they derive their energy and being.

Elemental Harmonics of Resonance

Throughout all creation there is a step-down in vibration from the inner spiritual essence of the Life Field outward to material manifestation. This step-down occurs through phases that reflect the proportional harmonics defined by the Golden Spiral.

Every level of macrocosm and microcosm reflects the polarity dynamic of energy radiating outward from a nucleus and being received again into that neutral center. Fields of force are organized in a quantum process of keynotes of entrainment. Thus, in the cosmology of the ancients we picture quantum zones of force emanating from the center of the universal energy field, each zone representing a specific resonant frequency of the step-down from Spirit into materiality.

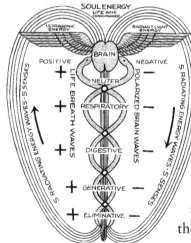

Universally, energy currents spiral in an octave of eight nodes that sustain seven steps. The seven chakras, where ida and pingala undulate around sushumna, personify this cosmic process.

The ancient wisdom describes these fields of resonance emanating from the center as *elements*. The elements are the vibratory archetypes of Cosmic Intelligence that underlie creation. Ether, Air, Fire, Water, and Earth. These five patterns of vibration are at the core of the sciences of the ancients. Dr. Stone writes: "*Patterns are mind energy fields.* Each type of energy has a vibratory speed and wave length which determines its function and affinity to other similar units of energy functioning in the body, or outside in the cosmos."[17] They are the fundamental resonances that are the inner vibrational basis of all phenomena. The elemental harmonics are understood as four phases of radiation (Air, Fire, Water, and Earth) from one center (Ether). Each of the elements represents a progressively denser, more material radiation of a single force, a harmonic of the underlying singular harmony. The elements are archetypal harmonics corresponding to the phases of materialization in the Polarity evolutionary cycle.

Energy Fields radiate from a center and step-down in quantum fashion from higher to lower rates of vibration from the core to the periphery. The radiation of energy fields is universally ruled by the Golden Spiral.

Like wheels within wheels, these whirling vortices of force emanate from the center of every energy field in a quantum fashion of discrete zones of resonance that provide the step-down mechanism from Spirit to matter. The Etheric center radiates a series of fields in a quantum step-down mechanism through which Spirit crystallizes into matter. The elements represent progressive stages in the cycle of movement of this energy from Spirit into matter. They form the outward, involutionary cycle from within to without, from above to below, from Ether to Earth, from subtle Spirit to the crystallization of consciousness into matter. This movement then becomes an inward-moving evolutionary cycle from below to above and from without to within, as energy cycles back through matter to Spirit.

In the Tantric *darsan* (a vision of God), nature (*prakriti*) is differentiated into five forms of motion. Ether (*Akasa*) fills space with the "Hairs of Shiva," nonobstructive motion-radiating lines of force in all directions, sustaining the space in which the other forces operate. Air (*Vayu*) is a transverse motion and the source of locomotion in space (from the sanskrit root *Va*, "to move"). Fire (*Tejas*) is an upward motion giving rise to expansion. Water (*Apas*) is a downward motion giving rise to contraction. Earth (*Prithivi*) is a motion which produces cohesion and obstruction, the opposite of the nonobstructive Ether.[18]

All vibration is entrained, through sympathetic vibration, with the resonance of these universal fields of force. All solidity, regardless of the phenomenon, resonates with the Earth harmonic; all fluidity with Water; all heat with Fire; all movement with Air; and all space with Ether.

The ancient seers articulated a distinction between the universal elements prior to manifestation, which they called "subtle elements," and the "gross" pentamirus combination of the elements that sustained manifestation. Thus the subtle *tanmatras,* or essences, combining and recombining, produce the five gross elements of the external universe the *Mahabhutas* or *Panchabhautikas.*

It is important to remember that this is a sacred model of creation. While a scientific perspective would hold that "mythology" points to a realm that is less than "real," sacred "mythology" asserts a transcendental dimension that is more than "real." The elements are not things, but are Primordial Being. The elements are aspects of the Divine Logos and are the root archetypes out of which Cosmic Intelligence manifests creation.

Vibration in nature spirals in fields goverened by proportional harmonics defined by the Golden Spiral.

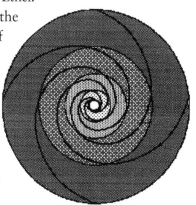

Cross section of logarithmic spiral pictured above.

Elemental Fields
Ether
Ether + Air sustain Air
Ether + Air + Fire sustain Fire
Ether + Air + Fire + Water sustain Water
Ether + Air + Fire + Water + Earth sustain Earth

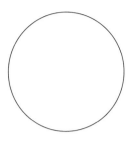

The Ether Element

The Ether Elemental Harmonic of Resonance

The center of the spiral of every energy field resonates with the neutral, Etheric harmonic. Ether is the first elemental resonance to evolve from the oneness of universal intelligence (*Mahat*). Ether is often referred to as the element of "space": "The characteristic of the Ether element is all-pervasiveness. It provides the space in which further creation takes place."[19] It provides the space that limits and contains the energy field. "Air is completely mobile, whereas Ether is completely inert. Air always wants to move and expand; Ether contains it."[20] Ether relates to the sonic, vibrational dimension of Cosmic Intelligence, which is the medium of creation. The overall space of the body, the sonic environments of the body cavities, the organ of hearing and sounds are formed from the space element.[21]

The classical glyph for Ether in the West is a circle. The circle defines a space. Ether defines the fields of the body. Manifestation is an appearance in an empty space within the infinite fullness, the plenum of cosmic being. Ether holds this space. The whirling atoms of manifestation sustain a vortex of emptiness. "Ultimately the seemingly solid body is composed of subatomic particles and empty space. What is more, even the subatomic particles have no real solidity; the existence span of one of them is much less than a trillionth of a second. Particles continuously arise and vanish passing into and out of existence."[22] Creation is a play of *unconsciousness* within the omniscient consciousness of cosmic Being.

Ether is the matrix through which thought forms are brought into etheric patterns which then crystallize into physical expression. Ether resonates with prana, the life breath of the cosmos and sonic medium of the Divine will to create. Ether is the principle in consciousness that draws from the universal medium of prana, cosmic breath, to project into physicality the intelligent patterns that sustain matter. The other elements radiate outward from within this all-pervasive subtle inner medium as they step-down into more and more solid matter. The potency of the emanation drops as the energy moves away from the center and steps down from Spirit to matter. The etheric body is a high vibrational matrix which steps energy down from the subtle universal elements into materiality. Ether is the womb through which the subtle inner quintessence of the universal elements combine into the pentamirus combination of the five gross elements which sustains matter.[23]

The Air Element

The Air Elemental Harmonic of Resonance

The potency of the Etheric core sustains the field of the Air elemental harmonic. In Air we move from the "o" of oneness into the polarity of the in- and out-breaths, the positive and negative fields. S. Radhakrishnan explains: "Vibration by itself cannot create forms unless it meets with obstruction. The interaction of vibrations is possible in air."[24]

All *movement* in nature is ruled by Air. All energy in manifestation moves in a double helix of centrifugal and centripetal orbs pulsating from a center in a whirling vortex of a fountain spray of manifestation. On the atomic level fields are a balance of proton and electron, positive and negative charges, centrifugal and centripetal vortices. The undulating interference pattern of these spirals is the basis of all vibration/movement in nature. All movement takes place within a balance of wholeness as an expression of universal law. Air rules all movement in manifestation as an expression of universal law. Air rules the gestalt of the pentamirus combination of the elements.

Air rules the relationship between the periphery of the body and the core, entraining the energy field with the infinite wholeness at the core. Thus Air fills space with the harmony of universal law. In the Chinese cosmology the Air element is known as Metal, pointing to the strength of universal law. Air is understood as the servant of Fire and Water, as Air always holds a balance of wholeness between the centrifugal and centripetal vortices, the periphery and the core.

The glyph for Air is the equal sign (=), the double parallel lines representing a balance of Fire/Heaven and Water/Earth. Air rules the east-west currents, which define the body from side to side and balance the periphery with the core. It is the bridge between the unified field and the polarity of involution/evolution. Air is the physical carrier of the prana, the universal life breath, in its dual polarity.

In the Pythagorean system the pentagram symbolizes the Air humor. The pentagram embodies the pentamirus relationship amongst the gross elements and the reciprocity of the play of the elements in wholeness. The pentagram is the crystallized expression of the relationship amongst the elements. It also represents the reciprocity between Fire and Water, Air and Earth, which vibrate in polarity relationships. The resonance of the elements are harmonious relationships within wholeness. The resonance of each element takes place within the profound pentamirus balance of these relationships. The pentagram is the

Air radiates from the central axis to unite the core and the periphery. (Purce, The Mystic Spiral)

crystallized expression of a spectrum of elemental attunement that universally undulates through every spiraling cycle of vibration.

In the Air element there is a *lengthening* of the arc of vibration from the stillness of the supraphysical unified field of space into the elliptical orbit (the cosmic egg) which characterizes all phenomena. The Air resonance is sattvic (*sat* means "absolute being"), *receptive* to the Absolute, the Unitive Field. In Air, both fields of force, centrifugal and centripetal, are equal and in a dynamic balance, and thus sustain a neutral field that resonates with the source. The parallel lines of the astrological glyphs for Air (♊︎♎︎♒︎) point to the dynamic energy fields of positive and negative, whose interference pattern underlies all vibration in nature. Air sustains the fulcrum, the bindu, around which Fire and Water emanate. Air is the bridge between the infinite sky and the finite world, the unity of space and diversity of time.

We can understand Air by contemplating spring, the season that is Air-predominant. In spring the Earth moves into resonance with the centrifugal spiral of Spirit. All forms that resonate with life are entrained through sympathetic vibration with this spiraling of life force from the center. Through sympathetic vibration with the omnipresent center, life springs out of the "emptiness" of space into manifestation in time.

Air sustains the spiraling transverse motion between the fields of Ether and Earth. In "re-spir-ation" the force of the Air element re-spires the more material elements through its potent inner resonance of attunement with the Life Field. All movement, breath, and the organ of touch and physical sensation are formed from the wind element, Air. The essence of Air is movement as an expression of living Cosmic Intelligence.

The Fire Elemental Harmonic of Resonance

The radiation from the nucleus of the Sun predominates in the involutionary currents which rule the step-down from Spirit to matter of the positive phase. As an oscillating wave moves through the spiral, there is a point where the spiral's fields of force sustain an arc in harmony with the solar influence—the arc of *radiance*. This is the Fire phase, where the positive, centrifugal, ascending fields come into entrainment. The Fire arc predominates, and the field is influenced by the universal positive harmonic. Fire is a radiation of the field of force of the Sun. It is a positive resonance in harmony with the universal electric harmonic of solar force. The classical alchemical glyph for Fire (△) captures its essence as a radiation of the Sun's field of force, as Fire always

The Fire Element

Element Attributes[25]

Element and Qualities	Attributes and Functions	Quality and Poles of Mind	Quality of Emotion	Quality of Body
Ether				
Stillness	Aesthetic sense	Tranquility	Governs emotions generally	
Harmony	Neutrality	Peace	Creates space for emotions	
Universal love	Emptiness	Neutrality	Ego—pride or humility	
Spaciousness	Core energy	Stillness	Grief	
Air				
Movement	Conscious emotions	Thoughts and ideas may dominate	Desire	Light
Mental activity	Heart-felt feeling	Cut off from feelings	Greed or aversion	Wiry
Attention	Compassion	Detached	Desireless	Drawn out
Thought	Rational thought	Thinks before acting	Charitable Compassionate	Thin Underweight
		Anxious, worried, fickle		
Fire				
Directed movement	Impulse behind movement	Enthusiastic, excitable	Anger	Moderate weight
Intelligence	Fire of metabolism	Willful	Resentment	Lean-muscular
Vitality	Drive	Focused, self-centered	Forgiveness	Medium build
Insight	Temperature	Honest, direct	Letting go	Muscular without exercise
Quickness of mind	Control	Expresses emotions quickly		Intention
Water				
Receptivity	Unconscious emotions	Intuitive, sensitive	Attachment	Moderate-stout
Intuition	Irrational pole	Patient,	Holding-on	Fully developed
Nurturing	Seeks lowest level	Fluid mind	Lust or chastity	Earth Goddess
Creativity	Grounding energy	In touch with feelings	Flowing	Padded look
Generation-seed	Potential	Acts intuitively	Moderation	Overweight
Earth				
Foundation	Field of completion,	Steady mind	Fear	Moderate
Support	Field of crystallization	Down to Earth	Courage	Stout, muscular
Stability	Final manifestation	Enduring		Block-like
Boundary	Inertia, heaviness	Routine		Heavy build

burns upward in tune with its source. The combined radiance of Ether and Air sustains the ring of Fire. Warmth, clear coloration, the organ of sight, and directed energy, are ruled by the Fire element.

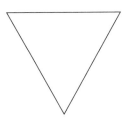

The Water Element

The Water Elemental Harmonic of Resonance

As the wave undulates through the spiral, there is a phase that resonates with and is energized by the lunar harmonic of influence. This phase is Water-predominant, and the arc is centripetal, descending, magnetic and attuned with the Moon's influence, which sustains fluidity and form. The combined potency of Ether, Air, and Fire sustain the resonant frequency of the field of Water.

Water gives birth to form. The glyph for Water is a downward-pointing triangle (▽). Water always sinks downward. This triangle is the receptacle of Spirit as the field of force drops in vibration to assume substantial solidity in this more material, fluid phase of the cycle. Water and Earth are centripetal fields of force.

The Earth Element

The Earth Elemental Harmonic of Resonance

The potency of all four inner fields of force sustain the gross Earth elemental resonant frequency. This next phase in the process of materialization is one that becomes dominated by the Earth's energy field, which sustains crystallization and elimination. At Earth the surface tension and resistance at the periphery of the energy field sustain a boundary. Earth is a most material expression of currents emanating from the nucleus. The glyph for Earth is the square (□). In Earth we have the crystallization of the holomovement of Ether and the play of the five forces, "the squaring of the circle." The pull of gravity materializes these forces.

Dr. Stone writes: "Energy is a living thing when approached from the Center flowing outward as currents and circuits. On the circumference it loses these qualities and becomes a sedimentation of energy whirls known as matter, through surface resistance and tension."[26]

At the Earth phase of resonance, the vibrations crystallize into matter and gradually lose their ability to resonate with the universal spiral of vibration. The energy field falls out of harmony with the universal keynote that sustains life. It drops to a linear energy field which resists the spiraling of vibration crystallizing into form. Eventually it breaks down, unable to reflect and draw sustenance from the universal harmonies. This is the "dust" of the Earth phase of the cycle. At this phase, elimination and decay take place, resistance is released, and the life potency is pulled back to the center by the centripetal return current of the evolutionary cycle, to step back up to Ether. All matter, by its nature, eventually crumbles to dust, releasing the Spirit to return to its source to continue in the eternal cycle of evolution. Gravity is the Earth's "prana," flowing from Heaven to Earth. The atmo-sphere of the Earth can be understood in terms of spheres of force emanating from the nucleus of the Earth, representing the centripetal return current of the evolutionary cycle. Dr. Stone called these crystalline emanations of force from the Earth "Gravity Lines."

Combinations of the Elemental Harmonics of Vibration Underlie Physiology and Anatomy

	Ether	Air	Fire	Water	Earth
Element	Emotion	Movement	Function	Liquid	Solid
Ether	Grief	Lengthening	Sleep	Saliva	Hair
Air	Desire	Speed	Thirst	Sweat	Skin
Fire	Anger	Shaking	Anger	Urine	Blood vessels
Water	Attachment	Movement	Luster	Semen/ovum	Flesh/fat
Earth	Fear	Contraction	Laziness	Blood	Bone

Creation is woven of the spectrum of vibration. The vibrational gestalt of the five elements underlies tissue and processes of the body. In bold are the qualities which result when an element predominates in the body/mind: Ether's essence of space rules grief, Air's quality of movement rules speed, Fire's directed energy rules anger, Water's life-giving essence underlies semen/ovum. The essence of the Earth element is the solidity of bones.

The Four Directions

All life undulates through the spiraling helix of creation in a cycle of phases of resonance: Air, from above to below; Fire, from within to without; Water, from without to within; and Earth, from below to above. All vibration entrains with the four primordial archetypes of the four directions of force.

The elements are the vibrational warp and woof weaving the resonant fabric of the body. Dr. Yeshi Dhondhen, physician of the Dalai Lama, summarizes the relationship between the keynotes of the elements and the harmonics of vibration that underlie the body:

> The sense consciousness arise from the mind. The flesh, bones, organ of smell and odors are formed from the earth element. The blood, organs of taste, and liquids in the body arise from the water element. The warmth, clear coloration, the organ of sight are formed from the fire element. The breath, organ of touch, and physical sensations are formed from the wind element. The cavities in the body, the organ of hearing, and sounds are formed from the space element.[27]

Midday
Summer
Solstice
Maturity
Fire Element

Dawn
Spring Equinox
Birth
Air Element

Dusk
Fall Equinox
Aging
Water Element

Midnight
Winter Solstice
Death
Earth Element

As above, so below—the universal cycle undulates through every cycle of creation.

Element	Sense	Food	Taste
Ether	Hearing		
Air	Touch	Fruit, Nuts, Seeds	Sour
Fire	Sight	Grains, Legumes	Bitter
Water	Smell	Sesame, Sunflower	Salty
Earth	Smell	Root Vegies, Tubers	Sweet
	Pulse		
Air	Snakelike, Fast-irregular, Feeble-faint, Moving from place to place		
Fire	Froglike, Jumpy-regular, High volume, Like beating drum		
Water	Full-bounding, Rolling-wavelike, Slow, Moderate full volume		

Keys to the Elements

The step-down takes place through the interaction of two poles—the Sun and the Earth. Life on Earth is a subsystem of the cosmic life of the solar system. The elements are phases of resonance reflecting the interaction of these two fields of force in the evolution of the universe. Creation takes place in an essentially "neutral field" between these two poles. This neutral ground is the medium of creative consciousness and the life processes. The Earth emanates an energy field. We refer to this field as the *atmosphere* of the Earth. A series of spheres of energy emanates from the nucleus of the Earth field. All growth on Earth aligns itself with the harmonics of these spheres of resonance. The word "atmosphere" comes from the Greek *atmos* meaning "breath." Life exists in the spheres of breath, in the respiration of the Earth as a living being.

Growth on Earth aligns with the elemental harmonics of resonance, the "atmo-spheres," sustained by the interaction of the energy fields of the Sun and Earth. The body grows within the energy field of nature. The body is aligned with the proportional harmonics of the atmo-spheres of resonance of the Earth's energy field. (Doczi, The Power of Limits)

The Atmo-Spheres of the Earth

The five spheres of emanation from the Earth represent the evolutionary return currents of the circulation of energy between the nucleus of our solar system, the Sun, and our "chakra" in that system, the Earth. The Earth emanates the spheres or fields of force that are fundamental to the organization of life. The surface of the Earth is itself Earth-predominant. Foods that resonate with the Earth element and are grounding grow within this sphere of resonance. The next emanation is Water-predominant, extending two feet above the Earth. Foods that grow in this Water-predominant sphere are cooling and cleansing, for example, vegetables and melons. The next sphere of resonance is Fire-predominant, focused at the level of an adult's solar plexus. This zone is where the warming, building grains grow. Above this are the spheres of Air and Ether, where more extreme foods such as fruits (very expanding) and nuts (very concentrated) grow. All life on Earth takes place as a subsystem within the spheres of force emanating from the Earth, interacting as subsystems of the solar forces. The evolutionary return currents rule structure and predominate in the musculature and skeletal systems of the body.

The Body Personifies the Evolutionary Process

In the Sanatana Dharma the world is a moment-to-moment creation of the Cosmic Being (Spirit, Sat Purusha). Spirit is omnipresent as unchanging Being. Nature is always changing, becoming, evolving. Nature is the sacred, moment-to-moment creation of Spirit. Nature is an emanation of Spirit, sustained by Spirit's animating force. Spirit is the One Life of Cosmic Intelligence which vivifies nature. Nature crumbles into decay and dust without the animating force of the Soul. The quintessence of the ancient wisdom is the realization of the mystery of eternal life and the sacredness of all creation as a moment-to-moment emanation of the presence of the One Life.

In the ancient wisdom the medium of creation is *the Word* of God. Everything in creation is understood as a microcosm of the cosmic breath of sacred sound vibration. In "sacred sound" the ancients were honoring the profound mystery of Cosmic Intelligence which is at the heart of all becoming. The Creative Intelligence of Brahma is omnipresent. All creation steps-down from the *pranava,* the unified field of the song of the cosmos, into the cycles of evolution. As transpersonal theorist Ken Wilber makes clear, "God" as consciousness, Satchitananda (being-consciousness-bliss), is wholly present as every level of creation: "God is the very stuff, the actual essence, of each and every stage/level. God is not the highest level, nor a different level itself, but the reality of all levels."[29]

The transpersonal perspective of the sacredness and unity of all life is the constant theme of the cosmology of the ancients. The chakras are the step-down mechanism from the Unitive Field of Cosmic Intelligence into its temporal channels of expression.

The lower chakras whose colors are red, orange, and yellow are ruled by Jupiter, *Brihaspati* the Guru. His unceasing faith, fire, and blessings illuminate the evolutionary process.[30]

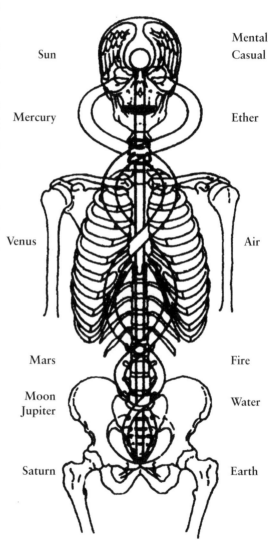

The chakras in our body are a microcosm of the solar system. (Stone, Energy)

The Chakras

The whole Vedic system is very comprehensive. It begins with the sound current or word of God as the first emanation. This forms the central core and axis of the Universe and of the cerebrospinal nervous system of the human body, and distributes the energy impulses upon the central axis as the motive power of the finer energies in man, from within outward, as the Life flows. First it is the psychic wireless energy and then precipitates into the physical life energy and currents.[31]

The body is understood as fields of resonance emanating from a central ultrasonic core. Dr. Stone writes: "The central core is the ultrasonic energy current of the Soul. It is the primary energy which builds and sustains all others. It flows through the sixth ventricle of the brain and spinal chord ..."[32] This core *(sushumna)* is located anatomically within the central canal of the spinal cord and resonates with "The Eternal Sound Current," the highest vibrational essence of Atman/Brahman, the Unitive Field of Cosmic Intelligence.[33] In Sanskrit the spine is called the *Brahmadanda*, the "stick of Brahma."[34] "The ultrasonic core is the central pathway of motion along which the five primary centers are created, located anatomically within the central canal of the spinal chord. Two intertwining currents of pranic energy (*ida* and *pingala*), each a polarized opposite of the other, pursue a double-helical pathway along this core."[35] Each point where the two intertwining currents intersect is a pulsating orb of force—a chakra (circle, whirl). Chakras are centrifugal and centripetal whirling vortices of force radiating from their ultrasonic core. The chakras are the primary centers of the pranas—the life currents—that sustain the body.

The chakras are the step-down mechanism from the ultrasonic core into the lower vibratory fields of the body. Reporting on clairvoyant research, Theosophist Charles Leadbeater writes:

> The chakras or force centers are points of connection at which energy flows from one vehicle or body of a man to another ... they show themselves as saucer-like depressions or vortices ... as small circles about two inches in diameter ... in the ordinary man ... when awakened ... they are seen as blazing ... whirlpools ... resembling miniature suns. All these wheels are perpetually rotating, and into the hub or open mouth of each, a force from the higher world is flowing ...[36]

Each chakra is the nucleus of an elemental energy field. The five chakras are located anatomically within the central canal of the spinal chord: Ether in the center of the throat, Air behind the heart, Fire between the second and third lumbar vertebra, Water at the lumbosacral junction, and Earth at the sacrococcygeal junction.

Leadbeater shares his clairvoyant vision of the subtle body, explaining:

> Though the mouth of the flower-like bell of the chakra is on the surface of the etheric body, the stem of the trumpet-like blossom always springs from a center in the spinal chord … In each case an etheric stem, usually curving downwards, connects the spine with the external chakra. As the stems of all the chakras thus start from the spinal chord, this force naturally flows down these stems into the flower-bells, where it meets the incoming stream of divine life, and the pressure set up by that encounter causes the radiation of the mingled forces horizontally along the spokes of the chakra.[37]

The streams of *primary life force* of the involutionary centrifugal currents and the *serpent power* of the centripetal evolutionary currents commingle in an undulatory circular motion. Leadbeater explains:

> The wave lengths are infinitesimal, and probably thousands of them are included within one of the undulations. As the forces rush around in the vortex, these oscillations of different sizes, crossing one another in this basket-work fashion, produce the flower like form … of wavy iridescent glass … of shimmering … petals … like mother of pearl.[38]

The Chakras as Centers of Consciousness in the Body

The elemental archetypes comprise a spectrum of consciousness. In the involutionary cycle where one's attention is directed downward and outward into the world to identify with the ego, the five chakras (Ether, Air, Fire, Water, and Earth) are associated with the passions, respectively: attachment, lust, anger, greed, and pride. In the evolutionary cycle back to the source, where the focus is directed inward and upward, in gratitude acknowledging the Great Mystery of life and action, the passions are transformed into the five virtues, respectively: detachment, continence, forgiveness, contentment, humility.[40]

Ten long line currents of prana emanate from the chakras.[39]

The Evolution of the Five Passions and Virtues

	Involution	Evolution
Ether	Attachment	Detachment
Air	Lust	Continence
Fire	Anger	Forgiveness
Water	Greed	Contentment
Earth	Pride	Humility

As we mature we have the potential to move from resonance with the involutionary passions to attunement with the virtues of the evolutionary phase.

Prana

Pranic currents emanate from the chakras to sustain the force fields of energy which our senses perceive as the physical body. While prana is usually translated as "life force," it is useful to point out that by prana we mean fields of living Cosmic Intelligence—not merely force or energy, not merely information or organization, but the consciousness of the omnipresent Spirit. Sir John Woodroffe writes:

> *Prana* is *Paramatma, Antaratma,* that is *The Supreme Being,* beyond and in bodies as their Controller and Director ... Another form of the mental principle is *Prana* or Life. Though not specifically called mind, it is nevertheless the aspect of mind which is wholly immersed in matter as the *directing* consciousness of the material energies of the body.[41]
>
> The cosmic *Prana* which pervades and vitalizes all breathing creatures (*Prani*) is the Brahman ... the source of the individual and collective life. Breathing is a microcosmic manifestation of the Macrocosmic Rhythm to which the whole universe moves.[42]

The "energy" of the body is a microcosm of the Sacred Word.

Prana is the life of a living universe, a form of divine energy that animates all living things. "Prana is a guiding, directing, and to such extent intelligent principle, which organizes matter into living forms with increasing degree of freedom and greater display of consciousness."[43] Prana links the living universal being and its subsystem, the living body. Five pranas (from *pra,* "forth," and *an,* "to breathe")[44] are the life currents which vivify the body to sustain the life of the body. The spiraling prana currents undulate through five phases of resonance: in-breath, out-breath, diffused breath, up-breath, and middle-breath. [45]

Prana is the vehicle of five stages of attunement to the cosmic forces and is the general term for "breath," in or out. *Apana* is the downward stage in the cycle of the breath. *Vyana* is the bond of union of prana and apana. It is the breath that sustains life when there is neither inspiration nor expiration. *Samana* is common to both inspiration and expiration. *Udana* leads the Soul in deep sleep to the central Reality or conducts the Soul from the body at death.[46] Prana the Sun-breath, is the supreme breath, the pure life force. It controls breathing and enables us to draw in Universal Life breath. *Udana* is

The Fivefold Phases of Prana: The Vital Airs[48]

	Resonance	Function	Movement	Seat
Prana	In-breath	Appropriation of Universal Life forces from the cosmos.	Inward movement, works ideo-motor-verbal processes and respiration	Heart
Udana	Up-breath	Uplifting, connects spiritual and material. Uplifts Spirit at death. Rules utterance, singing.	Upward movement, for example, erect posture	Throat
Vyana	Diffused breath	Distribution Circulation of life force. Rules separation and disintegration to resist entropy. Extension, contraction, and flexion of muscles, stores energy.	Union of Prana and Apana Life force/blue print of Cosmic Intelligence that holds space.	Throughout Body
Samana	Middle/ equalizing breath	Separates things: rules assimilation, digestion, and equilibrium of body chemistry and metabolism. Maintains organic life.	Kindles bodily fire, digestion, and assimilation	Navel
Apana	Out-breath	Excretory, rejection, casts out waste. Lower body death force.	Outward movement. Pulls against prana.	Anus

All vibration spirals through five phases of cosmic attunement. The breath is a microcosm of this universal holomovement.

an upward breath that links the physical to the spiritual. *Vyana* rules circulation. It keeps the body in shape and resists the forces of disintegration. *Samana* controls digestion and assimilation and promotes equilibrium. *Apana* is the downward and expelling force which rules elimination.[47] Dr. Stone writes:

> The life energy-Prana is a step-down current from the Eternal Sound Current which supports all living things. The origin of life is in the pattern-field

of the mental plane where it splits into potential sensory and motor currents. Its action is in the etheric field (akasha), and its center is that thousand rayed whirling disk of energy that supports all visible creation. In the body the heart chakra is the microcosmic life center which draws energy from within its center as the individualized life Prana. Externally the Sun is the source of Prana and life. In combustible matter it is stored as heat units.[50]

The index finger of the left hand and the second toe of the left foot deal with the heart center and the bulk of the "Pranic" life force in circulation.[51] Prana stimulates the fine nerve endings in the Schneiderian Membrane in the nose, on each side. These impulses are carried to the hemisphere of the brain, on each side ... [the] air has a positive or negative function in its flow to the brain.[52] Prana inlet is controlled by the sympathetic nervous system, in conjunction with the parasympathetic function. The artery rules in the tissues because it brings the positive pole or life energy or "Prana" to all parts of the body. . . Mind and "Prana" ARE TWINS IN FUNCTION, even as "Prana" vital force needs oxygen as a stepped down energy conveyor in order to function in matter, so does mind need "Prana" as an energy agent in order to connect with the body and the universal supplies of the four elements ... [53]

Cosmic Intelligence manifests the body to experience the world. Five pranic currents emanate from the chakras as centrifugal motor fields. These rule the operation of the five senses. Five centripetal, sensory currents rule the body's building and maintenance processes, as well as motivate the body in its expression in the world. Thus universal mind as an all-pervasive neutral field whirls through sensory (negative) fields and motor (positive) currents of pranas to sustain the body to experience the world through the evolutionary cycle. Positive motor currents flow outward from each chakra as the nucleus of oval fields of force. Negative sensory currents flow inward, to each chakra as the nucleus in the center of each field (see additional discussion in Chapter 12).

Dr. Stone writes:

The mind expresses itself through its five stepped-down agencies—which split into five sensory [currents] and five motor [currents]. These express themselves also as the mechanical leverage of the ten fingers, for motion and expression of skill ... and as ten toes and energy waves flowing through them for body motion through joints and leverage ... These waves of

Breath and the Elements[49]

Element	Purification Breaths
Earth	In nose out nose
Water	In nose out mouth
Fire	In mouth out nose
Air	In mouth out mouth

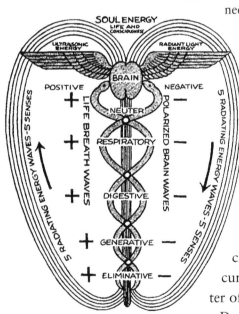

Five sensory and five motor pranas are formed from the two currents from the entry point of the ultrasonic core at the (third) "eye center."[54]

energy give the delicate sense of touch to the fingers … Such delicate patterns and finer energies waves link our mental patterns to the outside world, to experience sensations through contact with matter in its five modes of resistance. The five senses are modes of expression of the mind in matter.[55]

It is important to remember that this is a sacred model—not a godless scientific model. The miracle and mystery of life is only possible because of the animating presence of the Cosmic Person, Sat Purusha, the Soul acting as the principal of vitality through its prana. "This vital layer of wireless rays of energy waves is the breathing and living form of energy within the gross one, just as a hand fits within the glove and animates it."[57] The Soul is the hand which animates the glove of the body.

The Primary Heart-Shaped Fields[59]

Energy currents radiating from the core sustain chakras, whirling vortices of prana. Two currents, the centrifugal and centripetal, form the five primary heart-shaped fields of the head, neck, chest, abdomen, and pelvis, resonating with the five elements of Ether, Air, Fire, Water, and Earth, respectively. These fields are in a theomorphic relationship, and in each an elemental harmonic of resonance predominates which defines the qualities of the field (see additional discussion in Chapter 11).

The five fields of prana currents are fundamental to the makeup of the physical body. The resonance of the Earth chakra predominates in the bodily structures and in processes that sustain boundaries. The vibration of the Water chakra sustains the life medium and the processes that create, cleanse, and renew the tissues of the body. The radiant force of Fire predominates in the organs of assimilation and in the motive force of the muscles. The resonance of Air predominates in the nervous system and the functions of circulation and respiration. "As the sensory and motor pranas commingle, they establish an energetic physiology. From the five primary chakras emerge currents of subtle electromagnetic energy that fill the space of the 'etheric body' with etheric energy waves. The pathways of these etheric energy waves create a 'wireless' circuitry that interconnects every point of this subtle form with every other point in this energetic system."[60]

The centripetal currents of prana emanating from the chakras sustains five sensory pranas, to allow the perception of physical experience. The centrifugal currents predominate in structure and motor activity.[56]

Oval Fields

Dr. Stone introduces the concept of oval fields. He explains:

> All energy must have a circumscribed field in which to act, as a pattern design, a field of operation ... Without such a field of limits, even the vast energies in space would loose themselves in exhaustion and to no purpose ... The human skull has the outline and shape of an egg, a miniature planet-like shape. It is the individual microcosms where the pattern of all things to be in the body are cast or woven into the substance called brain tissue, which is the positive pole of the being. All things are represented here as patterns of mind energy and ideas, with a rhythmic wave length of their own. The whole body is but a duplication of these patterns in a more dense form and lower vibratory key of action.[61]

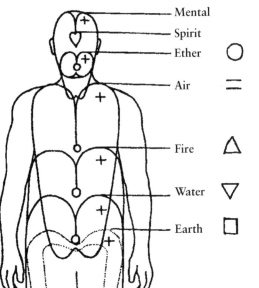

As above, in the brain, so below in the body. Oval fields of force arise from and surround the chakras. All energy resonates relative to a field, cycling from source to periphery and back to source. It is important to distinguish between the centrifugal and centripetal currents and the functioning of the sensory and motor pranas. The pulsating motor oval fields sustain a medium within which energy functions. The influence of the chakras is more sensory and psychological. In Dr. Stone's model, the oval fields' influence is more motor and functional.

In his book, *Polarity Therapy,* Dr. Stone points out in his discussion of the oval fields that he is choosing a "western viewpoint of anatomy and physiology" of "external relationships" to "avoid confusion" in presenting the functions described in the oval fields. He explains that other viewpoints, like that offered in the Ayurvedic system (which looks from within to without), describe the same functions differently.

In the model offered by Polarity Therapy, the centripetal sensory currents are identified with the chakras. The centrifugal motor currents sustain "oval fields." The chakras are ruled by the elemental archetypes. The oval fields are named for their dominant quality of motor activity or function. The head, ruled by Aries, is the Fire oval field. The eyes have the moving power of vision, without the light of vision there can be no direction for movement.

The second oval field is Ether, the neck, which rules the etheric body to

Chakras[58]

Earth	Chakra	Location	Consciousness/Desire
	Muladhara	Sacrococcygeal junction	Boundaries security issues, grounding, physical comforts, survival, shelter
Water	Svadhishthana	Lumbosacral junction	Nourishment, bonding, emotional attachment, procreation, lust, sexuality, fantasies
Fire	Manipura	Navel	Directed force, power, dealing with "reality," immortality, fame, will power, individuation, control
Air	Anahata	Behind the heart	Desire, aversion, ambition, hope, love, devotion, sharing, compassion
Ether	Vishuddha	Center of the throat	Knowledge, service, humility, discrimination, expression
Mahat	Ajna	Third eye	Self-realization, Unitive Consciousness, God-realization

provide the space for movement. Air rules the thoracic cavity as an oval field of the rhythmic movement of the lungs. The oval field at the pelvis is called the Water oval field because it is the center of the process of cleansing. Dr. Stone refers to the oval field at the navel as the Earth oval field because it is the center of the digestion and elimination functions.

It is well to be aware of the Western functional viewpoint, from *without to within*, which Dr. Stone offers as his point of reference. Yet we choose the viewpoint revealed in the ancient wisdom, which looks from *within outward* and expresses the elegance and parsimony of universal law. In Ayurveda the nerve and brain functions

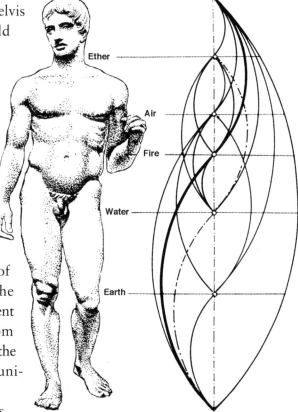

The electromagnetic spectrum vibrates as a unity aligned through the Golden Spiral. The cosmic octave underlies all vibration on every level of creation. Like wheels within wheels, vibration is embedded as fields within fields ruled by universal law through the harmony of the Golden Spiral. The chakra system of the body is a subsystem of the vibratory fields of nature and the universe. The chakra system is the nucleus of the energy field of the body. Its sevenfold vibratory field provides the step-down mechanism between the energy fields of nature and the cosmos and their microcosm, the physical body.

are associated with the Air element, which is ruled by the all-pervasive Cosmic Mind, "the one *polarized neuter center* within from which all external forms originate and emanate."[62]

Thus in the paradigm of the ancients the order of creation of the universal holomovement moves from: causal plane, to mental plane, to Ether, Air, Fire, Water, Earth. Thus the head in our model is ruled by *mind* and personifies the "mental plane" and the vision or identification which is the seed of our manifestation. The solar plexus, circulation, and respiration are ruled by the warmth and energy of the Fire element. The pelvis and the process of cleansing and regeneration predominate in the Water element. The colon, which rules the boundary between that which contributes to the life process and that which is eliminated, is ruled by Earth. Thus the prana currents emanating from the chakras sustain seven primary heart-shaped fields: causal at the third eye, mental at the brain, Ether at the throat, Air at the thoracic cavity, Fire at the the solar plexus, Water at the pelvis, and Earth at the colon and bones.

CHAPTER NINE

Purusha: The Omnipresent Cosmic Person

*In this universe, from the largest stars and planets
to the microscopic basic elements, all things
conform to cyclical spiralic form. In actual fact,
this form is established by the interaction of two
circular energies which are antagonistic to each
other. To our limited senses, whether these be
great suns or tiny electrons this whirling
movement appears to be undulating orbits of
diffraction of electrons or the elliptical orbiting of
fixed stars. In actual fact, however, all these
bodies are moving in logarithmic spirals.*
—*George Ohsawa*[1]

Here we will consider the fundamental principles in the universal mind that
underlie the stages in the manifestation of consciousness in form. A foundation of the Sanatana Dharma is the cosmology known as *Samkhya,* which
means "to enumerate." Samkhya, the oldest of the six orthodox schools of
Hindu philosophy, is said to be "the philosophical foundation of all Oriental
culture, the measuring rod of all Hindu literature, the basis for all knowledge
of the ancient sages [*rishis*] . . . and the key to Oriental symbolism . . ."[2] Through
careful reflection it enumerates twenty-five tattwas, or fundamental forces

that underlie creation, charting the stages of the evolution of consciousness into mind and matter. (*Tat* or *tad* imply the essential being of anything.) "This exposition is no mere metaphysical speculation, but is a purely logical account based on the scientific principles of conservation, transformation, and dissipation of energy."[3]

Samkhya postulates two ultimate realities, Spirit (Purusha) and nature (prakriti):

> The first principle postulated by the Samkhya system, *purusha,* is used to mean the soul of the universe, the animating principle of nature, the universal Spirit. It is that which breathes life into matter; it is the source of consciousness. Purusha is postulated to account for the subjective aspect of nature. It is the universal Spirit, eternal, indestructible, and all-pervasive; it is pure Spirit without activity and attribute, without parts and form, uncaused, unqualified, and changeless. It is the ultimate principle of intelligence that regulates, guides, and directs the process of cosmic evolution; it accounts for the intelligent order of things—why the universe operates with such precision, why there is cosmos and not chaos. It is the efficient cause of the universe that gives the appearance of consciousness to all manifestations of matter; it is the background that gives us the feeling of persistence; it is the static background of all manifest existence, the silent witness of nature.[4]

Purusha is a unified field of consciousness which is the ground state for all manifestation. All of the intelligence in creation is a reflection mirrored from this consciousness. In the phenomenology of our own personal experience, Purusha, the "soul witness," is what each of us refer to as the "Self." Purusha is human consciousness, the consciousness of an omnipresent God who lives within you, as you. Dr. Stone called Purusha "Soul," and he writes: "Soul is the inner energy field, the Doer and Witness in the field of consciousness; it is also the Perceiver."[5]

The second principle in Samkhya is prakriti, the cosmic substance, or primary matter. In the understanding of the sage Sri Aurobindo, prakriti is the will and executive power of the Purusha, the activity of Being. Quoting Aurobindo, Tyberg writes, "Prakriti is that producing element out of which springs the universe with all its various spheres and bodies."[6]

In Samkhya there is always an unbridgeable gulf between the unchanging Soul witness of the universal consciousness as the Self (Purusha) and its emanation, the evolution of nature (prakriti). Samkhya-Vedanta considered everything other than Intelligence—Purusha, the transcendental Self—to have arisen in the course of cosmic evolution. Purusha is immutable, an unchanging field of Ultimate Intelligence. The unchanging oneness of this field of living consciousness unites all beings. Prakriti is the ever-changing cycles of nature emanating from the One Life.[7]

It may be useful to think of the tattwas as the program in the computer of the One Mind. These quintessences are the primary forces in the universe, which sustain the appearance of differentiation of the One into the manifestation of phenomena.

The Samkhya teaches that the world-order is reason and is an expansion of the highest kind of intelligence; that there is no part without an assignable function, a value, a purpose; that there is always an exact selection of means for the production of definite ends; that there is never a random combination of events; that there is order, regulation, system and division of function.[8]

In the Sanatana Dharma a profound intelligence is consciously creating every atom of nature. In Samkhya the omnipresence of the omniscient Cosmic Consciousness is the innermost essence of everything. All creation is an emanation of this all-pervasive Being. This profound intelligence is the immediate cause of everything. Nature (prakriti) is a moment-to-moment emanation of Spirit (Purusha). All life is a moment-to-moment creation of this Ultimate Intelligence.

The Gunas

The rishis tell us that desire is the cause of creation. Desire generates a disturbance in the equilibrium of Mahat, the cosmic

The Twenty-five Tattwas of Samkhya Philosophy[9]

1. *Mulaprakriti—Root-Nature, Primary Matter, Cosmic Will. (sometimes Avyakta: The Unmanifest, the Primordial Element)*
2. *Mahat Tattwa—"The Great Principle," undifferentiated Cosmic Intelligence, the basis of Buddhi, the intellectual faculty by which objects are distinguished and classified.*
3. *Ahankara—The conception of individuality, of ego, of separate self.*

The five Tanmatras (tad, "that," and ma, "to measure") are the rudimentary or subtle elements.

The subtle Tanmatras, or "essences," combining and recombining, produce the five gross elements of the external universe, the Mahabhutas. From the five Tanmatras are derived the five Mahabhutas, or Panchabhautikas.

Five Tanmatras:	*Five Mahabhutas:*
4. *Sabda—Sound*	9. *Akasa—Ether*
5. *Sparsa—Touch*	10. *Vayu—Air*
6. *Rupa—Sight*	11. *Tejas—Fire*
7. *Rasa—Taste*	12. *Apas—Water*
8. *Gandha—Smell*	13. *Prithivi—Earth*

The five Jnanendriyani or Buddhindriyani are the five organs of perception or sense. An indriya is a force or power of Indra the god of the higher mind, and hence any exhibition of power—a faculty of sense or organ of sense.

The five sense organs, Niyantry:
14. *Srota—Ear*
15. *Tvak—Skin*
16. *Chakshus/Akshu—Eye*
17. *Jihva—Tongue*
18. *Ghrana—Nose*

The five Karmendriyani are organs of action:
19. *Vach—Voice or Larynx*
20. *Pani—Hand*
21. *Padas—Foot*
22. *Payus—Anus, organ of excretion*
23. *Upastha—Organ of generation*
24. *Manas—Mind*
25. *Purusha—The Spirit, The Divine Person*

substance. This initiates the play of the gunas and the process of cosmic evolution. Over the millennia, advanced yogis in their microclairvoyant research observed the electromagnetic spectrum and the vibrational basis of creation. They witnessed three universal forces which they likened to a rope of three strands: a unchanging core, sattva guna; a radiant centrifugal field, rajas guna; and a centripetal field, tamas guna. All vibration was the spiraling of these three forces.

As previously discussed in Chapter 7, the tri-gunas are the three "Universal Modes of Nature." All creation takes on its quality through its entrainment with these three archetypal essences, the three universal harmonics of force.[10] The English word *universe* is from the Latin *uniuersus,* "turned with a single impetus—hence, turned wholly—towards a single unity, turned into one …"[11] *Universe* comes from roots which mean "one coil." This derivation points to the ancient understanding that all creation and all subsystems of nature and the cosmos are part of one vast spiral harmony (from the Greek *harmos,* "a joining or fitting together," from roots meaning "a bodily joint").

The Archetypal Trinity

The tri-gunas are the three great archetypes whose modes give their characteristic qualities to all of nature. All three states of consciousness are present in everything, but one resonance always predominates. For example, sattva predominates in a innocent child; rajas, in an erupting volcano; and tamas, in a block of granite.[12]

In the Sanatana Dharma, all phenomena in nature are understood as manifestations of these three archetypes, or "types of arcs." The wave fronts of these archetypes make up the holomovement which is the sonic foundation underlying the harmony and integration in all of nature. Everything in nature takes on its distinctive qualities through its attunement with one or more of these archetypal forces. The predominant resonance defines the qualities of all phenomena. All matter, emotions, mind substance, and energies move by the three modalities of 0 (neuter), + (positive), and – (negative) polarities, the three gunas or tri-gunas.

Sattva Guna: Sattva (from *sat,* "Absolute Being") is the *superconscious* force of equilibrium in nature, and manifests as stillness, peace, lucidity, goodness, charisma, purity, harmony, balance, happiness, virtue, sympathy, and light. In the Sanatana Dharma this aspect of the Cosmic Person is personified as Lord Vishnu. *"Vishnu"* means "all-pervasive." Lord Vishnu refers to the all-pervasive Divine Self that is the one life of a living universe.

Rajas Guna: The rajasic centrifugal archetype (from the root *raj,* "to glow, to be excited") is the *conscious* force, which rules all energy and motion in nature and manifests as passion, action, effort, thirst, and desire. Rajas is ruled by Lord Brahma, the Creator. Lord Brahma is the personification of the creative potential of the Cosmic Person, the force of evolution and of conscious becoming, the power of growth reflected in all natural processes. Brahma is an ever-expanding pulsation of conscious breath, of infinite Creative Intelligence, whose force of creative potency encompasses all energy in creation.

Tamas Guna: Tamas (from the root *tam,* "to perish") is the centripetal force of inertia and *unconsciousness,* which rules all form in nature and manifests as solidity, resistance, crystallization, ignorance, fear, and death. Lord Shiva, the destroyer, is the personification of tamas guna, which rules the process of perishing and the dissolution of manifestation back to the Cosmic Person.

Every cycle in nature has a neutral, positive, and negative phase. All energy in nature is sustained by its attunement through sympathetic vibration with these elemental harmonics of force. Polarity energy balancing aspires to restore the integrity of this evolutionary cycle (or, Polarity cycle) by releasing resistance in the energy fields of the body to reconnect the fields with their source. In this way the harmonics of an individual's life force can be brought into harmony with the forces of nature and the cosmos which sustain life.

In Polarity Therapy the resonant frequency of the sattvic, neutral force is termed the Air principle; the wavelength of the positive, rajasic vibration is termed the Fire principle; and the negative, tamasic resonance is termed the Water Principle.

Dr. Stone writes:

> The idea of three bodies or fields of expression may seem new, but it is not an invention of mine. It is as old as mankind. The Trinity Principle of every religion bares evidence to this fact. If Deity is triune, man must be, since he is an exact image of his Creator."[13]

> This is the cosmic picture of energy modality, demonstrated by the orbits of planets around a central sun. The microcosmic picture we find in the atom and in the cell: A nucleus or neutron in the center, which is the "satguna" type of energy; the "rajaguna" energy, which is a positive, centrifugal energy of heat and expansion, flying upward and to the right, with a repelling quality in its red fiery Sun-like nature; and the tamas energy, called the "Moon energy," which is the negative, centripetal energy of moisture (like a precipitate of the positive pole), settling downward and to the left. Tamas is the attractive power of the negative pole of energy

emanation, the cooling, green rays, the feminine, centralizing toning and quieting principle. The central, sattva guna (or Satguna) neuter pole is the axial around which the other energies revolve. The human body is a magnified cell, the expansion of the brain pattern of the Airy, neuter, all-present Life Energy Principle, which has its base in the cerebrospinal fluid—in the brain and in the spinal chord and throughout the entire nervous system. The cerebrospinal fluid flows between the pia mater and the arachnoid sheets, and follows the nerves all through the body.[14]

Three Types of Touch

Polarity Therapy utilizes three types of touch, which resonate with the three universal harmonics. Sattvic touch is a gentle touch of stillness and presence. Sattvic currents arise from the core of the body, the heart of the being. In sattvic contacts the practitioner's intention touches the intelligence which is the medium of creation. Sattvic contacts promote a receptivity to the Soul. Sattvic touch corresponds to the Air-predominant harmonics which rule the sensory feedback mechanisms of the body and nervous system.

Rajasic touch is stimulating and moving, directive, and penetrating. It resonates with the solar plexus, muscles, and organs of the body. We use our muscles in an energy-balancing through movement and strength, in a Rajasic contact which releases tension in the muscles, directs movement, and vitalizes the organs of the body.

Tamasic touch is either deep and dispersing, or sometimes forceful with pressure, stimulating the client's resistance. In evoking resistance, emotional and structural blocks can be brought to the surface and eliminated from the body. Tamasic touch resonates with the more crystallized negative predominant fields of the body at the psoas, soleus, and achilles tendons.

Life in Helix

All of nature and the cosmos is an expression of the Polarity cycle of evolution. Nature is a manifestation of the involutionary cycles of conscious energy projecting into materiality and the evolutionary cycles returning to their source. This dynamic underlies the movement of all energy in the cosmos. We experience this universal force as growth. Growth as an expression of cosmic evolution is the central organizing principle of all of nature.

The evolutionary holomovement can be conceptualized as a spiral helix.[15] All vibration undulates as a spiral helix entraining with the centrifugal and centripetal forces of rajas and tamas guna. Polarity is the supreme ultimate principle of attraction and repulsion which universally underlies energy fields. The dynamic between the involutionary and evolutionary fields of force sustains the ubiquitous spiraling double helix that underlies all energy fields in nature from the DNA to the galaxy. In every cycle of creation, energy undulates in a dance of rajas and tamas gunas, stepping up and down in a helix of currents, coupling long waves to short waves and little worlds to big worlds.[16]

Nature is woven of a fabric of two spirals—centrifugal and centripetal—whose polarity universally forms the underlying structure and process of everything in nature. This polarity relationship has been labeled "dinergy" by Gyorgy Doczi. The spiraling double helix is the foundation of all organization in nature.[17]

In the sunflower, the seeds grow at the intersection of the centrifugal and centripetal radii, which may be seen to correspond to rajas, the centrifugal involutionary force of manifestation; and tamas, the centripetal evolutionary force of dissolution and return to the source. Everywhere that these two spirals intersect, a resonance of sattva (equilibrium) is sustained. This sattvic point resonates with the Absolute and is a *bindu*, a center of manifestation stepping-down into the next cycle of creation. Thus, in the sunflower the seeds grow at the bindus.

Dinergy in the structure of a sunflower. (Doczi, The Power of Limits)

All energy moves in cycles of vibration which can be better understood three-dimensionally as spirals. In this world all energy fields take the form of whirling vortices of energy. See the discussion of Dr. E. D. Babbitt's research in Chapter 11. Dr. Stone writes:

> So all our created objects are really whirls in space, like bubbles in an ocean.... Curvilinear motion or vibration operated in space and created sphere after sphere until all was manifested, *and the same energy maintains this creation everywhere.* Motion is always in the shape of a curve, from the path of the planets in the sky to the very atoms of matter, and spinning electrons.[18]
>
> Nature's patterns seem to arrange themselves in circles of ovals for action and function even in minute structures like spinning wheels of energy, like reflections from the central whirling chakra-pattern of energy.[19]

Nature is woven of a fabric of two intersecting vortices—centrifugal and centripetal—whose polarity universally forms the underlying structure of all energy fields.

In nature all systems of vibration are subsystems within the circulation of energy between the Sun, as the nucleus of the centrifugal field, and the Earth, as the nucleus of the centripetal field.

Like wheels within wheels, energy fields emanate from the nucleus as whirling vortices of electromagnetic resonance. In quantum fashion each whirling field of resonance is a step-down from the innermost highest vibrational fields to completion in physical manifestation.

As discussed in Chapter 8, in the ancient wisdom the step-down in vibration from Spirit to matter was ruled by five elemental harmonics of resonance: Ether, Air, Fire, Water, and Earth. These archetypes are the universal keynotes of conscious vibration that underlie all phenomena in nature. All energy in nature is entrained with these great Beings.

The Four Phases Undulate Through Every Spiral of Creation

Undulating through the spiraling helix of the evolutionary cycle are four phases of attunement. The primordial energy helix is made up of four phases in two opposing fields, as well as the sattvic bindu. The centrifugal phase includes a movement from above (higher vibrational) to below, and a movement from within to without. The centripetal phase includes a movement from without to within, and below to above. Waves spiraling through the helix entrain it with quantum points of universal resonance. On every level of macrocosm and microcosm energy entrains with these ubiquitous archetypes. These keynotes of universal resonance are the elements, or elemental patterns of resonance, which define phenomena. Ether, Air, Fire, Water, and Earth are the universal archetypes or "types of arcs" which are the elemental building blocks of creation. All manifestation as vibration is entrained by these inner forces of vibration.

There is always a balance of wholeness in the energy field. The centrifugal phase of Air and Fire and the centripetal phase of Water and Earth always counterbalance each other. This dynamic underlies the synergistic nature of the Polarity manifestation cycle. Everything in nature is spiraling. All movement in nature is in the form of arcs, helices, or tubes. The arc formed between the source and any point in the cycle has a unique resonance and wavelength or "type of arc." These are the root elements of differentiation in our paradigm. Differences in types of arc are expressed in the resonances of the four universal keynotes and reflect the four elemental phases of resonance (Air, Fire, Water, and Earth) in the three polarities (+, 0, –) whose harmonies sing the fabric of creation.[20]

Energy fields pulsate with undulations spiraling from within to without and without to within at the same time.

Jill Purce, writing in *The Mystic Spiral: Journey of The Soul,* points out these profound properties of the spiral:

> The simple two-dimensional spiral has a number of remarkable properties. It both comes from and returns to its source; it is a continuum whose ends are opposite and yet the same; and it demonstrates the cycles of change within the continuum and the alternation of the polarities of each cycle. It embodies the principles of expansion and contraction, through change of velocity, and the potential for simultaneous movement in either direction towards its two extremities. On the spherical vortex these extremities, the center and periphery, flow into each other; essentially they are interchangeable.[21]

(Purce, *The Mystic Spiral*)

All vibration embodies these profound dynamics. The same current/field of vibration is an avenue for both centrifugal and centripetal forces. Both undulate through the serpentine vibratory movement with energy both coming from and returning to its source in a balanced polarity of forces. There is always a balance of opposites in the undulating fields of alternating polarity, expansion, and contraction. A profound octavian process links fields nested within fields, following the universal harmony of the Golden Spiral. All vibration is a three dimensional vortex with fields embedded within fields braiding the standing waves of resonance which define phenomena. Every level of vibrating microcosm and macrocosm is the Golden Spiraling song of Brahma.

All vibration spirals as a reflection of these paradoxical principles. The vast harmony of the electromagnetic spectrum is unified through the mysterious properties of the spiral. All creation on the vibratory level of the electromagnetic spectrum is made up of spirals. Spirals in nature undulate in phase dynamics, which move from within to without (involutionary force) and at the same time from without to within (evolutionary force). Both directions of force, rajas and tamas, undulate through every spiraling helix of vibration in a profound balance of wholeness. This balance of wholeness is sattva guna, the Tao. The centrifugal pulsation manifests as two arcs, Air and Fire, and the centripetal manifests as Water and Earth.

In Tantric cosmology nature (prakriti) is differentiated into five forms of motion. Ether (Akasa) fills space with the "Hairs of Shiva," nonobstructive motion, radiating lines of force in all directions and sustaining the space in which the other forces operate. Air (Vayu) is a transverse

The Breath of Brahma
A spiral has the unique ability to move in two directions at the same time. All vibration is serpentine in motion, reflecting this dynamic. The gunas universally embody this ubiquitous force, the "breath of Brahma." Living vital fields are pulsating with life. The pulses undulate through the field in oval spirals from within to without and at the same time, from without to within. (Doczi, The Power of Limits)

All vibration following these key principles nests as spirals within spirals braiding the fabric of creation. (Doczi, The Power of Limits)

motion and the source of locomotion in space. Fire (Tejas) is an upward motion, giving rise to expansion. Water (Apas) is a downward motion, giving rise to contraction. Earth (Prithivi) is a motion which produces cohesion and obstruction, the opposite of the nonobstructive Ether.[22]

Vayu, the Air element, is the breath of Brahma, the motive force of all *movement* in creation. It is from the sanskrit *Va*, "to blow, to move." Air is the principle of motion, and its function is impact. It rules the sense of touch. Tejas, the Fire element, which rules *directed force,* comes from the Sanskrit, *Tij*, "to be sharp." It is the principle of luminosity and the function of expansion in nature. Fire rules the sense of vision which is the source of our direction. Apas, the Water element, is the principle of liquidity and the function of contraction. It comes from the root *ap*, "water." Water rules taste and the nourishment that we take in. Prithivi, the Earth element, is from the Sanskrit *plete*, "broad, flat, spread out, extended." It is the principle of solidity, and its function is cohesion.[23]

The four elements are four specific patterns of resonance—four moments in the evolutionary cycle. Air is a lengthening of the wave form from the sphere of Ether to the ellipse of Air as energy steps-down in a helix from above to below. Fire is a radiant expansive force from within to without. Water is a contractive flow from without to within. Earth is a release of Spirit from matter as the cycle finds completion in form and then returns from below to above. The four elements are four phases in the ubiquitous evolutionary cycle which underlies creation in nature.

The body tissue is very fluid and elastic. The body cavities, organs, and cells breathe through cycles of attunement with the cosmic forces. Each cell of the body pulsates through cycles of expansion and contraction, which are a microcosm of the order of creation. As the breath undulates through the medium of the body, the atoms, molecules, and cells are drawn in quantum fashion through phases of attunement with the holomovement of the forces of nature and the cosmos.

Through sympathetic vibration the elemental forces of nature and the cosmos entrain with the elemental energy systems of the body. Each phase in the cycle is qualitatively defined by its harmonic relationship to the source. Differentiation is determined by variations in the wavelength and degree of arc, positive or negative phase, and distance from the core. The quality and power of any aspect of physical manifestation is a reflection of its attunement to the vibrations of the core of the force field.

It is very important to understand that the four elements are moments in a cycle of wholeness. They seem differentiated, but they are only phases in a single process, relationships within a whole. What we are viewing is a single universal helix, a whirling vortex of bipolar energy. Within this gestalt are the four harmonics, four phases of the primal Oneness. These four elements are attuned to each other as opposite waveforms, balanced harmonics working together to reflect a fundamental Oneness as the basis of the universal synergy. They exist only through their dynamic relationship with each other.[25]

The rhythms of breath, heartbeat, vibration, and oscillation of life within every atom, molecule, cell, organ, and body cavity are profoundly lawful as the embodiment of universal law. The rhythms of these finite systems are organized by the spiral helix of in-breath and out-breath. The breath is our link between the macrocosm of the Air element as the life of a living universe and the microcosm of the finite systems of every atom, molecule, and cell of our living bodies. Our energy fields are entrained in a fabric of "spiration"—inspiration, expiration, respiration—with the universal spiral helix of Spirit, the life breath.

Standing Waves

On the atomic level, standing waves define phenomena. All phenomena are sustained by inherent patterns of resonance. The definitive pattern of resonance of any phenomenon is its standing wave. Standing waves are the pattern of resonance that define an energy system. Standing waves are the "natural" patterns of resonance which define phenomena. Standing Waves "appear whenever waves are confined to a finite region, like the waves in a vibrating guitar string, or in the air inside a flute."[27]

Standing waves appear whenever waves are confined to a finite region. Standing waves appear within the "finite system" of every unit of wholeness

The symbol of the Tao embodies the supreme ultimate principle that underlies all vibration emanating centrifugally and centripetally from a center, reaching the limit of the field, and reversing polarity to return to the source. (Purce, The Mystic Spiral)

of nature—every atom, molecule, cell, and organism. Our body is an example of a finite system which sustains patterns of standing waves. Each body cavity, organ, cell, molecule, and atom is a finite system sustaining standing waves which are the central organizing principle of the energy field. Standing waves always manifest as a logarithmic spiral which can be described through the Golden Ratio. As discussed in Chapter 8, this foundation of universal law is called the Golden Spiral. The Golden Spiral describes the rhythmic generation of the universe embodied in the ubiquitous breath of Brahma. Every level of macrocosm and microcosm manifests from a bindu radiating as the Golden Spiral. Standing waves align themselves with the fields of force of nature and the cosmos. By examining standing waves we can make transparent the underlying "ultrasonic" fields of force of nature and the cosmos.

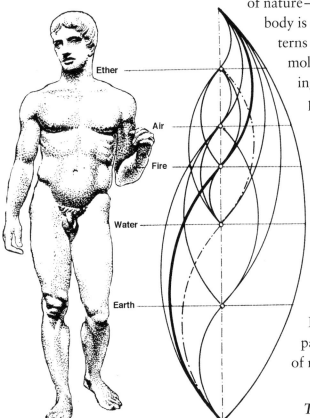

Ether

Air

Fire

Water

Earth

Standing waves are one of the fundamental organizing principles of a cosmology based upon resonance. (Doczi, The Power of Limits)

The Experience of Vibration

The science of the ancients, as expressed in the Sanatana Dharma, is based on the experience of sacred sound.[28] The ancients understood all of nature as woven of the fabric of the four resonant frequencies of vibration that emanate from the ultrasonic core of Ether. They viewed all phenomena as made up of four universal waveforms which sustain the boundaries (Earth), media (Water), energy (Fire), and lawful organization (Air) of all things. They understood that our senses are tuned into these elemental harmonics of vibration, and they based their science on the empirical experience of these four elements. In this model we are not dealing with an abstraction but are describing the living, breathing bodies of humans, nature, and the cosmos. "Energy" in this model is the breath of the cosmos—a living, sentient force. The "elements" describe phases of attunement with Brahma, the Unitive Field of cosmic being, the Creative Intelligence which is the all-pervasive ground of evolution.[29]

All vibration in nature entrains with the elemental archetypes. All creation is an undulating spiral helix. Every level of macrocosm and microcosm undulates in harmony with the spiral dance of this universal holomovement. From above to below (Air), within to without (Fire), without to within (Water), and below to above (Earth), this spiraling breath of God undulates through all creation.

These are the four directions held in reverence by Native Americans. The ancient Hebrews worshiped the sacred mantric utterance of יְהוָה (Yahweh) symbolized by the Lion, Eagle, Bull, and Human, who ruled the four quarters of the universe. The four humors of Greek cosmology throughout Europe, and the four elements which the Sanatana Dharma spread throughout Asia reflect the same ancient paradigm.

Our experience of boundaries is ruled by the Earth frequency of vibration, and throughout our dimension all phenomena sharing this resonant frequency are experienced by us as "solid." The experience of fluidity expresses the Water harmonic. Heat is attuned to the Fire element throughout our system, and movement itself predominates in the Air-resonant frequency. Our senses are attuned to these kingdoms of vibration: the ears to Ether, touch to Air, the eyes to Fire, taste to Water, and smell to Earth. The senses and mind weave the fabric of our experience through their attunement with these elemental frequencies of vibration, much as a television set constructs an image from the frequencies to which it is attuned.

Crossculturally and transhistorically, the sciences of the ancients are based on the five elemental keynotes of energy. The ancients understood that all phenomena, as energy fields, are made up of a combination of all the elements. It was understood that phenomena, seen as energy fields, are composed of Earthly boundaries, Watery media, Fiery functions, and Airy lawful vibrations, radiating outward from an suprasensory Etheric core. The element predominating in the energy field gives each phenomenon its characteristic qualities of solidity, fluidity, force, or movement.

Sri Krishna Prem (Ronald Henry Nixon, Ph.D.) insisted in reference to the elements: "We reiterate that no matter how much we use, and legitimately use, the symbols of cosmology and of daily life to evoke images in illustration of our contentions, we are in fact speaking of psy-

		Summary	
Gunas	Principles	Elements	Chinese
Sattva	Air	Ether/Air	Wood/Metal
Rajas	Fire	Fire	Fire
Tamas	Water	Water/Earth	Water

chic movements in consciousness and of this alone."[30] Dr. Prem, a university professor teaching in India, was a lifelong student of yoga. He personally realized the Self, the Himalayan truth that all creation is the play of consciousness of the universe as one living being.

Thus, though every physical phenomenon contains all of the elements, one elemental keynote of resonance gives the form its predominant qualities.

Particular structures and processes in the body and in nature are characterized as "Air-predominant" (nervous system), "Fire-predominant" (vital organs and muscles), and so on, and embody the qualities of those particular elemental archetypes.[31]

Air-Predominant Harmonics

The Air element resonates with sattva guna, the superconscious force of equilibrium in nature. Air, the first aura of emanation from the stillness of the Etheric core, is an energy field whose force embodies Cosmic Intelligence. Air is characterized by movement—not ordinary movement, but the intelligent movement which underlies the self-regulating fields of nature and cosmos. The qualities of lawfulness and of the profound harmony and integration of all movement in creation are expressed in the Air element, which fills the field of manifestation with the spiraling breath of universal law. Air is the fulcrum, or bindu, between the Unitive Field and its creation.

The Air element predominates in the thoracic cavity, lungs, ectoderm, nervous system, and the sattvic rhythm of the breath, which is always filling the space of the body with the re-spiration of universal law. Throughout the body this resonant frequency of parasympathetic vibration forms a fabric of wholeness and communion with the sattvic harmonic. Air rules the neutral, fulcrum field of the diaphragm and solar plexus and the balance of movement between the centrifugal resonance of the thoracic cavity above and the centripetal resonance of the pelvic cavity below. Air predominates in the colon and rules the rhythms of peristalsis throughout the body. Air (Libra) governs the kidneys, which purify the sacred waters of life. Air predominates in the ankles, which are the structural foundation of movement and direction. The fulcrum for Air and the center of our "spin" is the heart. The Air element arises "energetically" from the heart center, or Air chakra, and rules our desires in life, our passions, and our aversions as the basis of our movement in life.

The Air element predominates in a light touch and on the surface of the body. Air "inspires" and "respires." Air predominates as inspiration and high spirits and our enthusiasm for and love of life. Air rules mind with its incessant mercurial movement. Its characteristic type of arc is a bipolar spiral of prana. Air governs transverse motion and locomotion. Its field of force is a lawful balance of centrifugal and centripetal forces in a spiral, a helix of attunement with the one source. Air fills space with the breathing, living balance of universal law.

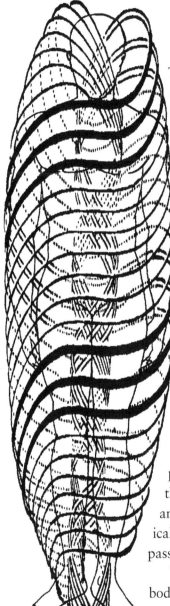

Air predominates in the transverse currents which integrate the periphery of the body with the core.

The thoracic cavity is the home of the Air harmonic of resonance in the body. Air predominates in the astrological harmonic of Gemini, with the spine and shoulders as the positive field; Libra, with the kidneys and renals as the neutral field; and Aquarius, with the ankles as the negative field. The Air element rules the sense of touch, the heart, and the breath of life, prana. Air-predominant touch resonates with the Air-predominant heart.

The currents of energy that spiral out from the ultrasonic core at the crown of the head and move from east to west around the surface of the body acting as sensory and feedback mechanisms, returning to the core, are Air-predominant. Air is a transverse force between the polarity of the two fields of the bilateral symmetry of the body and the ultrasonic core in the spine. Our nervous system resonates with the keynote of Air. A gentle touch, soothing movements, and balancing contacts activate the keynote of sattvic vibration, Air. Sattvic touch contacts the heart of Being. Sattvic touch fosters a receptivity to the healing presence of the Soul.

Fire-Predominant Harmonics

The second phase of emanation in the Polarity cycle is the resonant frequency of Fire. While Air rules the lawfulness of all movement as an expression of the superconscious mind of sattva guna, Fire rules the direction of movement as it resonates with rajas guna, the conscious mind, and the sympathetic system. Fire rules directed force in energy systems. Fire is a radiant, positive, electric, pushing, expanding arc of centrifugal force. The waveforms of Fire radiate outward and upward. Fire always burns upward, resonating with the Sun. Throughout our solar system all force is a step-down of radiation from the Sun. The solar force is embodied in the expansive alignment of the in-breath. Throughout all creation the solar force is the centrifugal force of radiance that rules, regulates, moves, and transforms. The initiative of Aries, the reign of Leo, and the integrity of Sagittarius express the nature of Fire as directed energy. All creation is aligned with the nuclear potency functioning through the solar harmonic of the Fire archetype.

Luminosity, heat, and caloric energy are sustained by the vibrational frequency of Fire. This electric pattern of energy predominates in the fields of

Fire radiates spirals from its center at the Umbilicus and Solar Plexus to provide warmth and vitality to the whole body.[32]

force which manifest warmth and function, as in the vital organs and muscles. Fire is the yang, directed, working, passionate, vital energy in the body. Its radiant working energy rules the organs, the processes of metabolism, and the endoderm, and is the motive force of the musculature. Any stimulating or rajasic contact resonates with Fire and promotes its attunement. Fire is centered in the solar plexus, two finger widths below the umbilicus.

Fire radiates vital force and defines the body from within to without.

> This type of energy flows mostly over muscle tissue to act and express itself. Any therapy designed to assist this must be active, vibratory, and positive in nature. All stimulating applications are in this field … [using] vibratory directional force over the muscles involved … blending into the energy current in purpose, speed and direction like two rivers joining into one. This has the greatest therapeutic value … The impulse must have momentum, but no compulsory force which causes a reaction.[33]

Water-Predominant Harmonics

The third field of emanation manifests waveforms that predominantly resonate downward, through the magnetic pull of gravity, toward the negative pole at the nucleus of the Earth. Water always flows downward, creating contraction and an arc of *centripetal* force. This receptive phase of the creation cycle sustains the keynote of Water and corresponds to the contracting outbreath of downward motion, cleansing, and elimination.

It is through the cohesive force of water that energy precipitates into substantial "solidity." In Water the field of force emanating from the source is reduced to a more material frequency that we experience as life forms. Our experience of fluidity predominates in the Water wavelength, embodied in the liquid medium and in the cycles that purify and sustain life, such as blood flowing through veins, or rivers flowing to the ocean.

Water rules cohesion, attachment, and love. Water rules the ways we connect to the world around us for nourishment. Cancerian love, the deep bonds of Scorpio, and the Piscian flow from Heaven to Earth nourish and connect us. Water rules the sense of taste. It rules the ways in which we are emotionally connected to others and the relationships that are emotionally nourishing.

The Water element predominates in the pelvic cavity. As Water is the mother of life, reproduction is centered on the Water level of vibration. The life-giving seed of semen and ovum is in the form of water. Water rules the cycles of cleansing, purification, and elimination in the body, it predominates in the mesoderm, the fluids, secretions, white tissue, and tendons. These magnetic negative fields of the body resonate with tamasic qualities, and are characterized by receptivity, materialization, contraction, resistance, inertia, and crystallization into form.

Water-predominant frequencies sustain the liquid medium that receives and embodies more subtle energy fields. As the field of force radiates outward and downward, it interacts with the more material atmosphere of the Earth's energy field and it loses its liquidity and resilience. Thus it loses its ability to resonate with the One Life. The vibrations flatten and become more and more crystallized into an Earth-predominant phase of resonance. The surface tension of the force field then predominates, creating an impenetrable boundary.

Earth-Predominant Harmonics

Throughout all of nature the phenomenon of solidity is sustained by a pattern of energy that we call "Earth." Earth is the force that produces boundaries and is the polar opposite of the unobstructed vibration of Ether. The crystallizing, linear tendency of the Earth archetype, at the limits of the potency of the core resonance, sustains boundaries and "solid" phenomena. This wavelength is the predominant resonant frequency in all fields of phenomenal force that have the quality of solidity, such as the bones in our bodies. Taurus rules the boundary of the etheric body which defines the overall space, Virgo operates the boundary between body and waste, and Capricorn is the foundation of the skeletal structure.

The colon and the bones are the home of the Earth element, which rules elimination. Earth is the final stage in the crystallization of Ether, and the colon and bones resonate with a phase dynamic that cycles between waves and form, Ether and Earth. The archetypal alchemical symbol of Ether is a circle of balanced energy where all forces are equal. The alchemical symbol of Earth is a square, in which all forces have crystallized into form. The relationship between Ether and Earth parallels the relationship between plasma/marrow as the Ether-predominant core and its crystallized emanation of bone.

At Earth the rajasic radiant spiral loses its ability to vibrate sympathetically in attunement with the One and tamas, the sinking spiral, predominates. The material form eventually crumbles into dust and the life-giving spiral of Spirit is drawn back to the One. Form manifests only as it is sustained by its harmony with the source and as its resonance with the "tune-i-verse." The Earthly level of vibration is force that is slowing in vibration from spiraling towards linearity and crystallization.

The Elements as Archetypes

Dr. Stone writes:

> The cosmic forces are vast reservoirs in which these myriads of units
> float and recharge their external field by energy from the cosmos through
> its spinning chakras and nerve centers to its own keynotes and use. When
> the central energy of the soul cannot toil and spin its wheels and shuttles
> effectively any more to extract what is needed from the outside reservoir,
> then its link is broken and the pitcher goes to the well no more. The Earth
> elements go back to Earth; the air to dust; the fire to ashes . . .[34]

All phenomena are made up of the pentamirus combination of these
wavelengths of vibration—the *boundaries* of Earth, the *cohesion* of Water,
the *directed force* of Fire, the organizing, *lawful movement* of Air—all radiating within the Etheric *space*. In every instance, energy in nature is an expression of these five harmonics of vibration. These are the underlying keynotes
of resonance that sustain all phenomena. Our senses and minds interpret these
vibrations as the experience of life.

The five elements are archetypes, founts of Creative Intelligence whose
cycles weave the splendor of evolution. The five elements are five stages in the
manifestation of consciousness. The essence of the elements as consciousness
is: Ether, the emotional space that seeds manifestation; Air, thoughts and ideas;
Fire, action; Water, feeling; Earth, completion. For example, desiring a home
is Ether-predominant. Formulating plans for construction is Air-predominant.
Taking action and building the home is Fire-predominant. The feeling of being
home is Water-predominant, and the fruition of the desire in form is Earth-
predominant.

Ayurveda, the ancient East Indian science of life and healing, works with
the image of all creation as energy fields of sacred sound vibration. All of
nature on every level is understood as a vast harmony of cycles of resonance.
All of creation reflects this harmony in a profound integration; every level of
wholeness is a microcosm of the harmonies of the whole.

On the physical plane this harmony is understood as being woven of the
five harmonics whose fields of force are the keynotes of a universal harmony.
Nature, our bodies, and our senses are woven of these universal patterns of
resonance. In our dimension, our five senses and the range of experiences
which they reveal to us are reflections of these five fundamental keynotes of

vibration that weave the fabric of our being in nature. All of nature is woven from the five elements of universal law:

Element	Sense	Food		Taste
Ether	Hearing	Amrita		Divine Taste
Air	Touch	Fruit, nuts, seeds		Sour
Fire	Sight	Grains, legumes, sesame, sunflower		Bitter
Water	Taste	Leafy greens, melons, squash		Salty
Earth	Smell	Root vegetables, tubers		Sweet

For the ancients, all phenomenal experience is a reflection of these five archetypes of energy. They believed that the totality of our experience of solidity, regardless of the particular phenomenon, can be understood as a reflection of the harmonic of the Earth element. All of our phenomenal experience of liquidity is ruled by the keynote of Water. All of the waters which sustain life are understood as a single harmonic. Igneous and caloric energy—light and heat—are ruled by the Fire element, and our experience of these phenomena is a reflection of a single pattern of energy. The ancients believed that all movement is lawful, taking place as an emanation from the One Universal Life.

The Law of Sympathetic Resonance

In his brilliant book, *The Principles of Light and Color,* E. D. Babbitt, the nineteenth-century father of color therapy, offers an understanding that parallels the ancient wisdom (see Chapter 11). Babbitt experimented with color in his quest for an understanding of "the harmonic laws of the universe." In his model, sound and sympathetic vibration are the key to the process which underlies the polar cohesion of atoms. He explains:

> Why are ethers drawn from spirals of one atom to the same kind of spirals in a contiguous atom, and why does a certain grade of ether exactly harmonize with, and seek out, a certain size spiral? For the same reason that a tuning-fork or the chord of a piano will be set into vibration by a tone made in its own key. In the case of the piano, a chord vibrates to tones of its own pitch, or in other words, to tones whose waves synchronize with its own vibrations. Let us apply this principle to atoms. The vibratory action of the red spiral throws the current of ether which passes through it into the eddy-like whirl which just harmonizes in size and form to the red spiral of the next atom above it with

Babbitt's Atom. (Babbitt, The Principles of Light and Color) which it comes into contact, and which must necessarily draw it on. This second atom passes it on to the red spiral of the third, the third to the fourth, and so on through millions of miles, so long as there is a spiral of the right grade to conduct it onward.[35]

Describing the same phenomenon, Itzhak Bentov explains rhythm entrainment, the process where: "periodic events that are close in frequency occur in phase or in step with each other." He goes on to write: "Our biological rhythms are entrained by light and to a certain extent by gravity … Magnetic, electromagnetic, atmospheric, and subtle geophysical effects influence us in ways that are not presently well understood." This process is significant in the macrocosm as well, where "the big and mighty asteroids and planets are rhythm entrained and develop resonances in their orbits as they rotate about the sun."[36]

The law of sympathetic vibration describes the tendency of vibrations of the same wavelength to resonate together. Sympathetic resonance may be the glue that holds creation together. Waveforms of the same frequency will entrain (synchronize their vibrations) to energize and influence each other.

The ancients understood our earthly sound to be an octave within a universal system of resonance. The "elements" of the ancients were the keynotes of earthly manifestation. All vibration is attuned to these archetypal harmonics of force. In the Sanatana Dharma a distinction is made between the *subtle* elements which are timeless universal archetypes, and *gross* elements, which are the fivefold patterns of vibration that underlie manifestation. The gross elements resonate sympathetically with the subtle inner elements and thus draw sustenance from these cosmic forces.

In the ancient wisdom, all phenomena were understood as woven of a fabric of five archetypes of vibration. The science of the ancients was based on these five elemental resonant frequencies. These are the four directions of resonance, plus a fifth, a stillpoint of wholeness (Fire, Water, Air, Earth, and Ether). There are the tattwas, the quintessential or fivefold essences of creation. These five phases of cosmic attunement undulate through every cycle of vibration. This is the key to understanding this archetypal paradigm and the relationship between macrocosm and microcosm.

To summarize: vibrations attuned with the ascending, solar, centrifugal force, radiating from the nucleus of our system, the Sun, resonate with the Fire element. Vibrations attuned with the descending, centripetal, negative pole of the Earth resonate with the Water element. Vibrations attuned to the flow from within to without, from the finer forces of the Unitive Field, resonate with the Air element. Vibrations attuned to the flow from without to within, from the periphery of the energy field back toward the core, resonate with the Earth element. The pentamirus combination sustains a stillpoint which resonates with Ether. These are the five phases of cosmic attunement that undulate through every spiraling vibration in creation. These are the cosmic forces that are the keynotes of creation. (We will focus on this key, to the understanding of the ancient wisdom, throughout this book.)

All phenomena exist through their resonance with these ubiquitous archetypal forces. All phenomena draw their energy, consciousness, and being from their attunement with these archetypal forces. In Dr. Stone's view, man is understood as an integral part of the quantum fields of nature and the cosmos where there is a constant exchange of energy. He explains:

> That man must have air, food, liquids, clothing and warmth from outside is accepted in so far as solids, liquids and air are concerned. But the idea of atomic energy fields of exchange in these conveyors and by direct absorption, is still a brand new idea, awaiting acceptance. Man is considered more or less an independent unit. His dependence upon exchange of energy in the wireless field of nature, like a receiving set in a broadcast field is still a new idea. Even his grounding to the Earth is overlooked—plus his possibility of direct perception and absorption of energy . . . [the] constant exchange of the finer forces of nature with man is questioned and doubted.[37]

> [these are the elements] which are built into the constitution of man as receiving sets or fields for the operation of the five forces or broadcasting wave lengths in nature [and an] . . . intricate receiving set for the cosmic forces which sustain man.[38]

Dr. Stone explains: "Small units of energy fields draw their supply of energy from larger units or orbits to which they are attuned, because they are their home, *source or parent fields*."[39] In the Sanatana Dharma, everything in creation vibrates through this universal holomovement. The five phases breathe through every atom and galaxy of creation. Everything is entrained by this holomovement. Everything is vibrating sympathetically with these archetypes.

The radiation of the potency of the Etheric center occurs in cycles, of which our breath is a microcosm. This central potency radiates a field of force in a cycle of four phases, sustaining the four elemental harmonics to form the basis of all creation. It is echoed and embodied in the cycle of the breath. As an organism breathes, the totality of the organism is drawn through a cycle of four phases. This holomovement takes place on every level of wholeness,

(Doczi, the Power of Limits)

as every cell and atom of the body breathes through the cycles of attunement and entrainment with the archetypal forces of nature and the cosmos.

The four elemental resonances manifest in relationship to the three principles of energy movement—sattva, rajas, and tamas—to sustain the twelve great archetypes of the zodiac which rule manifestation. Their spectrum of creativity is expressed in the range of creation in each kingdom: mineral, plant, animal, human, devic, planetary, etc. Each moment of creation has a unique "signature" that expresses the archetypal resonance of the cycles of eternal creativity that intersect in that facet of evolution. The moment of manifestation, the first breath, places a "being" in the infinite cycles of creativity and defines its pattern of resonance. Through sympathetic vibration the "personality" of the person, plant, animal, or stone reflects the cycles that intersect in its being.

The Synergy of the Body a Microcosm of Universal Synergy

The concept of synergy describes systems in which the organization and functioning of the parts maximally sustain the whole, and the organization of the whole maximally sustains the parts. Nature and the cosmos are a synergistic system, where there is ultimate integration between the parts and the whole.

Throughout every process, structure, and function we see a profound level of synergy and integration in our bodies. On every level, along every conceivable vector, what promotes the well-being of the part promotes the well-being of the whole, and vice versa. The body is a model of ultimate integration, ultimate synergy—a microcosm of the ultimate synergy of all creation.

On the subatomic level, elementary particles follow a set of laws which cause them to group together in patterns known as atoms. We now know of over one-hundred different kinds of atoms. Each of these has its own unique grouping of subatomic particles, and the pattern of each requires the particles to group in that way and no other.

Following the same laws that govern subatomic particles, atoms group together in patterns known as molecules. Hundreds of thousands of different kinds of molecules are now known, and new ones are constantly being found, or created. The number of possible molecules may be infinite. Each kind of molecule has its own distinct properties and its own distinct pattern that requires atoms to group in one way and no other.

On the next level of organization, many different kinds of molecules link themselves together in complex and intricate patterns to form the smallest unit of life, the cell. In the human body alone there are thousands of different kinds of cells, each with its own unique molecular pattern, and each with a unique job to do. The pattern of each cell will allow it to accept only certain kinds of molecules, and requires that those molecules be organized and structured in a certain way and no other.

Cells with similar structures and functions group together to form tissues. The tissue is the lowest level of macroscopic organization in the body. All of the body's organs and gross anatomical structures are composed of tissue. The organs of the body, likewise, are linked structurally and functionally into systems.

Biochemically and energetically, the body is a delicately balanced system of interlocking feedback loops, where a change in the function of even one tissue quickly causes changes throughout all the other tissues, organs, and systems. Change one subatomic particle in one kind of atom, and you change that atom into something different. Change one atom in a molecule, and you change the molecule. Change the molecule, and you change the cell, the tissue, the organ, and the entire functional balance of the body.

Any change in even the finest level of our being will ultimately be reflected in every level, up to the most gross. Conversely, no change can take place at the gross level without causing changes to echo back to the finest level. In every dimension, along every vector of possibility, along every level of organization, the human body, the body of nature, and the body of the cosmos is synergistic. Synergy is a foundation of universal law, a law that is grounded in the Divine Intelligence that underlies creation. As the *Gita* reveals: "God *becomes* creation."

The Synergy of the Holomovement

In the Sanatana Dharma all creation is one life, a single field of cosmic evolution. In this paradigm the elements are universal harmonics of resonance, emanating from Brahman, the Unitive Field of consciousness. The ancient scientists used the concept of sacred sound where we use the concept of energy. All phenomena on every level of microcosm and macrocosm are aligned vibrationally with the elemental harmonics of resonance.

This vast creation of nature and the cosmos is a synergistic system, where there is ultimate integration between the parts and the whole through the synergy of a ubiquitous elemental holomovement. The harmony of this elemental holomovement emanates from the Unitive Field of Brahman that is the innermost center of all that is. The nature of its emanation as the systems and subsystems of nature and the cosmos is theomorphic.

Meditation: Who Am I? Contemplation
Sri Ramana Maharshi[40]

Who Am I?
The gross body
which is composed of the seven humors,
I am not.

The five cognitive sense-organs, viz.,
the senses of hearing, touch, sight, taste and smell,
which apprehend their respective objects,
viz., sound, touch, color, taste, and odor,
I am not.

The five conative sense organs,
viz., the organs of speech, locomotion, grasping,
excretion, and procreation, which have as their
respective functions
speaking, moving, grasping, excreting, and enjoying,
I am not.

The five vital airs—
prana, etc., which perform respectively
the five functions of in-breathing, etc.—
I am not.
Even the mind which thinks, I am not.

The nescience too, which is endowed only
with residual impressions of objects,
and in which there are no objects and no functioning,
I am not.

If I am none of these, then who am I?
After negating all of the above-mentioned as "not this," "not this,"
that Awareness which alone remains—that I am.

What is the nature of awareness?
The nature of Awareness is existence/consciousness/bliss.

When will the realization of the Self be gained?
When the world which is what-is-seen has been removed,
there will be the realization of the Self which is the seer.

CHAPTER 10

Sound is the Temple of the Spirit

*Life is the expression of love in sound waves and
energy currents, throughout the creation and in
man. Love is light, which crystallizes as beauty in
the spectrum, becomes color and gases as it is
reduced in speed of vibration, and also forms the
beautiful colors in the buds, the flowers, and the
fruits. It precipitates as the delicate pink color of
the lacework of tissues in the human form. Everywhere is the expression of love and beauty.*
—Dr. Randolph Stone[1]

In earlier chapters we introduced the importance of sound for an understanding
of the wisdom of the ancients. Their cosmology was based on the concept of
sacred sound vibration. Their science was focused on sacred sound vibration.
In this essay we look more deeply into sacred sound.

In the Sanatana Dharma, energy is understood as sacred sound. In contrast to contemporary thought, in which we use the concept of energy, the
ancients use the concept of sacred sound vibration. According to the Vedas,
the universe was sung into being. The *Mandukya Upanishad* teaches: "The
syllable ॐ (Aum), which is the imperishable Brahman, is the universe. Whatsoever has existed, whatsoever exists, whatsoever shall exist hereafter is ॐ"[2]
The *Taittiriya Upanishad* teaches: "Thou art Brahman, one with the syllable
ॐ ... the mother of all sounds." In the ancient wisdom all creation is but an

echo of ॐ, the Word. ॐ, the "Logos," or "Divine Word," is vast, it is the medium of creation. It voices the Creator. Its song is creation. A divine depth fecundates sound. Its vastness encompasses totality. Its subtlety consciously spins every atom of creation. Creation is the song of God. Creation *is* sacred sound.[3]

Contemporary musicologist Joachim-Ernst Berendt explains that *Nada Brahma* means "Sound is God," or vice versa, "God is Sound." "So *Nada Brahma* means not only: God the Creator is sound; but also (and above all), Creation, the cosmos, the world, is sound. And: Sound is the world."[4]

The Tantric View of Sacred Sound

In the Tantric view, sound as a vibration of undifferentiated Intelligence is the catalyst that sets into motion the unfolding of the manifest cosmos.

> The causal vibration is the *shabdabrahma* or undifferentiated sound-less sound. It is the wave length of the experience of God. *Para shakti* is its slumbering energy. It contains the three qualities of sattva, rajas, and tamas. A primal shudder disturbs the equilibrium of this para shakti and arouses rajas, the active principle that carries out the creation of the manifest universe. The great cosmic vibration splits shakti into two fields of electro-magnetic force, *nada* and *bindu*. The centrifugal, positive male force, bindu, is the ground from which nada operates. Nada, the centripetal, negative, female force, unfolds the manifest universe. This duality of poles in the substratum of manifested shakti actually provides the electro-magnetic force holding together the molecules of the physical world in a state of vibration.[5]

After the first differentiation, the vibrating mass of energy continues differentiating and expanding as wavelengths. By the fifth vibration, the energy has evolved as the gross plane with the creation of the fifty primeval sounds, or *varnas*. The varnas are defined as "phonetic molds" with natural significance.[6] Varnas also means "color," and all sounds have corresponding color vibrations on the transcendental plane. From the combinations and permutations of these root sounds, the universe of forms is created.

> Sounds as physical vibrations are able to produce predictable physical forms. Combinations of sounds produce complicated shapes. In order to produce a particular form, a specific note of a particular pitch must be generated. Repetition of the exact note results in a duplication of the form. Therefore, underlying all the forms of the manifest world are the oscillating

wave lengths of primeval sounds in varying combinations. There are considered to be fifty of these primeval sounds. The Sanskrit language is directly derived from these fifty basic sounds, and thus of all languages it is the closest approximation to the language of creation. Sound is potential form, and form is sound made manifest.[7]

The ancient seers, in using microclairvoyant powers, witnessed the vibratory basis of reality. Madame Alexandra David-Neel, in her classic explication of the Tibetan cosmology, writes: "All things . . . are aggregations of atoms that dance, and by their movements produce sounds. When the rhythm of the dance changes, the sound it produces also changes . . . Each atom perpetually sings its song, and the sound at every moment creates dense and subtle forms."[8] Frijof Capra, in *The Tao of Physics*, writes in reference to the above quote: "The similarity of this view to that of modern physics becomes particularly striking when we remember that sound is a wave with a certain frequency which changes when the sound does, and that particles, the modern equivalent of atoms, are also waves with frequencies proportional to their energies. According to field theory, each particle does indeed 'perpetually sing its song,' producing rhythmic patterns of energy in dense and subtle forms."[9]

Gary Peacock, a well-known musician, interrupted his career and undertook a four-year study of molecular biology and organ physiology. He found that the relationships in the periodic table of elements of which matter is formed mirror the overtone structure of music. Peacock avers: "It becomes clearer and clearer to me that the actual structure of tone in music and the actual structure of matter are the same."[10]

The Law of Three[11]

Modern physics reveals that the entire perceptible universe is composed of vibrations. Gravity, electromagnetism, light, heat, and even matter itself are wave phenomena. Waves are patterns in time, moving configurations composed of amplitude, interval, and frequency. Every living body vibrates: organic, elemental, molecular, and atomic matter vibrate as wave phenomena. The model of universal law revealed by the ancients is a science of sacred sound. In the ancient wisdom all phenomena are manifestations of the underlying substratum of the vibration of sacred sound. Bentov points out in *Stalking the Wild Pendulum*: "We can associate our entire reality with sounds of one kind or another, because our whole reality is a vibratory reality."[12]

Audible sound consists of three elements: pulse, wave, and form. Although we can analyze them separately, they always occur simultaneously. "Pulse" describes the generating aspect of sound. Pulse is the ubiquitous breath of Brahma, whose rhythm is the Unitive Field of cosmic breath—the electro-magnetic spectrum—which is the motive force of all creation. The force of expansion and contraction of this breath is the tri-gunas, the supreme ulti-mate principle in manifestation. The throb (*spanda*) of this pulse vibrates as sound (nada). The breath of Brahma sings the Unitive Field of the underlying ॐ of the cosmos. All of the systems and subsystems of creation are but an echo of the cosmic ॐ.

Brahma, the Unitive Field of the cosmic breath, manifests as a trinity of sattvic, rajasic, and tamasic forces that are personified as the Lords Vishnu, Brahma, and Shiva (neutral, positive, and negative). Vishnu the Preserver is the sattvic *pulse* of the Unitive Field. Brahma the Creator is the positive field of *waves* radiating into creation; and Shiva the Destroyer is the crystalliza-tion of the wave into *form*. With pulse, wave, and form we have the neutral, positive, and negative—sattva, rajas, and tamas of the tri-gunas—which characterize the holomovement underlying all manifestation.

Wave and form are simultaneously created from pulse. They are symbolized esoterically as the two snakes of the caduceus. Usually we picture sound as a wave. The rising and falling of the wave is in alignment with the expansion and contraction of the pulse. Dr. Stone called the pulse of Sacred Sound within ourselves the "ultrasonic core."[13]

The Russian mystic Gurdjieff refers to the relationship of wave, pulse, and form as "the fundamental law" that creates all phenomena.[14] "This is the 'Law of Three,' the law of the three principles or three forces. It means that every phenomenon, on whatever scale and in whatever world it may take place, from molecular to cosmic, is the result of the combination of the meet-ing of three different forces."[15]

The evolutionary cycle of the polarity between centrifugal and centripetal forces circulating through a center is the essence of theomorphic holomove-ment. Usually we think of sound as a wave.

A Swiss scientist Hens Jenny has researched the relationship between sound and form. He calls this process cymatics. Describing the cymatic process of sound generating form, he explains:

The various aspects of these phenomena are due to vibration, we are confronted with a spectrum which reveals a patterned, figurative forma-tion at one pole (form) and kinetic-dynamic processes at the other (wave),

The rising and falling of the wave, shown by the arrows, is in alignment with the expansion and contraction of pulse. The original pulse is always contained in the wave. In this diagram one pulse is equivalent to one com-plete cycle of the wave. Sound waves are mea-sured in cycles per second (cps). It would also be correct to say pulses per second ... Form is the more elusive component of sound.[17] (Winters, Alphabet of the Heart) the whole being generated and sustained by its essential periodicity. The three fields—the periodic (pulse) as the fundamental field with the two poles of figure (form) and dynamic (wave)—invariably appear as one. They are inconceivable without each other. It is quite out of the question to take away one or the other; nothing can be subtracted without the whole ceasing to exist. We cannot therefore number them one, two or three, but can only say they are threefold in appearance and yet unitary: that they appear as one and yet are threefold.[19]

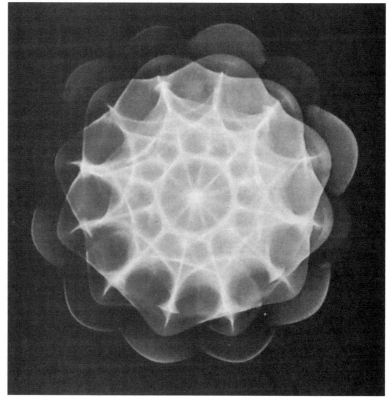

Cymatic: harmonic pattern created in liquid.[18]

As above, so below. Cymatics, the study of generating form from sound, offers images that parallel quantum particles. Both appear to be solid form, yet are also a wave; both are created and simultaneously organized by the principle of pulse. In each case, form that appears solid is actually created by the pulsation of vibration. Cymatic research suggests a parallel between macrocosm and the atomic microcosm. It appears that the ancients conceived of manifestation in ways that parallel the illustrations of cymatics. In the ancient wisdom all phenomena are manifestations of the underlying substratum of the vibration of sacred sound.

The Unitive Field

In the ancient wisdom we have a universe whose substratum is the all-pervasive creative potency of the ever-expanding breath of Brahma. This parallels the concept of the quantum field in the models of quantum physics. Capra explains: "One is led to a new notion of unbroken wholeness which denies the classical idea of the analyzability of the world into separately and independently existing parts . . . we say that inseparable quantum interconnectedness of the whole universe is the fundamental reality, and that relatively independently behaving parts are merely particular and contingent forms within the whole."[20] Capra further states that "The quantum field is seen as the fundamental physical entity, a continuous medium which is present everywhere in space."[21] The quantum field, like the Life Field of the breath of Brahma, is universal.

Standing Waves

Every shape no matter how complex
can be assembled or woven from a simple sum
of pure sine waves.
—Fourier Principle[22]

The unity and synergy of the electromagnetic spectrum together form a central organizing principle of nature. The universe is a vast *harmony* of waves of vibration. Capra writes: "The phenomenon of waves is encountered in many different contexts throughout physics and can be described with the same mathematical formalism whenever it occurs. The same mathematical forms are used to describe a light wave, a vibrating guitar string, a sound wave, or a water wave. In quantum physics, these waves are used to describe the waves associated with particles."[23] Gurdjieff concurs that:"the laws are everywhere the same, and that light, heat, chemical, magnetic and other vibrations are subject to the same laws as sound vibrations ..."[24] Modern physics parallels the ancient wisdom understanding that electromagnetic phenomena are subject to the same laws as sound vibrations. As above, so below—every level of macrocosm and microcosm is aligned with the harmonies of the electromagnetic spectrum.

Standing waves manifest as a logarithmic spiral.[25] (Picture this spiral three-dimensionally as an undulating spiraling vortex, spiraling both centrifugally and centripetally. (Doczi, The Power of Limits)

Resonance Is the Basis of Form

On the atomic level, standing waves are the "natural" patterns of resonance that define phenomena. Standing waves appear whenever waves are confined to a finite region. "Standing waves give the illusion of stability, segregation of momentum, and make possible the birth of matter."[26] Standing waves are the pattern of resonance that define a stable energy system. When a structure is in resonance (which means that it vibrates at a frequency that is natural to it and is most easily sustained by it), it implies the presence of standing waves.[27] On the atomic level the resonance of the standing waves *is* the basis of form.

Finite systems *are* patterns of resonance. Standing waves define finite systems. Standing waves appear within "the finite system" of every unit of wholeness of nature: every atom, molecule, cell, and organism. The illustration below points to the idea that the subsystems of the body are defined by the harmonic proportions of the standing waves. Our body is a finite system which sustains patterns of standing waves. The standing waves of the body align themselves with the standing waves of the fields of force of nature and the cosmos. The standing waves of the microcosm align with the standing waves of the macrocosm.

Standing Waves and the Golden Spiral

Standing waves always manifest as the logarithmic Golden Spiral, which is described by the Fibonacci progression (previously discussed in Chapter 8)

The Golden Spiral is a key to looking beyond the physical to the unified field which is the substratum underlying manifestation. All physical phenomena exist in alignment with this underlying field. The fundamental features, structure, and function of all organisms are reflections of the fact that they are subsystems within the larger energy systems of nature and the cosmos from which they derive their energy and being. The Golden Ratio and the logarithmic succession of the Fibonacci progression are fundamental expressions of universal law and describe the standing waves within every subsystem of nature and the cosmos. Such proportional harmonics of resonance from Spirit to matter are the basis of the unified field of nature and the cosmos.

Throughout all creation there is a step-down in vibration from the inner spiritual essence of the Life Field outward to a material manifestation. This step-down occurs through phases that reflect the Golden Spiral and sustain the theomorphic harmonics of force. The proportional harmonics of standing waves are fundamental to the organization of the body.

Dinergy: Proportional Harmonics in Nature

Gyorgy Doczi, in his illuminating book, *The Power of Limits*,[29] illustrates the role of proportional harmonics in nature. He argues that all growth in nature aligns itself with universal harmonics of force. All structures in nature reflect the proportions and patterns of these invisible forces in nature. In species after species of plants, insects, and animal forms, through almost two hundred

The resonance of the standing waves which define the energy fields of the body sustain a spiral helix whose center is the navel. Spiraling from this center are the proportional harmonics which define the resonant organization of the energy fields of the body.[28] (Doczi, The Power of Limits)

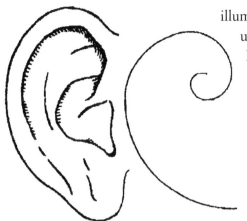

The harmony of the Golden Spiral underlies the radiation of energy in nature. The Golden Spiral is easily seen at the ears which resonate with the Etheric plane.

illuminating illustrations, Doczi clearly demonstrates the ubiquitous organization in nature along proportional harmonics. He demonstrates that nature is woven of a fabric of two spirals—centrifugal and centripetal—whose polarity universally forms the underlying structure of the growth process. He calls this polarity relationship "dinergy," and points to dinergy as the foundation of all growth in nature. All things grow through a union of complementary opposites. Structures in nature are formed by spirals that move in opposite directions—clockwise and counterclockwise. As a spiral unfolds from the center, the order of growth increases logarithmically at the same rate. The spiral contracts back to the center at the same rate. The proportional rate of growth of these expanding and contracting spirals is ubiquitous, found throughout all of nature and the cosmos. This proportion (1:1.618) is the Golden Ratio.

Following the Divine Proportion the radiation from center sustains a spectrum of harmonics of force. This spectrum of harmonics of force follows the same proportions throughout the universe. The same mathematical formula describes the curve of the galaxy as the curve of a chamber nautilus, the rhythms of an electrocardiogram, or oscillations of a tuning fork.[31]

All vibration is theomorphic, expressing the entire spectrum of harmonics of the creative process. Every level of creation has as its underlying "anatomy," a dynamic of vibration that spans the spectrum from the infinite to the finite. This dynamic sustains the structures that embody the process of creation. This spiral harmony of Polarity unites all energy fields. Thus, in our theomorphic model, every dynamic and structure in creation embodies the entire divine process of creation. All structures in nature are aligned with the ubiquitous creative potency of the proportional harmonics. All living things resonate with an infinite depth of resonance, in which everything is potentially enfolded within everything.[32]

Doczi is describing what we call the Polarity principle and the "elements," which can be better understood as elemental proportional harmonics, the vibrational steps down from Spirit to matter. All vibrations in nature align themselves with the proportional harmonics and resonate with the spectrum of harmonics of force between Spirit and matter.

In all natural phenomena energy radiates from a center in a fountain spray of centrifugal and centripetal spirals. The Golden Spiral describes the radiation of spiraling energy fields. In quantum fashion the spiraling energy fields structure proportional harmonics of force. This process is universal.

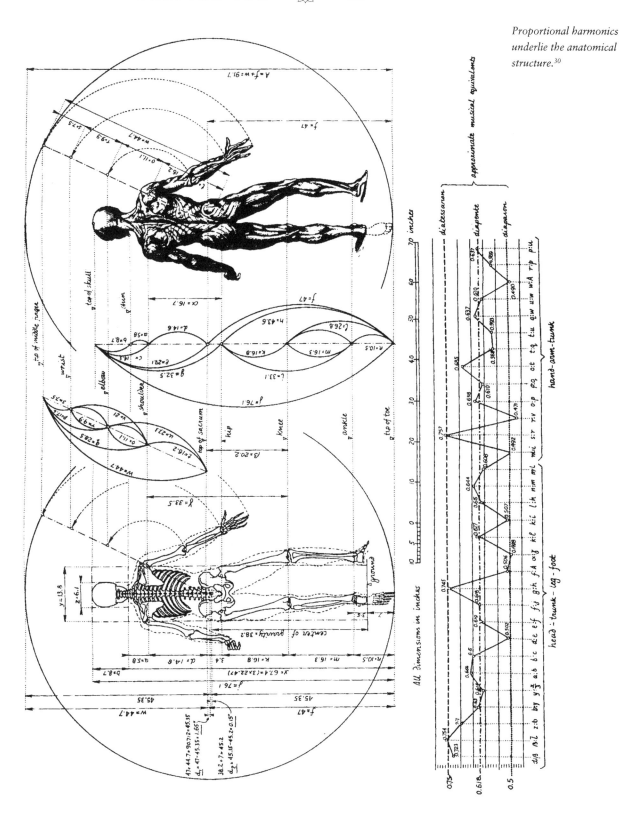

Proportional harmonics
underlie the anatomical
structure.[30]

Like wheels within wheels, energy fields are nested within energy fields in an infinite spiral. From the subatomic to the galactic the vibration of the universe is attuned to the Golden Spiral. At every undulation of vibration rajasic (centrifugal) and tamasic (centripetal) currents cross over to manifest a stillpoint of resonance of sattva guna. This stillpoint is the bindu or seed of creation. This bindu is the centerless center of the universe.

Again we are fascinated by the "supreme ultimate" principle, the "two hands of God," the principle of Polarity that underlies all creation—the spirals of yang and yin, rajas and tamas, positive and negative, centrifugal and centripetal: fields of force radiating out of and returning back to a sattvic center that is the ubiquitous theomorphic process. In our universe (from a root which means "one coil"), everything is tuned to one spiral that is reflected in the universal proportions of the Golden Ratio.

As we study the structures and process of nature through Doczi's profound illustrations, the underlying energy field of the cosmos is made transparent. We start to see patterns in the proportions of structure in nature which reveal the invisible underlying fields of force. We begin to picture how all biological structures take place within the larger suprasensory bodies of nature and the cosmos. We can finally see, through the Divine Proportion, how the spiraling pulse of the evolution of the macrocosm is expressed in the spiraling rhythms of the microcosm, and how our bodies are aligned with the fields of force of the universal body of consciousness, the One Life.

The Harmony of the Spheres

As early as Pythagoras and the ancient Greeks, acoustic proportions were understood in the West and were the basis of both music and physics. The significance of music among the ancients was intimately bound up with the paradigm of the "Harmony of the Spheres." Music in those days was seen as part of the terrestrial blueprint, as the foundation of the world, even as the world-soul itself.[33]

The monochord is a long, rectangular, box-shaped resonant instrument with a single string, used for demonstrating harmonic theory. The sounds obtained by touching the strings are called *overtones*. The sound of the full length of the string is called the fundamental, the lowest tone. The exact halving of the string produces the octave of the fundamental sound. Stopping the string at a third of its length gives a perfect fifth. A quarter of the string again produces the fundamental note, but two octaves from the original, and one eighth of the string takes us three octaves from the original. All tones are calculable as various fractions of the whole string.

A. Golden Spiral, B. Centrifugal spiral, C. Centripetal spiral, D. Union of opposing spirals, E. Illustration of dinergy from a sunflower. The seeds grow at the sattvic stillpoint of resonance defined by the intersection of the rajasic centrifugal and tamasic centripetal arcs. (Doczi, The Power of Limits)

By meditating and experimenting with the monochord, Pythagoras explored overtones and their relationship to nature, geometric proportions and sacred sound. The string of the monochord represents the potential of creation. Because it is not moving, it can be imagined as nothingness. The plucking of the string represents the beginning of creation: pulse, wave and form. The resulting overtones express the mystery of the One becoming many. All creation myths start with the Word (fundamental) followed by a series of events (overtones): "And a river (fundamental) went out of Eden to water the garden, and from thence it was parted, and became four heads (overtones)" (from Genesis 2).[34]

The overtone scale, which demonstrates acoustic proportions, is the "natural" tone scale of all music. The musicologist Joachim-Ernst Berendt points out: "What is astounding about this scale are the numbers [which show] the vibration of any tone on the scale will be higher than the preceding tone by precisely one whole number.[35]

The musical scale graphs the harmony of universal law. According to the seventeenth-century astronomer Johannes Kepler, who articulated the classical laws concerning the motion of the planets: "The fact that the planets move in elliptical orbits is indeed remarkable; but even more remarkable is the fact that ... they have chosen precisely those [orbits] which oscillate and sound in proportions of undivided numbers prevalent in our 'earthbound' music."[36]

The understanding of the proportional harmonics of the music of nature was used by astronomers to "discover" the planets Uranus and Pluto. In 1772, astronomers Titius and Bode formulated what is now known as "Bode's Law," in which one obtains figures very close to the actual mean distance of the planets to the Sun (measured in astronomical units) by taking four as Mercury's mean distance from the Sun, and adding to the series as follows:

Robert Fludd's monochord, illustrating the "Harmony of the Spheres" in the solar system. (Godwin, Robert Fludd)

$$4 + (3 \times 0) = 4$$
$$4 + (3 \times 1) = 7$$
$$4 + (3 \times 2) = 10$$
$$4 + (3 \times 4) = 16$$
$$4 + (3 \times 8) = 28, \text{etc.}$$

The series 4/7/10/16/28/etc. is a close approximation to the actual mean distances from the Sun of the planets:

Mercury, 0.387
Venus, 0.723
Earth, 1.000
Mars, 1.524, etc

Bode's Law[37]			
Planet	Mean Distance	Titus-Bode Distance	Tone (approx)
Mercury	0.387	4	c''''''
Venus	0.723	7	d flat''''
Earth,	1.000	10	g'''
Mars,	1.524	16	c'''
*Ceres (asteroid)	2.77	28	d flat''
Jupiter	5.203	52	e flat'
Saturn	9.539	100	e sharp
*Uranus	19.182	196	E+
*Neptune	30.055	388	A,
*Pluto	39.5	772	E,,

*Not known in 1772 (Chart in astronomical units, which equal the mean distance from Sun to Earth.)

Bode's Law predicted the placement of Uranus, the asteroid Ceres, and Pluto.[38] The Bode numbers represent the progressive approximation of a perfect musical octave.

Underscoring the musical octave as an expression of universal law, the contemporary musicologist Hans Cousto writes: "The octave as a unit of measurement can be applied to astronomic periods, to the frequency sequences of the Earth's atmosphere (spherics) and to vibrational microbiological phenomena ... In this way I will prove that it is not only valid for audible frequencies but is applicable on a truly universal level. Whatever the field, the uniting factor is always the octave."[39]

Cousto points out: "Apart from the prime or fundamental tone, the octave is the interval with the lowest degree of energetic resistance. All other tones vibrate with it ... " He explains:

> If you pluck a string on a guitar or sitar you will not only hear the fundamental tone but also a whole series of others which are whole number multiples of the basic tone. Timbre is the sum of the basic tone and the overtones, which together are known as the partial tone series ... The fundamental tone is the first partial tone; the first overtone (the octave tone) is the second partial tone. The second overtone, the fifth in the first octave, which is also known as the twelfth, is the third partial tone ... The second partial tone (octave tone) has exactly double the frequency of the fundamental tone, while the third partial tone (or twelfth) has three times the frequency, etc.[40]

Cousto underscores the logarithmic succession of proportional harmonics of universal law which are clearly revealed in music:

> The striking fact is that there is only one partial tone in the fundamental octave, while in the next higher octave there are two, with four in the second, and eight in the third. If the table is extended, there are sixteen partial tones in the fourth higher octave, thirty-two in the fifth, sixty-four in the sixth, and so on. In other words, the difference in frequency from one partial tone to another is a constant 64 hertz. From octave to octave, this difference in frequency is doubled, so that there are 64 hertz from the fundamental tone to the first octave (128 hertz), 128 hertz from the first octave to the second, 256 hertz from the third to the fourth and 512 from the fourth to the fifth.[41]

There is a logarithmic correspondence of the frequencies which reflect the universal proportional harmonics of the suprasensory fields of force of the cosmic body on nature.

> The conformity of the harmonic series to natural law pertains, on the one hand, to esoteric or mystical knowledge, and on the other, to that of scientifically demonstrable fact. Modern atomic physics and relativity theory support the monochord model and its demonstration of the derivation of the natural notes. It becomes clear that if the spatial variable (the length of the string) decreases, the temporal variable (the frequency count) increases, and vice-versa. Thus space and time stand here in inverse proportion—a situation reminiscent of the space-time relationship discovered by Einstein.[42]

Another basic law connected with the overtones or harmonics of a violin or cello string is closely related to Max Planck's quantum theory: "When the finger is moved up the string, only quite specific notes are sounded ... The notes jump from one position to the next—they are arranged in quantum fashion—and not as a continuum."[43] This is an important aspect of universal law. The elements operate in quantum fashion. Energy leaps from keynote to keynote in quantum fashion, thus maintaining the integrity of each elemental domain.

The partial tone series is an excellent demonstration of acoustic proportions as an expression of universal law. The tuned string on a guitar or sitar is a finite system, which produces standing waves. Standing waves align the energy in the finite system with the all-pervasive suprasensory fields of force of nature and the cosmos.

A string's attunement to a key allows it to resonate with universal law. The suprasensory fields of force of the universe entrain the guitar string, which then becomes a vehicle for revealing to the senses the underlying dynamic of the proportional harmonics of the fields of nature and the cosmos. The pluck of our tuned string is entrained with the "natural tone scale" of the holomovement of the universe, and manifests as the overtone series—a virtual "Jacob's ladder" of harmony to Heaven! The force of the pluck moves the string to double its frequency of vibration, again and again and again, exponentially, from the finite frequencies of audible sound to the infinite, limited only by the subtlety of the string's ability to resonate with higher frequencies.

The electromagnetic spectrum, the entire range of wavelengths of frequencies of vibration, is a unitary system. The laws which describe audible sound—as above, so below—apply to the entire spectrum and reveal the realms of vibration beyond the senses.

The study of sound provides a key to understanding universal law. The cosmos is a vast fabric of attunement. The harmony of the electromagnetic spectrum is an expression of attunement to the ubiquitous Golden Spiral. The proportional harmonics of the Golden Spiral generate the Cosmic Octave.

Attunement to the Standing Waves of Nature Through Joint Mobilization

Our word *harmony* comes from a root which means "to be joined together," like the joints of the body. The structure and function of the body always reflect the underlying process in which energy manifests through the theomorphic dynamic. This step-down is embodied in a movement from the subtle Ether-predominant core of the plasma in the bone marrow, through the Air-predominant movement of the joints, through the Fire-predominant functioning of the muscles, to the Water-predominant tendons, and out into expression on the Earth. Ether, Air, Fire, Water, and Earth are understood as fields of predominant resonance.

The mobility and range of motion of the joints are keys to releasing blocked energy and enhancing vitality. The joints, which step-down energy from the core to the muscles, are a fundamental regulatory system of the body and a nexus which allows the physical body to receive energy from the larger systems of nature and the cosmos.

The fundamental organizing principles of the physical body are a reflection of the sonic integrity of the resonant field. The physical body is organized around the proportional harmonics sustained by the standing-wave patterns of the body as a finite system, and the integration of these harmonics into the resonant fields of nature and the cosmos. Dr. Stone writes: "*Small units* of energy fields draw their supply of energy from larger units or orbits to which they are attuned, because these are their home, *source or parent fields.*"[45] The attunement of the standing wave patterns of proportional harmonics within the body sustains a harmonic pattern of wholeness that resonates with the ubiquitous implicit ultrasonic pattern of nature. For example, if you change the angle of any joint in the body, every other joint in the body compensates for this. Thus, change the angle at the ankle, and the sacrum, pelvis, and every vertebra and joint in the body compensate to sustain the *harmonic of wholeness.* You can focus on movement at any point in the body and witness the whole body compensating to maintain a harmonic of wholeness, adjusting the range of motion of the joints to accommodate the change.

If you watch a person or animal running, you can witness the harmony

Energy fields nest within energy fields aligned by the harmony of the Golden Spiral.[44] (Doczi, The Power of Limits)

and balance of the movement of the joints as a whole. This movement makes transparent our alignment with the energy fields of nature. The range of motion of the joints is the way the body reflects the standing wave patterns and aligns itself with the implicit order of nature and the cosmos. The entrainment of the standing wave patterns of the body is the way that the body receives energy, through sympathetic vibration, from the larger systems of nature.

Enhance the range of motion and resilience at the joints and you promote a receptivity in the body for resonance with a deeper level of harmony with the finer forces of higher intelligence. Much profound body therapy focuses on enhancing the range of motion and resilience of the joints, sacrum, spine, and cranial sutures. Dr. Stone referred to the sacrum as the keystone of the body, and felt that he had found the key to the healing process in perineal therapy, which enhances the range of motion and resilience of the sacrum. This allows the sacrum and the cerebral spinal system to resonate with a finer harmony of receptivity to Spirit. The body can then come into harmony with the finer forces of higher intelligence, within the standing wave patterns which open the door to health and happiness.

Growth Through Gnomonic Expansion

Gnomonic expansion describes one of nature's most common forms of growth, in which the old form is contained within the new. It is most easily pictured in nature, in the spiraling shell of the chambered nautilus or the florets of a sunflower, and is in fact the basis of all of the more permanent structures in animals, like bones and teeth, horns and shells.[46] In gnomonic expansion each successive cycle of growth spirals outward with the old cycles nested within the new, much like the rings on a tree. Gnomonic expansion in nature makes visible patterns of successive stages of growth and is the crystallization of the underlying forces.

Gnomonic expansion maps a sonic process that underlies growth. From a vibrational perspective, each finite system in nature is defined by the resonant frequency of its standing wave patterns. This sonic level of integrity is fundamental. As we explained above, phenomena have their being through the alignment of their standing wave patterns with the keynotes of the forces of nature. This allows them to draw energy through sympathetic vibration from the ever-spiraling cosmic forces.

One can picture the involutionary cycles of the day, the Moon, and the solar cycle of the year as a winding process that concentrates the cosmic forces from above to below, without to within in the seed. The process of growth

A chamber nautilus illustrates Gnomonic expansion.[47] (Doczi, The Power of Limits)

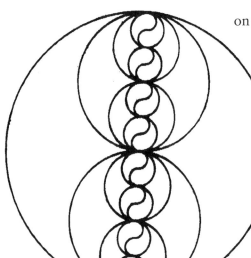

This illustration suggests the nesting of spiraling spheres within spiraling spheres which may be fundamental to the harmony of the vibrational realm. Centrifugal and centripetal currents universally spiral in a balance of wholeness.[48]

on Earth is the evolutionary unwinding and braiding of the evolutionary and involutionary resonances.

The cosmic involutionary forces of the subtle elements interact with the atmosphere of the Earth's electromagnetic field which shapes the evolutionary gross elements. The Earth's atmo-sphere (from the word breath) is made of spheres of force emanating from the Earth. These spheres of resonance influence the elemental harmonics of Earth, Water, Fire, Air, and Ether. These interact with the "music of the spheres" emanating from the Moon and Sun, whose cycles are fundamental to all growth. Phenomena in nature have their being through the braiding of the involutionary and evolutionary forces.

A molecule, cell, or organism grows through a succession of stages which are a microcosm of creation and of the sonic patterns underlying the cycles of creation. As it builds up the integrity of its resonance through sympathetic vibration with the harmonics of nature in quantum fashion, it spirals outward into entrainment with the next phase of nature's standing waves in a process of gnomonic expansion. Gnomonic expansion reflects the resonant process which underlies growth. Growth phenomena in nature *are* resonant structures—they grow through a process of attunement with the standing waves of resonance of nature.

Nature: The Ultimate Musical Instrument

All of nature is "a living string"—attuned to the music of creation. There is no musical instrument more harmonious than a "sound" body. There is no human-made instrument that can resonate as profoundly as the *living* "instruments" of nature and the cosmos. Each breath and heartbeat vibrates with an inner depth that unites Heaven and Earth. The "strings" with the ability to resonate to the vibrational rates of the Absolute are strings vibrating in the fabric of the fields of force of nature whose standing waves spiral from the finite to the infinite in every cycle of oscillation, vibration, heartbeat, and life breath.

All vibration is part of an infinite fabric of wholeness in which every wave is attuned to cycles within cycles within cycles, as emanations of the all-pervasive field of force of the Unitive Field that is the breath of the Cosmic Person, Brahma. Truly our "tune-a-verse" is "The Song of God."

Soul Communion through Union with the Sacred Sound Current

Dr. Randolph Stone's spiritual quest is described in a biographical article by his niece Louise Hilger. He was born to a Catholic family in Engelsberg, Austria, in 1890. His mother died when he was only two years of age. In 1903 he emigrated to the United States with his father. Dr. Stone was a lifelong student of Hermetic wisdom. As a young boy he was deeply immersed in the Bible. He taught himself English by comparing his native German Bible with an English translation. At nineteen he realized that what he was seeking was beyond conventional religion. He took the Nazarene vow and went on a strict diet with intermittent fasts, spending time in seclusion and meditation. He was a member of the Masonic Lodge and Manly Hall's Philosophical Research Society. He studied Rosicrucianism and Sufism. He was a student of Vivekananda's teaching as well as Swami Rama Tirtha, Yogananda, Krishnamurti, Swedenbourg, and Madame Blavatsky. He attended the lectures of the numerous spiritual teachers who visited Chicago and hosted many of them in his home. Over the years he accumulated a vast library of Eastern and Western esoteric, occult, and cabalistic teachings. He kept *The Bhagavad Gita, Light on the Path,* and *The Voice of Silence* under his pillow for constant contemplation.

Dr. Randolph Stone.

In later years he concluded that a living spiritual master was necessary to illuminate the truths of the ancient wisdom. In 1945, after decades of searching, Dr. Stone discovered the "Sant Mat" teachings through the book, *The Path of the Masters,* by Dr. Julian Johnson. The rest of his life was deeply dedicated to Surat Shabd Yoga and the living master of the Radha Soami Teachings, Maharaji Sawan Singh. Surat Shabd Yoga is the practice of Soul communion through union with the sacred sound current.[49]

Saint Kirpal Singh explains: "The audible Sound Current is in fact only one continuous creative life-principle which emanating from the Immaculate Pure One, steps down from plane to plane for the purpose of creating five regions below: pure conscious, great causal, causal, subtle and physical, and as It passes through varying degrees of density, peculiar at each place, It acquires a distinctive sound and hence has come to be known as the Music of the five melodies or Sounds."[50] Surat Shabd Yoga requires initiation from a qualified person.

Meditation: Nad Yoga

Nad yoga, the meditation on primal sound vibrations, facilitates Soul communion and Self-realization through attending to inner sounds. It is an extremely effective method for developing yogic perception.[51] Baba Ram Dass (Richard Alpert), and the Lama Foundation staff describe the practice of Nad yoga in *Cookbook For A Sacred Life*, a booklet included with the introduction to the Sanatana Dharma, *Be Here Now.*

This is a technique for climbing the ladder of divine sounds. You can understand these sounds as the vibrations of various nerves or as energies resonating with inner planes of vibration. In advanced Nad yoga one learns to associate each sound with specific visual and kinesthetic sensations that parallel the inner planes of consciousness.

With your head, neck, and chest in a straight line, find a comfortable position. You may lie down if necessary, as long as the reclining position does not lead to sleep. You may choose to use earplugs if there is external noise.[52] But it is better to find a quiet place in nature, or a quiet time of the night for practice. Keep your mouth and eyes closed.

Tune into any inner sound that you can find in your head. Focus on a sound till it is the dominant sound you are attending to. Let other thoughts and sounds pass by.

As you allow the sound to grow, filling your consciousness, you will ultimately merge with the sound and not hear it any longer. At this point you will begin to hear another sound. Tune in to the new sound and repeat the process. Depending upon the distinctions you make, there are seven or ten sounds. Descriptions of the sounds include the buzzing of bees, the sound of a crowd in a large hall, and drums.

Madame Blavatsky, the founder of Theosophy, writes:

> The first is like the nightingale's sweet voice, chanting a parting song to her mate. The next resembles the sound of silver cymbals of the Dhyanas, awakening the twinkling stars. It is followed by the plain melodies of the ocean's Spirit imprisoned in a conch shell, which in turn gives way to the chant of Vina. The melodious flute-like symphony is then heard. It changes into a trumpet blast, vibrating like the dull rumbling of a thunder cloud. The seventh swallows all other sounds. They die and then are heard no more.[53]

Do not concern yourself with the order of the sounds. Some of the sounds, like the flutes, are very attractive and you may wish to linger endlessly, fascinated by a blissful state of being. After a few days it is best to continue up the ladder of inner sounds. When you arrive at a high level of purification you may have a passing fever for twenty-four hours. Contemplating sacred sound can be a profound vehicle for healing, purifying the mind, and accelerating one's evolution through Soul communion.

CHAPTER 11

Babbitt's Atom: The Key to the Secret Teachings of All Ages

In this essay we look at the work of the father of modern color therapy, the brilliant nineteenth-century metaphysician Edwin D. Babbitt, M.D., L.L.D. In his profound book *The Principles of Light and Color,* Dr. Babbitt unfolds the "Harmonic Laws of the Universe." Like the Sanatana Dharma, Babbitt's atomic theory is based on sacred sound vibration.

Manly P. Hall has been called the "dean" of American esoteric studies. Centered on the very first page of his magnum opus, *The Secret Teachings of All Ages: An Encyclopedic Outline of Masonic, Hermetic, Qabbalistic and Rosicrucian Symbolical Philosophy,* is an illustration of Babbitt's atom. In the caption under this illustration he describes Babbitt as a genius. The placement of this illustration in the center of page one of his magnum opus, with Babbitt introduced as a genius, indicates that Hall felt that Babbitt's atom was the "Rosetta Stone" of Hermetic understanding—and the key to the secret teachings of all ages.[1]

Babbitt was reputed to have an expanded threshold of perception within which he could see energy fields and perceive different energies as different colors. Writing in the 1850s, he discussed left and right brain differences and theorized an "atomic bomb." Babbitt presided over The New York College of Magnetics, which had branches in California and abroad. The college offered courses in healing based on an exact science of "Magnetic, Electrical, Chemical, Solar and Spiritual Forces which underlie everything."[2] His home study courses in "Chromo Therapeutics" (color therapy) and natural healing had a large following. Through his influence color therapy was widely practiced

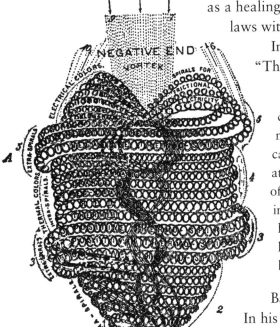

Babbitt's atom[3]: "atomic coil or helix."

as a healing modality until it fell under attack by antiquackery laws with the rise of "scientific" medicine.

In *The Principles of Light and Color,* Babbitt unveils "The Harmonic Laws of the Universe." He explains:

> The lines of atomic forces, are doubtless not in circles, this being contrary to the general untrammeled movements of nature, as the pathway of missiles, cataracts and planets is in the sections of a cone ... atoms are in *ellipsoids,* or rather in the modifications of this form in the *ovid* ... One thing in proof of this is in the fact that atoms combine and polarize better by having a smaller end, while, as we have shown, the law of positive and negative action forces one end to be smaller than the other.[4]

Babbitt's understanding parallels the ancient wisdom. In his model sound and sympathetic vibration are the key to the process of the polar cohesion of atoms. He writes:

> Why are ethers drawn from spirals of one atom to the same kind of spirals in a contiguous atom, and why does a certain grade of ether exactly harmonize with, and seek out, a certain size spiral? For the same reason that a tuning-fork or the chord of a piano will be set into vibration by a tone made in its own key. In the case of the piano, a chord vibrates to tones of its own pitch, or in other words, to tones whose waves synchronize with its own vibrations. Let us apply this principle to atoms. The vibratory action of the red spiral throws the current of ether which passes through it into the eddy-like whirl which just harmonizes in size and form to the red spiral of the next atom above it with which it comes into contact, and which must necessarily draw it on. This second atom passes it on to the red spiral of the third, the third to the fourth, and so on through millions of miles, so long as there is a spiral of the right grade to conduct it onward.[5]

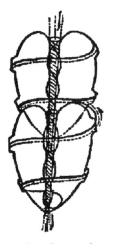

Polar cohesion of atoms.

Babbitt's atom is based on the principle of polarity. The positive fields are characterized by heat and expansion; the negative are cold and contracting. Babbitt's model portrays a hierarchy of energy fields.

This hierarchy includes the "main spiral" which passes around the atom as well as "sub spirals" which encircle the main spiral, called first spirilla or little spiral, second spirilla, and third spirilla, following nature's "law of trinal gradiation," to use Babbitt's terms. These all move in the same direction—as above, so below—just as the Sun, planets, and Moon move through space in the same direction. There is also a hierarchy between the more coarse, external spirals which Babbitt calls the "extra spirals," the finer spirals set farther in, called the "intra spirals," and the third level of the hierarchy, the "axial currents" or "atomic vortex" at the core. Babbitt pictures a gradation of three to seven spirals in each hierarchical field.

Thus, Babbitt's atom portrays a hierarchy of three fields following his law of trinal gradiation. The triad of resonance facilitates the alignment of atoms through polar cohesion to manifest solidity. In polar cohesion (figure 1) the ligo (male/protruding and female/receptive connectors) of the upper atom glides into the ligo of the lower, and thus the two atoms become riveted together by the vortex of the axial currents. The more gross extra spirilla keep the upper atom from being drawn into the lower atom.

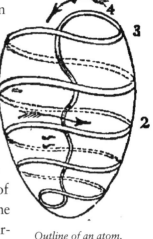

Outline of an atom.

Lateral cohesion (see figure) takes place through the eddies of the external positive spirilla aligning with the negative spirilla. Lateral cohesion is ruled by the thermal expansive positive polarity and is not as strong as the contractive cohesion of the axial current.

Babbitt continues:

Paraverse layers.

> We have ascertained ... that the spiral, itself the most beautiful of continuous forms, is the great leading law of motion in nature. Let us presume, then, that the spiral direction rules in atoms as well as in worlds, especially as, according to the great unity of law, we must judge the unknown by the known. In fact the spiral is a necessity if we are to get any continuous lines around the atom and have it progress regularly so as to cover its whole form and then to convey its force to the next atom. So far, then, we have the external atom clad with spiral lines of force, or rather, a spiral framework, and tube work through which, and over which, this force must vibrate and flow.[8]

Babbitt's research leads him to a paradigm which parallels that of the profound ancient Samkhya model:

Transverse cohesion of atoms.[7]

> But from the nature of things there must have been an almost infinitely subtle elastic, infrangible Intersoul, all-penetrating, all quickening, and filling the whole realm of being. This may be termed Spirit or the soul of things. If we should take these two great departments of the universe separately, we should have Spirit on the one side as a limitless, unformulated substance and matter on the other side, as an unpolarized and lifeless mass.[9]

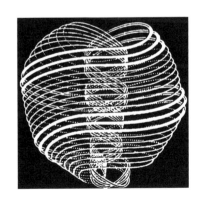

As below, so above. We can find valuable suggestions for research in energy systems through Babbitt's model. The law of trinal gradiation is fundamental to the dynamics of Babbitt's paradigm much as the three gunas are the foundation of cosmogenesis in the Sanatana Dharma.

Polarity and Babbitt's Atom

Synthesizing Babbitt's model with the Polarity model we find that in the torso, the shoulder and hips parallel the "extra spirals" and embody the Fire principle of directed force for external interactions. The internal focus of body building and cleansing parallel Babbitt's "inter spirals" and the Water principle as an energy system. The "axial current" parallels the ultrasonic core which Babbitt calls the "Intersoul." Thus, the evolutionary currents parallel the extra spirals, the involutionary currents parallel Babbitt's inter spirals, and the ultrasonic core parallels the axial current.

The extra spirals in Babbitt's atom parallel Polarity's Fire principle, the intra spirals parallel the Water principle, and the ultrasonic core parallels the axial currents.[10]

Human energy fields are subsystems of the energy of nature and the solar system. The poles of this system are the Sun and the Moon/Earth. Everything in nature grows in entrainment with these forces of rajas and tamas. During involution, superior fields entrain with the centrifugal solar forces, inferior fields with the lunar contracting pole. In evolution the fields reverse polarity with inferior fields expanding centrifugally and superior contracting centripetally. Physical energy fields take on their characteristic heart shape as subsystems within this energy dynamic.

The Heart of Our Perspective

Dr. Annie Besant and Bishop Charles Leadbeater's clairvoyant picture of the heart of the universe "the ultimate physical atom" seen as a manifestation of pure life force spinning perpetually through itself. This image is remarkably similar to Babbitt's atom.

The muscle fibers of the human heart are spiral shaped and contract in a spiral to twist the blood through the chambers.[12] Dr. Stone reveals that each chamber of the heart resonates with an elemental keynote, attuning the blood to the elemental sonic essences.[13]

Occult Chemistry is the report of clairvoyant research which took place between

1895 and 1933. A school of Theosophist scientists using microclairvoyant powers developed a "Periodic Table of the Elements." From carbon to uranium, they described and diagrammed fifty-seven chemical elements, the basic constituents of matter, and developed a formula for ascertaining their atomic weight. They found ninety-nine distinct elements including a number that were yet to be discovered. The book, *Extra Sensory Perception of Quarks* (Theosophical Publishing, 1980), demonstrates that the research of Dr. Besant and Bishop Leadbeater had indeed been accurate and has been largely confirmed by a half-century of scientific subatomic research.[14]

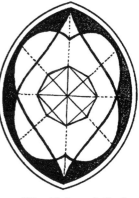

"Vita-life is symbolized as a radiant sacred flame of energy which is the heart and core of every living thing." The downward evolutionary and upward involutionary flames emanate from the neutral "diamond heart" of light and sound of the soul.[16]

Stone, *Polarity Therapy*

In his book, *Vitality Balance*, Dr. Stone depicts "Vita," the symbol of life as a heart-shaped, radiant sacred flame of energy. He goes on to say that the human heart itself is in the shape of the sacred flame. In the first and second charts in his book, *Polarity Therapy,* Dr. Stone offers diagrams of "the life circuit" in the shape of a heart. In the first chart the psychophysiological step-down of the Soul-energy to the physiological, sensory, and motor functions is pictured in a heart shape, focusing on the dynamic between the centrifugal and centripetal currents. In the "Diagram of the Pattern of Life Force and the Tissue Cell" the life circuit, in the shape of a heart, is illustrated.

The Heart of Creation

Dinergy, which is the foundation of all growth in nature, creates and sustains heart-shaped patterns which are whirling vortices of bipolar energy formed by the interaction of centrifugal and centripetal energy fields. For our purposes, the energy fields of nature are under the predominant electromagnetic influence of the Sun, the nucleus of the solar system, as a positive centrifugal energy field; and of the Earth as the nucleus of the negative centripetal field of this helix. Earth is a relatively slow-moving field. So the centripetal resonance is also driven by sympathetic resonance with the Moon, which parallels the Water chakra in the human energy system.[18] All energy in nature is a subsystem within the field of force of the solar system. The energy fields nearer the nucleus of the positive field resonate with expansive solar radiance, and the fields nearer the negative field of the bipolar helix are shaped by the contracting force.

Between these two poles, a spectrum of resonance is sustained that ranges from the subtlest innermost harmonics of resonance of the nucleus, which would tend to be perfect spheres, to the elongated linear crystallization in the

Form follows energy. The muscles of the human heart express a spiral. The heart is a center of resonance in the body—its four chambers are attuned to the four elements. Its beat is the fundamental for the overtones of cosmic attunement that sustain life in the body.[11]

Earth-predominant surface of the energy field. Thus we are picturing a spectrum of harmonics of resonance from the spherical Ether-predominant resonance at the nucleus, to the elongated, elliptical energy field of the Air element, to the upward-radiating field of the Fire-predominant harmonic, to the downward-radiating resonant force of the Water harmonic, to the more crystallized, linear fields of force of the Earth-predominant harmonics of resonance. As we move away from the nuclear resonance, the fields tend to be more and more influenced by the Earth's pull, becoming more and more elongated and linear.

The influence of this dynamic sustains a heart-shaped pattern, which, following Babbitt, we picture as a fundamental feature of the energy fields of the body. The heart shape of dinergy can be seen as a fundamental basis of structure in the human body. Each body cavity can be seen as a heart, with the system of "hearts" reflecting the theomorphic step-down from Ether to Air, to Fire, to Water, to Earth. The heart shape reflects the proportional harmonics of the Golden Spiral, which as we've seen, universally defines the radiation of energy from a center.

The heart shape is an expression of the cycle of emanation and negation, centrifugal and centripetal fields emanating from a neutral core of higher potential. It is a literal depiction of the essence of creation. This primordial type of arc of yang and yin force fields sustains the chakra, or whirling energy

A) line spiraling out from center following the Golden Ratio; B) centrifugal and centripetal spirals; C) the heart-shaped blending of opposing centrifugal and centripetal spirals radiating from the same center.

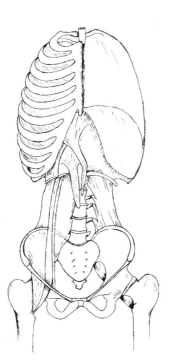

Energy fields emanate from a center in dinergy of centrifugal and centripetal forces following the Golden Spiral, sustaining heart-shaped fields of force.[19]

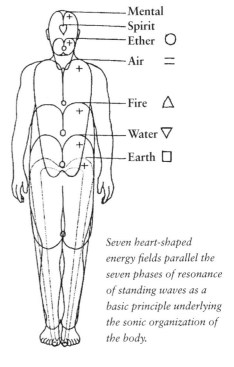

Seven heart-shaped energy fields parallel the seven phases of resonance of standing waves as a basic principle underlying the sonic organization of the body.

vortex, the primordial heart-shaped archetype which underlies all phenomena in the material "whirld."[20]

Looking down from above at the cross section of the thoracic cavity (see figure), you can picture the Air oval field emanating from the heart. In Air predominance, the centrifugal and centripetal founts of energy unite the core and periphery of the energy field. Air is a lengthening of the centrifugal energy field which predominates in the east/west currents defining the body from side to side. The opposite polarity of Air is Earth, which is a lengthening of the periphery through the resonant arc of the centripetal currents that define the boundaries.

From an anterior view of the body, picture the heart and sternum as the neutral center of the field. The positive centrifugal field radiates upward and outward as the thoracic cavity and shoulders and then falls downward, contracting into the negative centripetal pole at the symphysis pubis. This radiant flow of energy creates a heart-like shape. On the back we have the reverse energy flow, with the nucleus at the sacrum. Energy radiates downward and outward at the buttocks, and centrifugal then flows upward, contracting at the negative centripetal pole at the occiput (base of the skull). Again, a heart-shaped energy flow is created through the dynamic of expanding and contracting positive and negative energy fields.[22] There is always a blending of positive and negative fields, yang and yin. Centrifugal and centripetal fields balance and oppose each other in every gestalt. In this example, top and bottom and front and back sustain a phase dynamic of "perpetual motion" of tension and harmony, sustaining the balance of the overall energy field.

As the field becomes more material, it separates more and more from oneness into duality—from the oneness of the head as an oval, embodying the mental plane, into the duality of thighs and legs, feet and Earth, at the negative pole of the physical plane.

Energy fields only exist as patterns of wholeness within patterns of wholeness. Energy fields are patterns of resonance in which every part of the pattern is influenced by the whole and by the resonance of the rhythms of wholeness of which they are a part. Thus centrifugal and centripetal forces only exist in resonance with each other as part of a dynamic gestalt.

All patterns and structures in nature reflect the proportional harmonics of the Golden Ratio. This ubiquitous relationship reflects the underlying vibrational harmonics of universal law. In our model, the energy of creation is the omnipresent creative potency of the breath of Brahma. The waves of this breath pulsate as the cycles of universes, the same heartbeat throbbing through our bodies and the phase dynamic of atoms. All creation is a vast energy

Form follows energy. We see a striking parallel between the proportional harmonics of the spiraling hearts and the thoracic inlet.

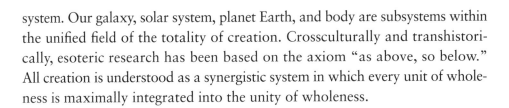

The caduceus or "staff of life," an ancient Egyptian and Greek esoteric symbol of life essence.

system. Our galaxy, solar system, planet Earth, and body are subsystems within the unified field of the totality of creation. Crossculturally and transhistorically, esoteric research has been based on the axiom "as above, so below." All creation is understood as a synergistic system in which every unit of wholeness is maximally integrated into the unity of wholeness.

The Caduceus

Dr. Stone writes:

> MAN KNOW THYSELF is the admonition contained in the symbolism ... of the Staff of Hermes. The wings of the Caduceus represent the two hemispheres of the brain. The knob in the center is the pineal body. The upper staff is the path of the finer energy which produced the center portion of the brain, noted by the rings and spinal cord below it ... The two serpents represent the Mind Principle in its dual aspect. The fiery breath of the Sun is the positive pole as the vital energy on the right side of the body ... On the left side of the body flows the cooling energy of the Moon essence of Nature ... These two currents cross over in each oval cavity of the body and change their polarity. Thus they flow in and out of each other constantly and produce alternating currents in their action ... The last open loop and the lower part of the staff is the Cauda Equina, The Tree of Knowledge of God and Evil. It is situated at the end of the spinal cord, proper, below the second lumbar vertebra.[23]
>
> The central polarized energy current of the Caduceus is given as the major step-down current from the ultrasonic *energy flow which is actually the soul of man* in this mortal body ... The Caduceus is the symbol of the airy life currents in the breath, which polarize one side of the body with the other and the above with the below in each major field as a positive and negative current. Thus the soul or Spirit energy is stepped-down through the mind and senses ...[24]

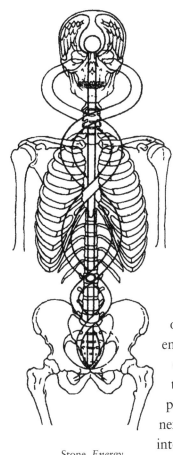

Stone, *Energy*

The caduceus is a representation of the ultrasonic core, the nucleus of the subtle energy fields of the body. All energy fields of the body are emanations radiating from this potent fulcrum. In the caduceus, pingala (positive, centrifugal) and ida (negative, centripetal) forces intertwine themselves in a serpentine pattern around a neutral sushumna. Sushumna, pingala, and ida are the channels of subtle energy (nadis) which are the nexus that steps-down energy from "the universal creative energy" (prakriti) into the physical body. Sushumna, pingala, and ida resonate with the universal vibrational quintessence of the gunas (sattva, rajas, and tamas).

The Body is Theomorphic, the personification of the process of creation.

The caduceus of sushumna, ida and pingala in the core of the spine, sustains fields of force which are a microcosm of creation. The fundamental organization of the body reflects this step-down of energy from spirit to matter. Whirling vortices of centrifugal and centripetal fields of force emanate from the chakras as the energy underlying form. The Air Field rules movement in the body. The beat of the heart is its neutral source, the rhythm of the lungs its positive field and the peristalsis of the colon its negative pole. The Fire Field rules directed force, its neutral source is the umbilicus as the center of vital force, its positive field the Solar Plexus, diaphragm and organs of assimilation. The negative pole is the thighs where force is directed into manifestation. The Water Field rules form. Its neutral source is the sacrum (and womb), the innominate is its positive field and the feet are its negative pole. The Earth Field has its neutral source in the ether predominant plasma of the bones, its positive field in the knees and its negative field in the colon which rules elimination.

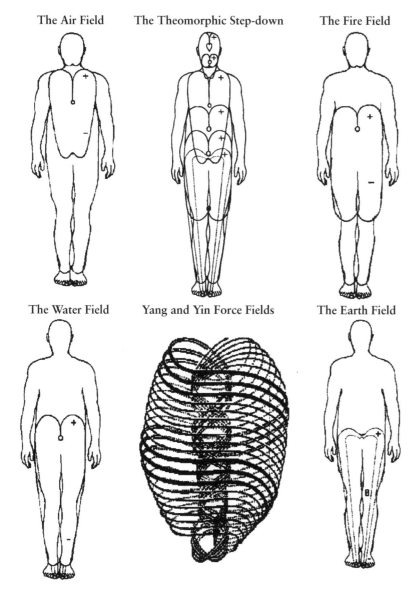

The Air Field   The Theomorphic Step-down   The Fire Field

The Water Field   Yang and Yin Force Fields   The Earth Field

The polarity of the Earth field with the colon negative and knees positive is a reflection of the fetal position in the womb.

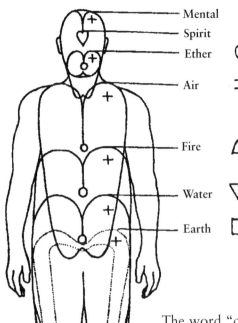

Mental

Spirit

Ether

Air

Fire

Water

Earth

Where pingala and ida cross in the caduceus, chakras are formed which spin radiating oval fields of subtle force. The chakras emanate whirling, throbbing fields of force. The currents pulsating from the chakras are the step-down mechanism from the primary energy of the unified field of nature to the subsystems of the subtle body. The whirling emanations of the chakras are the nexus between the subtle universal elements and the pentamirus[26] combination of the elements of the physical body. Like wheels within wheels, the pulsating currents of force emanating from the chakras at the ultrasonic core sustain the elemental step-down into the harmonics of resonance for the fields of Ether, Air, Fire, Water, and Earth.[27]

Welcome to the "Whirld"

The word "chakra" is usually translated as "wheel," and pictured two-dimensionally as a spinning hubcap of etheric energy. A chakra is better understood as a three-dimensional heart-shaped whirl of energy. This whirling vortex of energy embodies the underlying energy gestalt found universally throughout the cycle of manifestation and negation in this "whirld," where ultimately all phenomena are based on the spiraling microcosm of a field of bipolar energy sustaining heart-shaped whirling vortices of energy. From above to below, from third eye to coccyx, the chakras are the primary step-down mechanism from Spirit to matter. Each chakra is the center of a pulsating energy field attuned to its predominant elemental resonance.

The body is made of energy fields. Each major body cavity is sustained by the radiation of its whirling chakra of centrifugal and centripetal forces. These dynamic pulsations of force are perceived by the senses as the physical body, but it is actually the physical emanation of subtle inner forces. In much the same way that a television set is tuned into a particular spectrum of vibration which it then organizes into an image, our senses receive a range of vibrations from which they build our subjective experience of a "solid" phenomenal world.

Through dinergy the three pulsating fields of spiraling energy of the caduceus—the sushumna, pingala, and ida (corresponding to sattva, rajas, and tamas)—define the body in three axes, or dimensions: side to side, back to front, and top to bottom.

The Three Principles and Three Axes of the Body

The Air Principle

The transverse or east-west current relates to the sattvic quality of neutrality, to balance the left and right sides of the body and unite the core with the periphery. Its field of force sustains the Air principle, which emanates from the neutral sushumna current of the core energy system. It spirals transversely around the surface of the body, positive at the top of the sushumna and negative at the bottom, functioning in communication and coordination, integrating the periphery of the body with the core, and carrying sensory input from without to within. It is thus associated with the parasympathetic nervous system, which performs these functions in the body.

The Fire Principle

The Fire principle, which Dr. Stone called the "spiral current," relates to the rajasic quality of energy. Its centrifugal (pushing, positive), and centripetal, (pulling, negative) currents radiate throughout the body, spiraling outward from its neutral center at the umbilicus to energize the whole of the body for warmth and movement. The Fire center may be pictured as a fountain spray of energy radiating upward from the navel and becoming most predominant in the solar plexus. It is the center of vitality and governs the distribution of internal vital energy throughout the body. It relates to the sympathetic nervous system, defines the body from posterior to anterior, and radiates from within to without.

The Water Principle

The Water principle defines the dimensions of the body from top to bottom. Water and Earth both resonate with the tamas guna, the principle of resistance or crystallization, which rules completion and form and predominates in the periphery of the body. Water is the receptacle of form and the medium of creation. It is at this step-down into matter that energy assumes substantial solidity. Water predominates in the pelvis and in the tendons. Water, which always flows downward into the Earth, carrying off impurities, rules the cleansing processes of the bladder and lymph system. Earth, which rules at the limits of the energy field, sustains boundaries and predominates in the solidity of the bones of the body and in the colon, which provides the boundary between that which sustains life and that which is to be eliminated.[30]

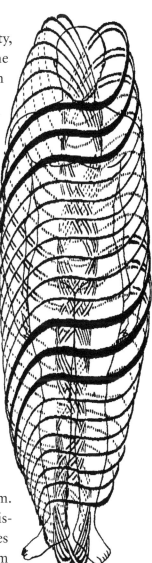

The Air principle unites core and periphery.

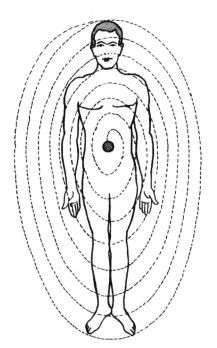

The Fire Principle[28]

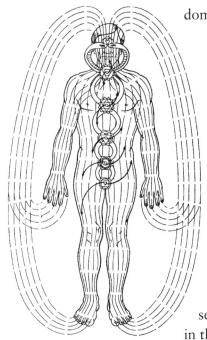

*The long line currents.
The five elements arise
from the chakra system
on the level of the Water
principle to function in
the body.*[29]

The long line or bipolar currents pulsate from each of the five chakras to sustain five zones of force radiating out from the core of the body. Each of these fields resonates with the resonant frequency of its source chakra, running vertically through the body in its positive phase and returning to its source chakra in its negative phase. These five fields of force resonate with the five elements—Earth, Water, Fire, Air, and Ether. Dr. Stone called these five fields of force the "Bi-Polar" or "Long Line Currents." "Energy waves on one side of the body balance those on the other side of the body by traveling in opposite direction . . . [these] wireless waves of the body . . . have their *polarity reflexes* to the *five senses*, the *five fingers* as touch and skill, and the *five toes* as motion and action."[31]

The long line currents define the bilateral symmetry of the body and function to regulate its physiology. Each element governs specific organ functions and tissues in areas of the body which correspond to analogous elemental harmonics. The elemental domains are also expressed in the five fingers and toes. The long line currents carry the energies of the mind into the body and govern the senses. They resonate with the negative, tamasic qualities of completion and form, and relate to the central nervous system.

Like wheels within wheels, the elemental currents emanate from the core and step-down in vibration in quantum fashion from Ether to Earth as fields of force. Each of these fields vibrates with the resonant frequency of its source chakra, running vertically through the body in its positive phase and returning to its source chakra in its negative phase.

A Theomorphic Perspective

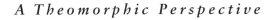

The basic recurring theme in Hindu mythology is the creation of the world by the self-sacrifice of God . . . "sacrifice" in the original sense of making sacred . . . whereby God becomes the world which, in the end, becomes again God.[32]

As discussed in Chapter 6, in a theomorphic perspective, all of the systems and subsystems of nature and the cosmos embody the process of creation. Every level of macrocosm and microcosm emanates from a single living conscious Unitive Field of Ultimate Intelligence. Every level of creation embodies the same energetic/vibrational laws. At the crossover of every undulating spiral of rajas and tamas gunas there is a stillpoint of sattva guna which is the omnipresent center of creation. Every spiraling vibration on every level of macrocosm and microcosm in every cycle touches the omnipresent creative center. All creation is a moment-to-moment emanation from the omniscience and omnipresence of the One. All life shares this omniscient center. This omniscience is reflected in holographic aspects of nature and the cosmos. The ancients understood this theomorphic principle and articulated it in the fundamental dictum of Hermetic research, "as above, so below."

Holographic Models in the Healing Arts

A holographic model of the body is a common denominator underlying many healing arts. Chinese medicine, for example, offers us the fruit of centuries of empirical practice and refinement. One aspect of this body of knowledge, which has grown rapidly in the West in recent years, is acupuncture. Part of the practice of acupuncture is auricular therapy, a system in which the body as a whole is represented in the form of a fetus superimposed on each ear. Stimulation of acupuncture points on the ear eliminates pain and promotes healing in analogous parts of the body. We also find holographic representations of the body on the face, nose, *hara* (abdomen), hands, feet, and tongue in other aspects of Chinese medicine.

There are many other examples of holistic perspectives based on holographic evidence. Iridology is a diagnostic practice based on a holographic view of the body. Every part of the body is represented, in microcosm, on its analogous place on the irises of the eyes. Foot and hand reflexology likewise map the whole body onto the feet and hands; the reflexologist works with the microcosm of the feet or hands to affect the macrocosm of the body as a whole. Colon therapists offer charts on which each organ of the body is represented by an analogous place on the colon. Chiropractic theory views the spine as reflecting the body as a whole: Each part of the body is represented somewhere on the spine, and chiropractors work with the spine to effect healing throughout the body. Though it is barely a hundred years old, the empirical effectiveness of chiropractic care has made it one of the most widely accepted healing arts in the United States.

These treatments have a synergistic impact on every level of wholeness throughout the body. Each of them addresses not just the body, but its harmony with the universe as reflected on every level of the body. The body is

"Primordial Mind Pattern" underlying the holographic body.[33]

merely one level, one expression, of the universal energy field which expresses itself over and over again throughout the whole of the cosmos.

In his book, *The Principles of Light and Color,* E. D. Babbitt wrote that the harmony of creation arises from the fact that each level of creation is a radiation within a larger field of radiation, with all creation ultimately sharing the same center. The leaf of a tree, for example, is a radiation of its stem— a more materially elaborated reflection of the energy pattern concentrated in the stem. In the same way, the stem and leaf are radiations from their nucleus in the branch, which is a reflection of the radiation of the trunk. The trunk is a reflection of the core energy of the Earth, and the Earth is a holographic radiation expressing the higher concentration of the potential of the Sun, and so on.

In Babbitt's view, all phenomena in nature can be reduced to energy fields that are microcosmic radiations of the macrocosm within which they are contained—emanations, in other words, from a center of higher electromagnetic potential. Thus, everything in creation ultimately shares the same center and has a holographic relationship to the same whole. Dr. Stone referred to the holographic dynamic in terms of a "primordial mind pattern" ultimately reflected on every level of organization in the body.

In the Sanatana Dharma, all creation is a moment-to-moment emanation from the all-pervasive holograph of the Unitive Field. The sages who experienced this level of consciousness described it as a "realm of jewels, where each jewel reflected the light of all the other jewels, and indeed each jewel was all the other jewels." This jewel is consciousness, the omniscience of the all-pervasiveness of the Self. All creation is an appearance within the Unitive Field of consciousness, Sat Purusha, Spirit, the One Cosmic Person. The Tantric axiom "What is here is elsewhere ... What is not here is nowhere," refers to the *present*. While the past and future are always changing, the present is immutable as the presence, the consciousness of Being. All creation shares this essence of the awareness of Being. Creation is a moment-to-moment emanation from the Unitive Field of awareness.

The Tri-gunas as Holomovement

To see a World in a Grain of Sand
And a Heaven in a Wildflower
Hold Infinity in the palm of your hand
And Eternity in an hour.
—*William Blake*

Again and again we build upon the axiom of the identity of all consciousness which is the root realization of the Sanatana Dharma. All creation grows from the same center, the Self. The emanations from the One universally follow the same laws. The ancient wisdom reveals that the first principle of the manifest universe is Cosmic Intelligence (*Cit*). Creation emanates from this Omniscience. The miracle of life is sustained by this living Presence. The synergy of the living fields of atoms, molecules, cells, tissue, organs, organisms, nature, and the universe is sustained by the depth of Cosmic Intelligence and the sharing of the One Universal Life Field. All life emanates from omniscience; this is reflected in the holomovement of the breath of Brahma which is the vibratory ground of all creation. The word of God is pregnant with Divine Being, Divine Intelligence, Divine Will (Sat-Chit-Ananda). As above, so below.

A hologram is produced when a single laser light (an extremely pure, cohesive form of light) is split into two separate beams. The first beam is bounced off the object to be photographed. Then the second beam is allowed to collide with the first. They create an interference pattern which is then recorded on a piece of film. When another laser beam is shone through the film, a three-dimensional image of the original object reappears.

There is a suggestive parallel between the above process and the dynamic of the trigunas. In the Sanatana Dharma, different qualities in nature arise from different phases of the cycle of creation (rajas) and dissolution (tamas), and the interference patterns sustained by these movements in the cycle and their relationship to the source (sattva). The holomovement of these forces sustains standing waves. The interference pattern between the outgoing and incoming wave fronts, interacting with the force of the nucleus, weave the vibrational fabric of creation. This balance of polarity between rajas and tamas, interacting with sattva, is the essence of a universal holomovement.

In a theomorphic model the stuff of creation is Divine Intelligence. The Creator is fully present in every atom, molecule, and cell of a living universe. "What is here is elsewhere, and what is not here is nowhere." Every vibration in creation is theomorphic, a holographic representation of the process of creation. On every level of macrocosm and microcosm the holomovement sustains standing waves of vibration. The interference pattern of the outgoing and incoming wave fronts of the standing waves weave the fabric of creation.

Thus, every level from microcosm to macrocosm is a reflection of the same underlying energy field on different octaves of contraction and expansion. Each octave of wholeness is sustained by its standing wave fronts. The arcs of every element are analogous in both atom and universe. The dynamic gestalt of each field of force is a universal balance of yin and yang. Each unit

of wholeness as a field of energy is balanced and attuned to the same spectrum of resonant frequencies. Following the Golden Spiral, energy fields are nested within fields—all sharing the One fulcrum, the One Source of the omniscience of the One universal center. Microcosm and macrocosm on every level are thus in a synergistic relationship, the energy fields of force of each level influencing and supporting every other level.

We must understand our fundamental unit of wholeness, therefore, as a radiant field of force—"radiant" because it is a radiation from its center, and "field of force" because it is a dynamic circulation of energy. The nuclear potency of force expresses itself throughout the totality of the field as the Polarity cycle. In our solar system, for example, everything is an expression of the potency of the Sun.[34] The profound harmony and synergy of all creation is an expression of the circulation of the holomovement of the supreme potency of this neutral force (*Vasudeva*) throughout the body of the universe.

Know this my "Prakriti" [nature, matter],
United with me, the womb of all beings.
I am the birth of this cosmos, Its dissolution also.
I am he who causes, no other besides me.
Upon me, these worlds are held,
like pearls on a string.[35]

Doczi, *The Power of Limits*

The spiritual is understood as an all-pervasive inner force whose radiant potency, a living intelligence, sustains the fields of force in matter. Spirit is the string upon which the pearls of manifestation are strung. In the human body, seen as a field of energy, the fundamental processes which underlie its structure on every level are based on this dynamic. Every unit of wholeness—every atom, molecule, cell, and organ of the body—is sustained by this underlying dynamic of Spirit moving into matter and back again. This involutionary emanation (rajas) from the nucleus (sattva), interacting with the evolutionary return currents (tamas), forms the interference pattern that weaves the fabric of creation. Each crossover of these two forces sustains a pattern of resonance which becomes the reference beam (sattva bindu) of a center of manifestation, and the holomovement manifests on another level of the microcosm.[36]

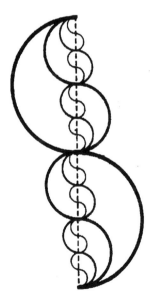

CHAPTER TWELVE

Breath and the Cosmos

Our lips are Shiva and Shakti. Our upper lip is attuned to a centrifugal vortex; our lower lip resonates with a centripetal vortex. Every breath of life attunes us with the Life Field of the universe without and within. Our every breath is a sacred microcosm of the breath of creation. Our every breath is a reenactment of the primordial mystery of creation, our every utterance a microcosm of the Sacred Word. As jurist and Tantric scholar Sir John Woodroffe wrote:

> On the physical plane, Prana manifests in the animal body as breath through inspiration (*Sa*), or Sakti, and expiration (*Ha*), or Siva. Breathing is itself a mantra, known as "the mantra which is not recited" (*Ajapa-Mantra*) because it is said without volition. The divine current is the motion of Ha and Sa. This motion, which exists on all the planes of life, is for the Earth plane (*Bhurloka*) created and sustained by the Sun, the solar breath of which is the cause of human breath with its centrifugal and centripetal movements, the counterpart in humans of the cosmic movement of the Hamsah or Siva-Sakti tattvas, which are the soul of the Universe. The Sun is not only the center and upholder of the solar system, but the source of all available energy and of all physical life on Earth. Accompanying the sunshine there proceeds from the orb a vast invisible radiation, the prerequisite of all vegetable and animal life. It is these invisible rays which, according to science, sustain the mystery of all physical life. The Sun as the great luminary is the body of the great Solar God, a great manifestation of the Inner Spiritual Sun.[1]

Like wheels within wheels, energy currents spiral from above to below to above in a double helix.

All creation is a spiral dance of attunement with the underlying oneness of the universe ("one coil"). The word "universe," which means "one coil," comes from the Latin *unio*, oneness, union, unity, and *versus*, turned with a single impetus—hence turned wholly—toward a single unity, turned into one.[2] The essence of creation is "spiration." Respiration is re-spiration—as we breathe, we re-spire. We come into attunement with "the cosmic spiral." The Sanskrit word *prana*, usually translated as "vital force," comes from the root *an*, meaning "to breathe," and refers to the universal breath or vibration. Our bodies are made up of spiraling energy fields. All energy vibrates in a double helix of wholeness which we call dinergy. All energy is the universal spiraling breath of the holomovement of the tri-gunas. All vibration is a spiral waving, a helix of force. The body as a whole is a spiral helix of force. The cycles of the breath move through the body much as a wave moves through an ocean as its fundamental organizing force.

It is valuable to understand the primacy of the breath and the profound reality of the building, balancing, and elimination phases of the breath. As waves move through an ocean, breath moves through the watery medium of the body. From head to toe we can experience the body's alignment with the radiant, building, solar, yang force of the in-breath. In the out-breath we can feel our alignment with the releasing, purifying, eliminating, contracting yin breath. These cycles shimmer through every body cavity, organ, muscle, cell, molecule, and atom of the body as the very breath and heartbeat of life.

In the ancient wisdom of the *Vedas*, we often find images that underscore the primacy of the breath. All of the energy of creation is understood as the breath of Brahma, the Creator (recalling that the word "Brahma" derives from the Sanskrit *brih*, "to expand"). It is said that the Lord "breathes creation into existence," so that breath is seen as the motive force of creation. Again, Atman, the universal essence in the form of the personal essence, is from the root *at*, "to breathe." We picture a world *sui generis* as the spiraling breath of the universe. Our world is one in which the fabric of phenomenal form is woven of spiraling threads of suprasensory vibration. All vibration is a subsystem of the spiraling breath of Brahma.

Breath is the foundation of life. From our first breath to the time we expire, it is breath that vitalizes the body. Without breath there is no life. It is the breath which gives us our place in nature and the cosmos. Our first breath defines our place in the cycles of infinity. The imprint of this breath defines the patterns of archetypal resonance of our astrological signature. Respiration is a process of re-spiration, the rhythms of our breath fostering our bodies' attunement with the all-pervasive spiral of Spirit that is the ground of all being. With inspiration, the spiraling rhythms of the breath radiate outward from every atom, molecule, and cell of the body. With expiration, the spiraling movement is inward. The breath thus moves through the body much as a wave moves through a body of water. Breath is the motive force of the body, fundamental to the rhythms of movement that sustain the body.

The *Vedas* portrayed prana, "the Supreme Breath," the pure life force, as a horse "whose various energies are depicted as the chariots of the Gods ... There are five life-breaths that sustain our vital and nervous systems ... [prana] controls our breathing and enables us to draw in the universal Life-forces into our physical being."[3]

Apana, the downward "breath" which pulls against prana, governs the excretory functions; *samana* kindles the bodily fire and governs the processes of digestion and assimilation; *vyana,* or diffused "breathing," is present throughout the body, affecting division and diffusion, resisting disintegration, and holding the body together in all its parts; and *udana,* the ascending vayu, (current of prana) is the so-called "upward breathing."[4]

As above, so below. The spiraling currents of prana undulate through five phases of resonance: in-breath, middle-breath, diffused breath, out-breath, up-breath. Prana is centered in the heart region; *apana* in the region of the anus; *samana* in the navel and abdominal region; and *udana* in the throat region. *Vyana* pervades the whole body.

Our bodies exist as energy fields within the energy fields of the bodies of nature and the cosmos. It is through the breath that we realign and attune with the vibratory forces of these energy systems. Our breath embodies the cycle of Polarity, and is the link between the microcosm of our body and the macrocosm of the universe. It is the living microcosm of the universal breath of creation. The universal polarity of expansion and contraction, electric and magnetic, yang and yin, birth and death, involution and evolution is the rhythm of this omnipresent force. Each cycle of respiration is a movement through a positive, radiant, building phase, and a negative, contracting, cleansing phase, and a neutral phase of entrainment with the source. Through in-spiration and ex-spiration, our own life-breath is the link between the macrocosm of the cosmic breath of Brahma and the microcosm of our life as the personification of the conscious energy of evolution. The Polarity Therapist releases tension and resistance in the body to foster the receptivity of the body to the regenerative processes of re-spiration. An emotional energy block is tension, which inhibits breathing.

The Spiral Dance

Spirit manifests through spirals. A synonym for creation is "spiration." All energy in nature is spiral in form. Dr. Stone writes: "Curvilinear motion or vibration operated in space and created sphere after sphere until all was manifested. AND THE SAME ENERGY MAINTAINS THIS CREATION EVERYWHERE. Motion is always in the shape of a curve, from the path of the planets in the sky to the very atoms of matter, and spinning electrons."[5]

Throughout all creation there is a primordial breath which in-spires and re-spires. To be "inspired" is to be in harmony with the ubiquitous spiral of Spirit. To breathe is to "respire," to bring our life-force into a spiral of attunement with the universal helix carrying all force in the cosmos. The fields of force of prana entrain the spirals of the physical phases of life energy. The force of all energy in nature is sustained through its spiral form and, through the law of sympathetic vibration, through its capacity to breathe with universal waveforms. To respire is to inspire. To respire is to recreate harmony and attunement with the archetypes and to be tuned into the subtle harmonics, which reflect deeper levels of the finer vibrations of the higher levels of intelligence.

We all breath the same air. This air is *Spiritus* (Latin, "life breath"), the breath of life. The presence of *Spiritus* gives life. Air is the presence of life essence. Air is one, unchanging throughout nature. Summer and winter, the Sun and Moon of Fire and Water are unceasing in their cycles of change. Bodies are born and die. Air, the life essence, is unchanging. Is the body merely a vehicle for *Spiritus,* the one unchanging life essence? Is a life a leaf on the tree of life? Is *Spiritus* the one life? the life of nature? the life of the universe?

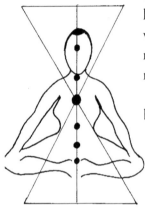

Purce, *The Mystic Spiral*

Breath is the living omnipresence of universal law. Respiration is the act of re-spiring. As we breathe in and out, we draw in spirals of air through which sympathetic vibration reestablishes the spiral harmonics of nature's fields of force throughout the body. The more sound and harmonious our bodies, the higher the level of Cosmic Intelligence we entrain with, and the more profound the clarity, joy, and inspiration that inspires our life. The words "health" and "whole," as well as "heal" and "holy" derive from the Old English root word *hal,* which means "sound, whole, and healthy."[6]

The Five Phases of Breath: A Quantum Process

> *Everywhere we look in Nature, we see nothing*
> *but wholes. And not just simple wholes, but*
> *hierarchical ones: each whole is a part of a larger*
> *whole which is itself a part of a larger whole.*
> *Fields within fields, stretching throughout the*
> *cosmos, interlacing each other and everything*
> *with each and every other.*
>
> *—Jan Smuts*[7]

The key to understanding the reality of the elements and of the living link between the macrocosm and the microcosm is in the phases of the breath. Standing wave fronts undulate through every body of wholeness as the foundation of integration in the cosmos. In nature we picture standing wave fronts as spirals of undulating energy. Earlier, in our discussion of sacred sound, we pictured the dynamics of the "Law of Three," where the ubiquitous breath pulse of Brahma sustains wave and form in the cycle of vibration. As the undulations of vibration move along a spiral, the spiral's shape changes, and aspects of the spiral come into attunement with the universal archetypes of resonances. Thus there are, in quantum fashion, phases of attunement waving through all the spirals of creation.

Air

Fire

Ether

Water

Earth

As the waves phase into attunement, they are energized through sympathetic vibration and entrainment with the elemental archetypes of resonance. The root archetypes resonate through the spirals in five keynotes of materialization. First is Air, a lengthening of the emanation from the Etheric core, toward an oval field of resonance. Second is Fire, a radiant, positive, centrifugal arc of vibration. Third is a pentamirus stillpoint between Fire and Water, which is a double helix of resonance, Ether-and-Earth-predominant. Fourth is Water, a negative, sinking, centripetal, eliminating resonance. Fifth is a crystallization of vibration in tune with the Earth's radiance of gravity lines of evolutionary currents that releases Spirit, to complete the cycle. That brings us back to a sattvic stillpoint between the breaths, in tune with the unified field.[8] Air and Earth as harmonics of resonance are in a polarity relationship, the lengthening of Air predominating in an inner undulation of attunement and receptivity to "Heaven" (Ether), and the lengthening of Earth, an outer undulation of crystallizing resonance attuned with planet Earth's force field of resistance and dissolution. Air and Earth resonate in the vertical phases of the undulating holomovement. Fire and

Fire Fire Fire Fire

Earth Air Earth Air Earth Air Earth

Water Water Water Water

(Doczi, The Power of Limits)

Water resonate in the centrifugal and centripetal entrainments, and in the ascending and descending horizontal phases of the holomovement.

All vibration entrains with the elemental archetypes, through sympathetic vibration through sympathetic vibration with the quantum processes of these universal fields of force. All solidity, regardless of the phenomenon, resonates with the Earth harmonic, all liquidity with Water, all heat with Fire, all movement with Air, and all space with Ether. On every level of macrocosm and microcosm energy spirals as the Cosmic Octave. In quantum fashion, it entrains in five steps from Spirit within to matter without. Everything in creation is cycling, breathing, beating, vibrating, oscillating in entrainment with this theomorphic holomovement. As we breathe in and out, a dynamic rhythm is sustained which is reflected in the rhythms of the breath, heartbeat, peristalsis, oscillations, and vibrations of every level of the body. These rhythms are the living, breathing presence of universal law. The helix of the thoracic cavity and pelvis expands and contracts with each breath. This cycle of expansion and contraction waves through the body on every level. These waves affect the shape of the body's cavities, organs, cells, molecules, and atoms. These patterns of resonance are our link to nature and the cosmos.

A major focus in Dr. Stone's work is the concept of the "oval field," a form which emphasizes the dynamic of polar opposites and forces of attraction and repulsion emanating from a neutral field that underlies all electromagnetic processes. Dr. Stone writes:

> The oval or ellipse is the Creator's design of curved lines of energy travel in the cosmos. It rises from one latent point or neuter center [bindu], like a stream from one fountain and then flows in opposite directions, away from the Center into space. When the greatest point of expansion has been reached, it falls into graceful curves of an oval or ellipse, creating universes with planetary systems and solar systems within its sphere. In the miniature world of individual units, this is also reproduced in microscopic dimensions as electric lines of force, as the dew drop, the cell, the egg, magnetic lines of force in the bar magnet, in the molecule and in the atom with electronic orbits. The lines fall gracefully to the right and to the left, as male and female energy, equally balanced by the neuter center. The Great Oval was called. The Cosmic Egg which brought forth all created things.[9]

The oval field is an expression of the quantum step-down mechanism of the elemental resonances. Living vital fields are pulsating with life. The pulses undulate through the field in oval spirals from within to without, and at the same time, from without to within. The interaction of the phases of the cycles of pulsation in quantum fashion sustain the attunement with the elemental resonances. There is a movement undulating through the spiral fields of force which changes the curve of the arc, and thus the attunement, through sympathetic vibration with the archetypal harmonics. Oval fields are made up of the arcs of the five phases of manifestation. In quantum fashion the moving wave entrains with the keynotes of the ubiquitous elemental harmonics.

The movement of energy through the universal cycle of phases of manifestation and negation is the root archetype and foundation of all manifestation. All vibration undulates through these phases of cosmic attunement. This is the key to understanding the link between macrocosm and microcosm in the "tune-i-verse." The breath is a microcosm of this holomovement.[10]

Cosmic Attunement in the Quantum Cycle of the Breath

The first phase of breath is the in-breath, a lengthening of the torso, body cavities, organs, muscles, cells, molecules, and atoms into an oval field with an inner sattvic receptivity to the core. This involves a quantum leap as a step-down from the ultrasonic core into the harmonics of the physical body. The thoracic cavity, lungs, and nervous system, which are Air-predominant, entrain with the Air element of resonance.

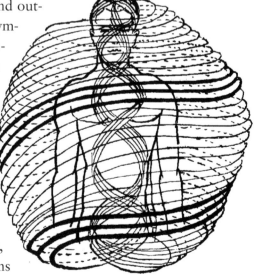

(Purce, The Mystic Spiral)

The second point of entrainment in the quantum cycle of the breath resonates with the Fire element, as an expansive, electric, radiant, building, rajasic field of force. As we inhale, the body cavities expand and the molecules, cells, and atoms are drawn into a radiant harmony with the universal Fire elemental keynote of vibration. The fabric of the body as a resonant structure entrains with the universal Fire resonance to energize the Fire-predominant fields of force in the body—the diaphragm and organs of the solar plexus.

The third phase is the pause between the in-breath and out-breath. This phase is pentamirus and is entrained through sympathetic vibration with the standing waves of the atmosphere of the Earth. Here we have a lengthening of the fields of force to a double helix, with an outer receptivity, "to the surface tension at the limits of the pattern field ..." This point in the balance of yang and yin also resonates with the innermost core as an Etheric stillpoint, as there is a quantum crossover from centrifugal to centripetal in the wave.

The fourth phase resonates with the Moon, as a downward-flowing pulsation of purification, elimination, completion, and cleansing. In this rounding of the out-breath, the diaphragm, body cavities, molecules, cells, and atoms entrain with the cleansing lunar influence. The pelvis and perineal diaphragm resonate tamas guna with this centripetal Water-predominant force.

(Stone, Evolution, Energy Charts)

The fifth phase makes the quantum leap back to the ultrasonic core. It is the magnetic, contracting, eliminating attunement, a pulling inward of energy in alignment with the crystallized radiation of the Earth's gravity lines, the return currents of evolutionary force. The coccyx and mysterious sacrum resonate with this Ether-predominant force.

Appreciate for a moment that life is always vibrating. A live body is throbbing with movement. This movement is profoundly lawful, expressing universal law, and is cyclical, as is all movement in nature. Each breath of our lungs, beat of our heart, rhythm of peristalsis, pulsation of the cells, and oscillation of our atoms reflects the same lawful harmonies of creation. The breath attunes the body to the universal forces of nature and the cosmos which vivify all life.

The organism as a whole—each body cavity, organ, cell, molecule, and atom of the living being—cycles through the phases of breath. These cycles are the living, breathing presence of prana, the living intelligence of the universe. Thus, as we breathe, the body as a whole is drawn into attunement with the keynotes of the five elemental harmonics.

The standing wave patterns moving as the energy fields of the body align with the harmony of the standing wave patterns of the suprasensory wave fronts of nature and the cosmos. The cycling of the breath is the fundamental force that aligns the standing wave patterns of the microcosm of the body with the standing waves of nature and the cosmos.

When we strike chords of universal attunement, harmonics are created with each oscillation of the chord which instantly bring us to supraphysical rates of vibration. Thus, it sustains harmonics which bridge Heaven and Earth. All harmonious vibrations in nature sustain wave dynamics that ultrasonically phase from the physical to the supraphysical. In any whole system, standing waves are generated that fill the whole with a spectrum of vibrations that are a bridge to the infinite. In the sound human body, as in all subsystems of nature, all vibration embodies this theomorphic music of Heaven and Earth.

Polarity Therapy and the Ultrasonic Core Radiance

The focus of Polarity theory and Therapy is the relationship between Spirit and matter. The work of the Polarity Therapist is to release tension and resistance in the body, to bring the physical into deeper harmony with the Soul. The Life Field is understood as an all-pervasive inner force whose radiant potency sustains the fields of force in matter.

While the source of Spirit is eternal and unchanging, its ever-evolving potency, or shakti, is the force which sustains creation. We work with a cycle of harmonic phases—of involution and evolution—in which the spiritual potency steps down in vibratory rate to the resonant frequency of physical expression and then returns to the spiritual.

The Body Personifies the Theomorphic Process

The structure of the body as a whole is theomorphic, the personification of the process of Spirit emanating into matter. Adding two nonmaterial phases— Spirit and mind—to the five elements described above gives us seven phases of energy step-down from Spirit into matter. It is an expression of the "Law of Seven" and is organized as a field of resonance by standing waves whose wavefronts sustain the seven body cavities that emanate from the chakras. The seven core chakras personify the seven planes of consciousness from Heaven to Earth. The body cavities reflect the theomorphic pattern of seven phases of Spirit stepping down into matter. Each cavity is a unique resonant environment which predominates in its attunement to an archetypal force. Each body cavity parallels one of the seven planes of the quantum step-down of Spirit to matter in a progressively more material resonance.

The pineal gland is the link that resonates with the primary energy of the source. The energy steps down from the causal plane through the pineal and radiates upward as the brain. The brain is a microcosm and personification of the mental plane. These mental plane images are the inner seeds of our manifestation. Our desires begin to manifest when they are voiced at the throat, which is a microcosm of the Etheric plane. The Ether-predominant throat steps energy down into the physical body for expression. In the Air-predominant thoracic cavity the heart, lungs, and diaphragm establish the appropriate emotional rhythms to bring the desire into physiological and physical expression. Air stimulates Fire, directed energy. In the next step-down to the Fire-predominant solar plexus, these impulses are empowered with the force to move into the musculature. The impulse is birthed through the Water-predominant pelvis, the level of form and elimination, to complete the process of manifestation through the colon and bones, as the expression of our desires manifesting on the field of the Earth.

The fingers and toes also reflect the theomorphic spectrum of five elements in the physical expression of the evolutionary cycle. The big toe and thumb resonate with the Etheric frequency in the core of the body. The index finger and second toe resonate with the Air element, which predominates in the thoracic cavity. The third finger and toe resonate with the body's Fire harmonics, which predominate in the organs of the solar plexus. The ring finger and fourth toe predominate in the Water element and resonate with the cleansing and reproductive organs in the pelvis. The little finger and toe resonate with the Earth element and the processes of elimination. This underscores the organization in the body of Spirit moving into matter.[11]

The energy fields of the front and back of the body interact as a unity, creating a helical pattern of energy flows between the thoracic and pelvic cavities.[12] The body expresses the symmetry of the double helix.

As the centrifugal rajasic resonance of the thoracic cavity entrains with the in-breath, the centripetal resonance entrains with the contraction of the pelvic cavity in a balance of wholeness. The entrainment is reversed with the out-breath. There is always a balance of wholeness between rajas and tamas, above and below, front and back.

The five-pointed star is the crystallization of the play of the phases of the elements which sustain manifestation on Earth, and it is the personification of the sacred geometry which is the quintessence of the principles behind manifestation.[13] Through movement from above to below and within to without, our bodies are the embodiment and personification of sacred geometry. The five-pointed star describes the gestalt through which energy moves from within to without, from Spirit to matter, and above to below, and from Heaven to Earth. In the five-pointed star we are pointing to the pattern of resonance embodied in the atomic structure and patterns of oscillation of the molecules and cells. The cosmic breath, on all levels of our bodies—organic, cellular, molecular, and atomic structures—aligns itself with universal law and vibrates as an emanation of the harmonics of the unified field of the elements. Through standing waves on every level of the macrocosm and microcosm, the cycles of vibration of the atoms and cells of our body are entrained with the energy fields of nature and the universe. The atoms, molecules, and cells of the quadrants of the body resonate in a cycle of yin and yang, positive and negative, Fire and Water, Air and Earth.

The five-pointed star is the crystallization of the gestalt of cosmic modulation that breathes through every cycle of vibration. The Air and Fire keynotes resonate in the thoracic cavity with the lifting of the diaphragm with the in-breath. The Water and Earth resonance is modulated in the pelvis by the con-

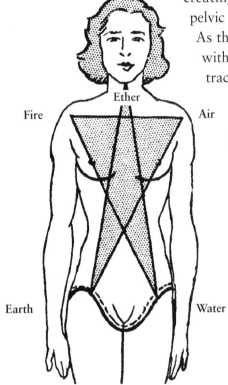

The Helix, Five-Pointed Star, and Breath
The ancient Hindus tell us that "God sings the universe into creation." the body, as all of nature, is a fabric of attunement. Every muscle and bone is a field of energy vibrating as an expression of attunement with the standing waves of universal law.[14]

tracting of the abdomen in the diaphragmatic lift. The reverse happens with the out-breath. The falling diaphragm contracts the thoracic into a centripetal resonance which is now Water, then Earth-predominant as the expanding abdomen modulates into a centrifugal Air and Fire keynotes of resonance. This gestalt is happening on every spiraling undulation of microcosm and macrocosm. Fire and Water, Air and Earth resonate in reciprocal relationships of balance and wholeness. There is a stillpoint in the movement of the diaphragm, where rajas and tamas intersect, which resonates with sattva guna and the keynote of Ether. The breathing process modulates the attunement of the body to the elemental cosmic forces.

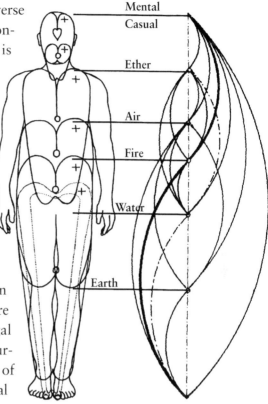

Each of the body cavities sustains a predominant resonance which parallels the planes of creation. (Doczi, The Power of Limits)

Archetypes are actually types of arcs—changes in wave length—and keynotes of resonance. Air and Fire are the step-down into manifestation through the centrifugal radiance. Air is a lengthening of the vibrations of the currents from the absolute sphere of Ether into the duality of the Hiranyagarbha (cosmic egg). Fire is a radiant centrifugal current of vibration entrained with solar forces. Water is a rounded arc of vibration resonating with the downward centripetal lunar force, and Earth is a flattening crystalline field whose resistance releases Spirit from gravity's grip to leap back to the stillpoint of Ether. The return currents step up to the source through the centripetal resonance. This is a quantum process though which all vibration is entrained with these four primordial archetypal keynotes. These keynotes are nested as fields within fields in a profound balance of wholeness and attunement. The ubiquitous cosmic breath undulates through these four phases of resonance in every cycle of vibration.

With the in-breath, the lengthening of the thoracic cavity sustains a predominant resonance with the Air element, and the barreling of the thoracic cavity resonates with Fire. With the out-breath, the barreling of the abdominal cavity resonates with Water, and the lengthening of the abdomen resonates with Earth. The stillpoint between the breaths resonates with Ether. From in-breath to out-breath this pattern is reversed from above to below, from front to back, and from side to side. The body is thus a microcosmic personification of the five elements and the five-pointed star. The phase predominance in each quarter of the body resonates in a dynamic unity with the gestalt of the energy field of the body. Again, the body *is* the embodiment of the ubiquitous dynamic

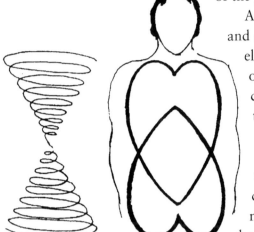

of the play of the elements and the five phases of manifestation. As we breathe, the body cavities, organs, cells, molecules, and atoms are drawn through a process of attunement to the elemental cosmic forces. The cells oscillate through a pattern of expansion and contraction of centrifugal and centripetal cycles which entrain with cosmic forces. All energy spirals through the elemental phases of cosmic attunement.

As discussed in Chapter 11, the muscle fibers of the human heart are spiral shaped and contract in a spiral to twist the blood through the chambers.[15] Each of the four chambers of the heart predominate in an elemental resonance and are a microcosm of the process of entrainment to the elemental archetypes of Cosmic Intelligence.[16] This process is cycling through every spiraling atom of creation.

As above, so below. Manifestation is a dynamic interplay of the five elements. We picture the universal energy field as a three-dimensional, heart-shaped field of force sustaining four predominant resonant frequencies. As the standing wavefronts breathe through this whorl, the four directions of motion phase into perfect harmony with the universal types of arcs and are energized through sympathetic resonance. The waveforms of the phases of the elements interact as a dynamic gestalt of influence, balance, and wholeness.

This is the driving force of manifestation. To understand this process of cosmic attunement through the life breath is to understand the validity of the elemental models of nature and the cosmos which prevail in traditional cosmologies.

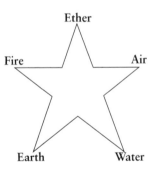

Centrifugal and centripetal vortices. As a wave moves through an ocean, the breath literally moves through every atom of the body.

Five-Pointed Star.[14] The body is a musical instrument played by the cosmic forces. The cycle of the breath modulates our attunement to the elemental keynotes of the cosmos. The five-pointed star is the crystallized expression of the harmonic relationship between the five elemental keynotes that breathe through every spiral of creation.

Meditation on the Sacred Breath

Throughout this work we have been a finger pointing to the Moon of Higher Knowledge. This is a unique realm of intuitive insight and *realization* that goes beyond the linear mind and perception of the five senses. This "knowledge" can only be made real, that is, "realized" through a profound process of personal purification to bring us into harmony with the finer, higher vibrational forces of Cosmic Intelligence. This process of Self-cultivation yields a unique epistemology and a category of knowledge *sui generis* called *Yogic Perception.* This body of sacred knowledge, *Self-Realization,* has been replicated crossculturally and transhistorically for millennia.

We already have what we are seeking. It is only the busyness of the impurities of our minds that clouds our perception of the sunshine of our being.

Quiet the mind and the Self illuminates our being.

The breath is often praised as the most powerful tool for quieting the mind. We have all experienced the way we hold our breath to still our minds in concentration. Perhaps you have been taught to take a deep breath or count breaths to ward off anger and regain equanimity.

Using the breath we can immediately break the unconscious yoke of identification with the mind and emotions and establish a Self independent of our thoughts or feelings. This can be a profound tool for cultivating detachment and equanimity and conquering the demons of the impurities of our minds. Most of us are deeply identified with the impurities of our minds. We blindly identify with our every thought, emotion, and passion. The profound fruit of Yogic psychology is the understanding that you are not your thoughts, you are not your emotions, you are not your passions. These are merely input, the contents of the mind. Yoga offers a myriad of techniques for cultivating inner peace to quiet the mind and reveal the deeper Self.

Vipassana Meditation[17]

The ancient Buddhist technique of *Vipassana* (Insight) Meditation begins with a process of breath control called *anapana-sati* (awareness of respiration). Breath control through anapana has a profound power to quiet and concentrate the mind.

Find a very quiet space where you will not be distracted. Sit upright and assume a comfortable posture with head, neck, and spine in a straight line. Close your eyes.

Turning away from the outer world, the most prominent activity of our inner world is breathing. Give your attention to the breath entering and leaving the nostrils. For as long as possible keep your attention on this movement. When thoughts arise, simply bring your attention back to your breath.

This is not a breath exercise. This is an exercise in awareness, a technique for concentrating the mind. The intention is not to control the breath, but to simply be conscious of it as it is naturally—long or short, heavy or light, rough or smooth.

As simple as this practice sounds, appreciate that quieting the mind is one of the most challenging things in the world. Discomfort, pain, and fantasy will inevitably catch your attention. As soon as you realize that you have been distracted, simply move your focus back to the sensation of breath entering and leaving the nostrils. Do not react to your loss of concentration or judge yourself in any way. Simply focus with renewed presence on the breath entering and leaving the nostrils. As you fail to hold your attention for more than a few breaths, simply, smilingly, in kindness, and without discouragement bring your attention back to the sensation of the breath at the nostrils. Anapana-sati requires repeated, continuous practice, patience, and diligence. It can be practiced fruitfully for years—even decades—in a process of purifying the mind. It can be practiced for ten to fifteen minutes daily, by itself to actualize the higher mind, or as a prelude to other forms of purification.

CHAPTER THIRTEEN

Intelligence Fixed in the Breathing Spirit Is the Hub of Being

In the east the primal image is of Narayana (Lord
Vishnu) who after the destruction of the world of
the previous cycle and before the creation of the
next is asleep on his bed of the great serpent
Ananta (time) which is floating on the ocean of
milk (eternity). From his navel a lotus springs
forth with the god Brahma, the creator seated on
it. The sleeping lord dreams creation.[1]

It is always valuable to remember that in the Sanatana Dharma all creation
is a play of consciousness. God as *Satchitananda* (being, consciousness, bliss)
becomes creation. The evolution of the universe is an "appearance" within
this consciousness, representing the unfolding of the infinite creative potency
of the Absolute. Every atom of nature is the eminent expression of this Intelligence, and is a center for manifestation of this creative potency.[2]

The *Upanishads* are revered as the pinnacle of the ancient esoteric teachings of the Sanatana Dharma. In the *Brhad-aranyaka Upanishad* (which is
considered to be the most important of the *Upanishads*), Yajnavalkya explains:
"As a mass of salt is without inside, without outside, is altogether a mass of
taste, even so, verily, is this Self without inside, without outside, altogether *a*
mass of intelligence only."[3] [emphasis added]

In one of the principal *Upanishads*, the *Kausitaki-Brahmana*, the God
Indra, the King of Heaven, is asked for a boon deemed to be the most beneficial

for mankind. Sri Indra exhorts: "Understand me only," and reveals the doctrine of "The Supremacy of Intelligence."

> For verily, without intelligence, speech does not make known (to the self) any name whatsoever. "My mind was elsewhere," he says, "I did not cognise that name." For verily, without intelligence breath does not make known any odour whatsoever. "My mind was elsewhere," he says. "I did not cognise that odour." For verily, without intelligence the ear does make known any sound whatsoever. "My mind was elsewhere," he says, "I did not cognise that sound." ... Verily, if there were no elements of intelligence, there would be no elements of existence.... For as in a chariot the felly is fixed on the spokes and the spokes are fixed on the hub, even so these elements of existence are fixed on the elements of intelligence and the elements of intelligence are fixed in the breathing spirit. The same breathing spirit is, truly, the intelligent self, bliss, ageless, immortal.[4]

Intelligence fixed in the breathing Spirit is the hub of Being around which the mind and senses turn. It is Intelligence that is the foundation of life and sentience.

The esteemed philosopher S. Radhakrishnan points out that: "The basis of the individuality of the ego is *vijnana,* or intelligence which draws round itself mind, life and body."[5] In the Hindu pantheon the gods are depicted typically with four arms which symbolize the first four tattwas, or the principal forces within consciousness. First is the foundation of creation: chit, divine consciousness, a self-aware force of existence. The essential nature of chit is unity, wholeness, and freedom. It is the omnipresence of being which is the fundamental ground of creation. Second is the divine counterpart to lower mind, manas (from the verb root *man,* "to think"), which is the mind, the thinking principle in general. Third, buddhi (from *budh,* "to know"), is the intellectual faculty, discrimination. And fourth ahamkara (from *aham,* "I," and *kara,* "action"), is the conception of individuality, ego, self—the "I-maker." The foundations of creation are these four states of consciousness.

Thousands of years ago, the ancients developed a sophisticated psychology of liberation. In this model, to quote the revered sage Shankara (AD 686): "The mind, together with the organs of perception, forms the *"mental covering."* It causes the sense of "I" and "mine." It also causes us to discern objects."[6]

Mind in its relationship to the input of the senses is called the "inner instrument" *(antahkarana).* Mind functions as manas as it receives impressions of the outer world through the senses. Mind functions as buddhi when classifying, relating, and distinguishing one object from another through the senses. Mind functions as ahamkara in relating to the incoming sense data as I, me, or mine.[7]

Shankara perceptively describes the relationship between the mind and desire. "The mental covering may be compared to the sacrificial fire. It is fed by the fuel of many desires. The five organs of perception serve as priests. Objects of desire are poured upon it like a continuous stream of oblations. Thus it is that this phenomenal universe is brought forth." Desire creates the world. Liberation is experienced as desire ceases. Shankara concludes: "Ignorance is nowhere except in the mind."[8]

Further expanding on manas, it is the Sanskrit for "mind," which comes from the root *man,* to think, believe, imagine, suppose, conjecture." The Indo-European word *man* shares the same Sanskrit root. In the philosophy of the Sanatana Dharma, mind *is* the basis of creation as *Mahatattva* or Mahat, which is the principle of Cosmic Intelligence. Mahat is the Great Principle, "the first product of the cosmic substance . . . the first stage away from the original condition."[9]

In the Western tradition of Hermetic philosophy, the principle of the mental universe is fundamental. In *The Kyballion,* in a chapter entitled, The Mental Universe, the Initiate writes: "The Universe is Mental, held in the mind of THE ALL . . . THE ALL IS SPIRIT! But what is Spirit? This question cannot be answered . . . Spirit is simply a name that men give to the highest conception of Infinite Living Mind."[10]

The Dalai Lama writes: "According to the Buddhist explanation, the ultimate creative principle is consciousness. There are different levels of consciousness. What we call innermost subtle consciousness is always there. The continuity of that consciousness is almost like something permanent, like space particles. In the field of matter, that is the space particles; in the field of consciousness, it is the clear light . . ."[11]

The Neoplatonic philosopher Plotinus attempted to describe his experience of pure consciousness:

> For There, everything is transparent, nothing dark, nothing resistant; every being is lucid to every other, in breadth and depth; light runs through light. And each of them contains all within itself, and at the same time sees all in every other, so that everywhere there is all, all is all, and each all, and infinite the glory. Each of them is great; the small is great; the sun, There, is all the stars, and everything is all the stars and sun. While some one manner of being is dominant in each, all are mirrored in every other . . . since everything contains all things in itself and again sees all things in

another. So that all things are everywhere, and all is in all. Each thing like-wise is everything ... There each part proceeds from the whole, and is at the same time each part and the whole. For it appears indeed as a part; but by him whose sight is acute, it will be seen as a whole.[12]

In the monastic establishment of Tibetan Buddhism the ancient Sanskrit texts have been well cared for over the ages. The essence of the science of Self-development of the Sanatana Dharma has been widely practiced. The Unitive Field of consciousness is called the *Dharmadhatu*. The Dharmadhatu is the boundless fundamental ground of all being. Garma Chang, quoted in Ken Wilber's *No Boundary*, explains: "In the infinite Dharmadhatu, each and every thing simultaneously includes all other things in perfect completion, without the slightest deficiency or omission, at all times. To see one object is, there-fore, to see all objects, and vice versa. This is to say a tiny individual particle within the minute cosmos of an atom actually contains the infinite objects and principles in the infinite universes of the future and of the remote past in the perfect completeness without omission."[13] The ancient seers realized the omnipresence and omniscience of the Self in a direct experience of the holo-graphic ultimate ground of Being.

Wilber, who has been called "the Einstein of the consciousness move-ment," continues: "In Mahayana Buddhism the universe is therefore likened to a vast net of jewels, wherein the reflection from one jewel is contained in all jewels, and the reflections of all are contained in each. As the Buddhists put it, 'All in one and one in all.' This sounds very mystical and far-out, until you hear a modern physicist explain the present-day view of elementary par-ticles: 'This states, in ordinary language, that each particle consists of all the other particles, each of which is in the same way and at the same time all the other particles together.' "[14]

The physicist James Jeans is widely quoted for his pronouncement, writ-ten in the 1930s, that: "Today there is a wide measure of agreement ... that the stream of knowledge is heading towards a non-mechanical reality; the universe begins to look more like a great thought than a great machine."[15] In his classic discussion of science and mysticism, *Nature of the Physical World*, Professor A. S. Eddington concludes: "The idea of a universal Mind or Logos would be, I think, a fairly plausible inference from the present state of scien-tific theory, at least it is in harmony with it."[16]

A similar conclusion is reached by physicist Sir Fred Hoyle, the founder of the Cambridge Institute of Theoretical Astronomy and the man responsi-ble for our current understanding of the origin of all the heavy elements in the universe. According to Hoyle, within the laws of physics there is not only mathematical evidence that the universe was designed by some sort of Cosmic Intelligence, but that that Intelligence is unfathomably old, billions of years older than the age of the known universe.[17] Thousands of years ago in the *Brhad-aranyaka Upanishad,* the sage Yajnavalkya revealed: "so verily, this great being, infinite, limitless, consists of nothing but knowledge."[18]

It may be valuable to remember the fundamental axiom of our cosmology, that God as Absolute Awareness becomes the universe. The Jewel whose light is so fascinating is the omnipresence of the Absolute, the Unitive Field of consciousness. Everything we are discussing is fundamentally Creative Intelligence. All of the forms and processes outlined above are states of consciousness. The forces are essentially principles of consciousness within the dynamic of the Universal Intelligence. Our thesis is that nature and the body are the life breath of Cosmic Intelligence.

Dr. Stone's Hermetic Conception of Mind

Again we find ourselves contemplating Dr. Stone's first principle of Polarity Therapy. "The Soul is a unit drop of the ocean of Eternal Spirit which is the dweller in the body as the knower, seer, doer; it experiences all sensation and action. It alone is the power in the body which reacts to any mode of application of therapy or action. *Consciousness and intelligence reside within the soul.*"[19]

Dr. Stone, who was a life-long student of Hermetic philosophy, writes:

Mind energy is the finest form of matter ... "Prana" ... is the mystic link or neutral common denominator between these two opposite poles of Mind energy as the Molder, and Space energy as resisting material, matter ... In the human body Mind Energy flows over the brain and the nervous system and becomes animated Intelligence, Feeling, Perception, Consciousness; the root of all the senses and the awareness of all sensations in and through the form of matter. SOUL IS CONSCIOUSNESS; "PRANA" IS LIFE; MIND ENERGY IS A NEUTER meeting ground between Soul and matter.[20]

Later in the book, *Energy,* he continues:

Mind is the negative pole of the Soul, its source. Spirit Energy, in its descent from Spirit to matter, creates a field or neutral pole in the middle of its travel, that partakes of the nature of both poles. And it is the mind which is this neuter pole. This mind, which is slowed-down Spirit Energy becomes the functioning factor or positive pole which rules the negative pole of matter. *RULE:* All opposites have an intermediary which partakes of the nature of both; like mind, reaches halfway to the *Polarity* of Spirit Energy, and rules everything below that level. *SOURCE:* Even as spirit energy precipitates all creation below itself, so does Mind condense and solidify its energy as various stratas of energy ... By this process it is of the same nature and substance as the created forms; and, being a finer energy itself, it can readily flow in and through all parts and forms. This is the secret of Mind and its rulership of all created things.[22]

Dr. Stone describes the currents of resonance of prana, the life breath of the vital body, as the positive pole throughout the body, and the physical body itself as the negative pole of the fields of action. He writes:

Now comes the third or mental body as the neuter pole of action, which can effect the whole from any standpoint because it is the center core of all neuter fields and forces in the body and in nature. Even as the mental body blends with the physical in every cell of being, function and structure, so does the mental body, as the finest essence and phase of matter penetrate every tissue cell of the entire body, in normal health.[23]

Every tissue cell has a mind, as a diffused particle of the central mind. The mind as the neuter pole animates the other two forms or bodies for its use and the purpose of expressing the mind in and through matter ...[24]

Mind energy waves do not need wires for conducting of its alternating impulses. Mind substance itself is a continuous media plus Ether. Concentration of mind energy is the directing power of its substance. It acts by preponderance of impulse, impact or mental weight.[25]

The cerebrospinal fluid of the brain, in the meninges of the chord and in the center of the nerves is such a medium for the conduction of the AIRY PATTERN ENERGY OF THE MIND over this intricate network.[26]

In Dr. Stone's writings all of the energies that weave the fabric of the body are a step-down from the primordial mind currents. He explains: "Mind energy is the first essence of matter ... It is the pattern energy of geometric proportions in the atomic fields of matter as the shape of things to be. As above, (in the brain) so below (in the body.) ... The whole body is but a duplication of these patterns in a more dense form and a lower vibratory key of action ... Diffused mind energy rules every cell of the body."[27]

"Evolutionary mind energy flows from the mind principle over the brain and nervous system."[21]

The five elements are five kingdoms of conscious vibration which form the step-down mechanism from Spirit-mind to matter-mind. Matter manifests through five keynotes of resonance: space, heat, gas, liquid, and solid, which predominate in Ether, Air, Fire, Water, and Earth elements, respectively. Each of these kingdoms of resonance is a denser, more material vibration of consciousness. "When mind currents are reduced in vibratory intensity, they also fall into the category of electromagnetic waves and functions."[28]

Dr. Stone reveals: "So the mind is the first step-down of the Soul's energy, as a neuter pole which expresses itself perfectly through five sensory currents as five negative circuits and through the five motor currents as five positive expressions in all five fields of matter ... Mind, being a neuter activity itself, is therefore capable of every capacity of action and sensation through matter ... The senses are the five modes of expression of mind in matter."[29]

Each of the senses is tuned into one of the elemental harmonics that sustain phenomena. Hearing resonates with the Ether element which rules the realm of sound vibration. Touch and movement as sensation resonate with the Air element. Sight, which resonates with light, is ruled by the Fire element. The experience of fluidity is Water-predominant, and the experience of solidity is ruled by the resistance of the element Earth. All of nature is woven from these five archetypal harmonics of consciousness. These five levels of resonance sustain the appearance of form in consciousness.

The body as consciousness is woven of the fabric of the five elements which pulsate from the chakras as motor and sensory currents, motor currents predominating in structure, and movement and sensory currents predominating in the senses. Dr. Stone continues:

> The mind expresses itself through its five stepped-down agencies, which split into five sensory senses and five motor senses ... These express themselves also as the mechanical leverage of the ten fingers, for motion and expression of skill ... and as ten toes and energy waves flowing through them for body motion through joints and leverage ... These waves of energy give the delicate sense of touch to the fingers ... Such delicate patterns and finer energy waves link our mental patterns to the outside world, to experience sensations through contact with matter in its five modes of resistance. The five senses are modes of expression of the mind in matter.[30]

Polarity Therapy is an understanding of the body as a crystallization of consciousness. The five elements are five phases in the precipitation of consciousness. Dr. Stone's model underscores the presence of consciousness throughout the body as he stresses that: "The mental body, as the finest essence and phase of matter penetrates every tissue cell of the entire body ... Every tissue cell has a mind, as a diffused particle of the central mind ... Diffused mind energy rules every cell of the body.... " Dr. Stone continues with this sage understanding:

> To learn to control our own mind is the real purpose of all experience, because the mind is the neuter agent of the very Essence of all matter in motion ... It is the most subtle and intangible form of matter, like an airy nothing that pervades and rules all things. All conditions and limitations are produced by mind essence and substance. The molds of pleasure and pain, health and sickness, etc. are formed by *our own thought patterns* of harmony, limitations or discord.[31]

CHAPTER FOURTEEN

Archetypes as Founts of Creative Intelligence

The Vedic tradition tells us that God created the universe from a desire to see himself and adore himself. The being of this inconceivable God may be considered as an all-conscious, all-containing, all-powerful, homogeneous, endless expanse of pure, formless Spirit. His desire to see himself created (or distinguished from himself) an idea of himself, called in Indian thought the "Real-Idea." This divine self-apperception, the "Creative Word" in Judeo-Christian thought, this event itself, is the cosmic human. And this cosmic human is what we, actual humans, call the universe.[1]

In our evolutionary paradigm the universe is understood ultimately as a unified field of awareness. The absolute, Brahman, is the unchanging source of unlimited consciousness. All things express the unfolding of the creative potency of the Cosmic Person. Brahman as illimitable intelligence emanates the ever-expanding field of Brahma, creative evolution. Brahman, the Unitive Field of awareness, is the innermost essence of all beings.

Our thoughts are likened to waves on the surface of the ocean of universal mind. Our awareness, our presence, our consciousness, is the ocean. If we can only sink deeper into the ocean of the One—past the obsessive, reactive mind—each of us can experience the Ultimate Intelligence of cosmic consciousness that is the omnipresence of the Self. God lives within me, as me. What it is that actually is consciousness within me—not my mind, not my feelings, not my body, but the unchanging ground of awareness—is the omnipresence of the One.

Archetypal images in the universal mind are projected into phenomenality. This centerless center of creativity is everywhere—the all-pervasive Unitive Field of consciousness is an omnipresent spring of creativity. Everywhere that tamas guna and rajas guna intersect, a bindu, a sattvic seed-point of equilibrium, is sustained, resonating with the superconscious Unitive Field. This stillpoint, in every wave, cycle, breath, and heartbeat of creation, is the omnipresent centerless center of the One. All creation is woven of the fabric of cycles which express the evolutionary journey of this limitless creative potency.

As light reflects off of every facet of a jewel, so does the Creative Intelligence manifest its creativity through the wellsprings of the archetypes. Archetypes are channels of creativity. They are the primordial patterns of Creative Intelligence, and express its differentiation. Archetypes are fountains of creative expression. Archetypes reflect the immortal Logos from the one mind of the Cosmic Person in the universal drama of the cycles of "time" and "space."

In our model, the substance of the universe is Creative Intelligence. The Creator is fully present in every bindu—stillpoint—of creation. The world is a moment-to-moment creation of an omnipresent God. As "Gods are immortal men, and men are mortal Gods,"[2] each person is the mortal expression of the archetypal realm. The Logos or Divine Word is the force of evolution and the life breath of intelligence which is the Sacred Self of each being.

In our model, where everything is sacred sound vibration, and resonance is the medium of creation, archetypes are understood as "types of arcs." Archetypes sustain resonant patterns in the vibrational fabric of creation. Archetypes are vibrational essences which link the ideal realms of the macrocosm of the One Eternal Being with the microcosm of the cycles of evolution in time.

Creation is *spiraling vibration*. In this universe ("one coil") everything exists by virtue of its attunement with the unity of the infinite *spiral* of the ever-expanding evolution of the One. All differentiation results from different qualities of attunement and resonance with the one omnipresent source— the centerless center of all that is. All physical manifestation is a radiation from this omnipresent field of Creative Intelligence.

Throughout this discussion we have been focusing on a universal holo-movement. We call this process the evolutionary cycle, and describe it as a universal pattern of theomorphic resonance. The fundamental organizing principle, both structural and function, of all processes in nature, in the cosmos, and in all dimensions (as above, so below) are patterns of resonance corresponding to the cycle of creative evolution. The discussion below is a rudimentary attempt at interpreting astrology as a key to the consciousness of the universal holomovement. Following the discussion of astrology are discussions of additional archetypal correspondences.

Astrology and the Holomovement

Consciousness is the image of The Divine One
and, as such, is present on all levels of reality …
—Plotinus[3]

Astrological archetypes are founts of Creative Intelligence. They are the immortal "logos," the "Divine Ideas" spoken of by Plato, through which the Universal Mind manifests its infinite creative consciousness into creation.

> These "Ideas" are not conceptual abstractions at all, but living Spiritual Powers which, as the Gita says, "stand" in their own nature eternally and are reflected in the flux of beings, giving to each its form and its essential nature, not abstracted from beings but formative of beings, the perfect types and patterns of all things below.[4]

In the understanding of many traditional societies, these archetypes were understood as "beings" and personified as demigods, nature spirits, and ancestors. Modern man in the myopic vision of scientism views this as mythology and relegates it to realms that are less than real. Crossculturally and transhistorically in the ancient wisdom, *mythology* points to something *more* than real. Mythology avows a transcendental realm of higher being, which is the source of our becoming.

The astrological "signs" can be better understood literally as *archetypal energy patterns* rather than as symbols, arbitrary carriers of meaning. We suggest that the twelve astrological glyphs actually depict twelve archetypal energy patterns—twelve moments (types of arcs or harmonic resonances), twelve processes in the creation cycle.

The twelve archetypes of the zodiac are the alphabet which spell out the infinite drama of creation.[5] They are the fundamental types of arcs in the holomovement of the ubiquitous spiraling vibration of the Unitive Field. The infinite spectrum of Creative Intelligence manifests its diversity within the unity of this system. The one spiral resonates universally through four great archetypes—Air, Fire, Water, and Earth—four universal types-of-arcs, four fields of force emanating from the unified field. These four universal patterns of resonance and intelligence undulate in a gestalt of wholeness which includes three phases of manifestation—positive, negative, and neutral—giving us the twelve great archetypal resonances of the Logos of the Cosmic Person, which we call the zodiac.

Astrology is the study of the cycles of harmonics of attunement with the cosmos. When these harmonics come into more harmonious attunement through conjunction, triune, and sextile aspects, the archetypes are energized to manifest more sattvically. When the harmonics come into disharmonious attunements, or into attunements with the keynotes of "malefic" planets, there is a tendency to express, attract, or experience stress and disharmony, providing the more challenging aspects of evolution.

The cardinal archetype of resonance is centrifugal radiating energy, which correlates with rajas guna and the principle of directed energy. The fixed signs resonate with centripetal energy and are associated with the inertia of tamas guna. The mutable signs resonate with the balance and harmony of sattva guna.

Our First Breath

Our first breath marks our place in the infinite cycles of evolution. It defines our relationship to eternity and imprints a signature of resonance that defines our relationship to the "immortal realm" of archetypal being. While the holomovement is a ubiquitous pattern of "wholeness," each bindu or stillpoint of manifestation resonates with a unique identity defined by its unique place in the cycles of evolution.

David Tame points out that the ancient Chinese understood this process. "According to the Chinese, the monthly changes from one sign of the zodiac to another indicated *cosmic modulations* in the pattern of celestial harmonics. With each new stellar configuration new Tones inundated the Earth, bringing with them new tendencies in thought, new moods, different behavior patterns, and different activities in the Nature kingdom."[6]

Our first breath defines our place in the cosmos and the cycles of vibration with which we resonate. If, at the moment of birth, the eastern horizon aligned with the force fields of the zodiac which we call Capricorn, the individual taking breath in that moment will manifest a persona characterized by groundedness and practicality. If it is in Gemini, the individual is more likely to live in a world of ideas and words. Each manifestation is said to be

the signature of a moment of creativity and is defined by a pattern of resonance or correspondence which links the personal life cycle to the immortal—
to universal Being.

Seasonal Cycles

Polarity, the play of dual opposites, is the fundamental archetype of manifestation. All experience in nature can be viewed as cycles of energy embodying the Polarity dynamic: nothingness and being, involution and evolution,
birth and death, sunrise and sunset, spring and fall, expansion and contraction, positive and negative. The oscillation between polar opposites—between
the phases of out-breath and in-breath, yang and yin, active and receptive,
pushing and pulling, male and female—is the fundamental pattern
of all creation. The resonance of this archetype of dinergy has
been called "the two hands of God."

 This universal cycle is manifested each day as the
Earth rotates on its axis, and in each year as the Earth
revolves around its nucleus, the Sun, as well as on every
other level of being. Our fundamental experience of
nature and of its cycles of light and darkness is played
out from dawn to dusk and from spring to fall as we
witness the birth and death of each cycle of life. At
every moment the Earth is changing its angle of relationship to the Sun. At every moment the Earth is changing its quality as an energy field in resonance with its
source—the Sun.

Midday
Summer Solstice
Maturity
Fire Element

Dawn
Spring Equinox
Birth
Air Element

Dark
Fall Equinox
Aging
Water Element

Midnight
Winter Solstice
Death
Earth Element

 The constant oscillation of the rajas and tamas fields of force
is the warp and weft of the fabric of creation. The dynamic of oscillation between these forces describes the order of creation on every level of the
macrocosm and microcosm of the cosmos. The integrity of the resonance of
the gunas is the force which defines the archetypes. This is also well illustrated
by the cycles of the year, as the Earth oscillates through solstices and equinoxes
in its cycles of attunement to the force field of our nucleus, the Sun.

 The planet Earth (as a field of vibration) moves through a pattern of resonance where all life *springs* into manifestation as we approach the center of
attunement with the rajasic harmonic resonance at the spring equinox in
March (in the northern hemisphere). As the Earth locks into entrainment with
the rajasic radiant solar force, all the rajasic subsystems of nature are energized through sympathetic vibration. The very word "spring" points to a spiral

essence and the renewal of our resilient attunement to the unseen inner fount of life. As the Earth cycles toward the fall equinox, it "falls" out of resonance with the rajasic archetype of creation, and all life contracts and sinks into the wintery death and crystallization of resonance with the centripetal tamasic archetype.

The Earth and all of the planets move in oval orbits around the Sun. These ovals are like sliding scales in the music of creation. These ovals chart the planetary breathing of the cycles of rajas expansion and tamas contraction. As the planets move through these cycles, in quantum fashion, they lock into the archetypal harmonics defined by the four elements in their three polarities. Dr. Stone writes: "Our Sun and other planets and their orbits have an oval or spherical shape in space. This similarity is shared by the humble cell, the dew drop, the molecules, atoms and whirling electrons. Our physical body is built on this plan..."[7] All creation on every level of the macrocosm and microcosm breathes through these cycles of polarity. All creation, through sympathetic vibration, entrains with the archetypal keynotes of the universal harmonics of the elements.

Signatures in Astrology

As we discussed for archetypes in general, astrological archetypes manifest the infinite creative expression of the Cosmic Person. The astrological archetypes are wellsprings of creative consciousness. The illimitable creativity of the universal being is differentiated by the astrological cycles. The astrological signature defines the uniqueness of each manifestation within the infinite. Everything in creation has an astrological signature and can be defined by its relationship to totality. The astrological signature defines the unique place of each manifestation within the infinite cycles of the evolution of Creative Intelligence.

The twelve archetypes of the zodiac are the embodiment of the four elemental harmonics—Air, Fire, Water, and Earth—resonating through the three polarities of Cardinal (+), Fixed (−) and, Mutable (0). Air and Fire are positive radiations of the involutionary cycle of manifestation, and Water and Earth radiate with the negative evolutionary return currents. The involutionary cycle describes consciousness in its childhood, expressed as the self-centeredness of Aries, the possessiveness of Taurus, the superciliousness of Gemini, the sensitivity of Cancer, and the pride of Leo. The archetypes which resonate with the evolutionary return currents are characterized by more maturity and wisdom. Thus, the archetype Virgo predominates in the consciousness of discrimination, Libra manifests harmony, Scorpio represents introspectiveness, Sagittarius predominates in sincerity, Capricorn predominates in mastery of the material plane, Aquarius focuses on universal ideals, and Pisces focuses on the mystery of unity in diversity. The modes of consciousness manifested in the twelve archetypes of the zodiac are defined in fundamental ways as the consciousness of the involutionary and evolutionary stages of the universal holomovement. The study of astrology allows us to begin to fathom the consciousness of this holomovement as cycles of infinite creative evolution.

The archetypal spectrum of creation manifests its essence on every plane and kingdom of creation. We can witness this spectrum, beginning with an understanding of the Sun as a fount of the consciousness and being of the Light of Eternal Life. Mercury manifests discrimination; Venus radiates love; the Earth resonates with strength, and the Moon predominates in attachment; Mars manifests an essence of initiative; Jupiter expresses expansiveness and faith; Saturn manifests limitation, crystallization, and form. On Earth, our metals span the same spectrum: gold resonates with the Sun, silver with the Moon, the mineral mercury resonates with the planet Mercury, copper corresponds with Venus, iron with Mars, brass with Jupiter, and lead with Saturn. In *Culpepper's Complete Herbal*,[8] a classic seventeenth-century herbal guide, you will see each herb described by its signature—for examples, basil is Mars in Scorpio, garden mint is an herb of Venus, and mulberry is ruled by Mercury. In every kingdom of creation—canines, felines, flowers, herbs, jewels—the archetypal signatures underlie the spectrum of creation. Books of such correspondences are available, which catalogue all the phenomena of creation by archetype. Transhistorically and crossculturally, astrological signatures were the *lingua franca* of the literati.

The Seven Days of the Week

The seven days of the week are defined by this same harmony of correspondences. The august and uplifting eternal quality of Sun-days; the blue, watery, emotional, cleansing of Moon-days; and the brash, fiery impulsiveness of Tuesdays, which is ruled by Mars. Wednesday, ruled by Mercury, is an excellent day for agility and communication. Thursday is Thor's day and is ruled by Jupiter, the great benefactor. It is characterized by wisdom, expansiveness, and generosity. Friday, ruled by Venus, is the day for lovers (remember the exquisite Venutian quality of Friday afternoons). Saturday, ruled by Saturn, is a day of resistance, usually very down-to-earth and practical. Notice the polarity of opposite qualities from ray to ray. The uplifting positive solar ray of Sunday cycles to the negative cleansing watery lunar ray of Moonday. Its opposite is the fiery martian Tuesday. The brash quality of Tuesday is balanced by Wednesday's Mercurial discrimination. Thor'sday casts discrimination aside for an unbridled jupiterian benevolence. Friday moves from Jupiter's universal love to Venus's personal love. Saturn's day is ruled by structure and limitations. Check out your experience of the week; see if these signatures ring true in your own experience.

Color Archetypes

The colors of the rainbow span the same quantum spectrum of the Law of Seven, or the Cosmic Octave. Out of the white light of the Sun emanate the uplifting violet, spacious blue, balancing green, illuminating yellow, cleansing orange, and grounding earthy-red. Numerous books on the psychology of colors describe these archetypal forces.[9]

Correspondences of Planets, Colors, Gems, and Metals.[10]

Sun	Red	Ruby, garnet	Gold
Moon	White	Pearl, moonstone, cloudy quartz crystal	Silver
Mars	Dark red	Red coral	Iron
Mercury	Green	Emerald, peridot, jade	Mercury
Jupiter	Gold	Yellow sapphire, yellow topaz or citrine	Brass
Venus	Transparent, variegated	Diamond, clear zircon, clear quartz crystal	Copper
Saturn	Dark blue, black	Blue sapphire, amethyst	Lead

Elemental Archetypes in Astrology

The Tao, the unceasing play of yin and yang.

The archetypes Fire, Air, Water, and Earth define basic *realms* of being. It can be quite revealing to analyze the elemental archetypes from the point of view of the spectrum of consciousness. The essence of the elemental archetypes is Creative Intelligence. Astrological patterns of resonance are described as exalted when there is attunement and the more sattvic qualities are able to manifest, or in detriment when the archetype is in a more tamasic and resistant pattern of resonance in the gestalt of the native's astrological signature, so that the characteristics manifested are less conscious.

The Fire element as consciousness is *radiance* in all its dimensions. Excitable, enthusiastic, its light brings color and warmth to the world. Stephen Arroyo, in *Astrology, Psychology, and the Four Elements*, writes:

> This element has been correlated with the dynamic core of psychic energy by C. G. Jung, that energy which flows spontaneously in an inspired,

self-motivated way ... self-centered ... The Fire signs exemplify high spirits, great faith in themselves, enthusiasm, unending strength, and a direct honesty ... [and an] unrelenting insistence on their point of view. Fire signs are able to direct their will power consciously ... Their will to be and to express themselves freely is rather childlike in its simplicity. They may come across as rather willful, even overpowering at times.[11]

For the Air signs, ideas are real. Arroyo explains:

The Air realm is the world of archetypal ideas behind the veil of the physical world, the cosmic being actualized into specific patterns of thought. . . The Air signs focus their energy on specific ideas ... By concentrating on these ideas [they] ensure that they will eventually materialize ... the Air signs deal with "experience in its concern over theoretical relations." The emphasis on theory and on concepts in the life of the Air sign people leads to their finding the most compatible mode of expression in art, words and abstract thought.[12]

The Water signs are focused on feeling, Arroyo writes:

The Water signs are in touch with their feelings, and in tune with nuances and subtleties that many others don't even notice. The Water element represents the realm of deep emotion and feeling responses, ranging from compulsive passions to overwhelming fears to an all-encompassing acceptance and love of creation. Since feelings by their very nature are partly unconscious, the Water signs are simultaneously aware of the power of the unconscious mind and are themselves unconscious of what really motivates them. When they are in tune with the deeper dimensions of life with full awareness, they are the most intuitive, psychically sensitive sign ... They are able to help others by means of an empathetic responsiveness to the feelings of their fellow beings. When, however, they are not fully aware of their own feelings, they find themselves prompted by compulsive desires, irrational fears, and great over-sensitivity to the slightest threat.[13]

The Earth signs are masters of the physical plane:

The Earth signs tend to rely more upon their senses and practical reason than upon the inspirations, theoretical considerations or intuitions of the other signs ... Their innate understanding of how the material world functions gives the Earth signs more patience and self-discipline than other signs. This element ... has strength of endurance and persistence that enable Earth signs to always look out for themselves. The Earth element tends to be cautious, premeditative, rather conventional and usually dependable.[14]

Theomorphic Resonance:
The Body and Astrological Archetypes

God dwells within you, as you.
—*Swami Muktananda*[15]

In a theomorphic perspective the body is the personification of the Creator and the process of creation. The astrological archetypes embody the steps in the cycle of manifestation. The fundamental organizing principles of the body are an expression of the theomorphic principles and the embodiment of these archetypal forces.

In this archetypal anatomy of the body we see the underlying genius of the theomorphic nature of universal law. Each of the twelve glyphs represents a step in the order of creation. Each is a step in the process of Spirit manifesting as matter and the evolutionary cycle. Each body area, predominating in a resonant frequency, personifies the stages of creation from Spirit to matter.

The astrological archetypes are the stages of manifestation and a microcosm of the processes of creation. Man is the personification of the creative process. We are made in the image of the Creator, a microcosm of the creative process. As above, so below.

Living Archetypes: The Astrological Body

The infant has a complete zodiac in its own make-up, an exact duplicate of the cosmos in which it lives ... This is woven by the four pattern threads of "the four rivers of life" ... [which] act in triple function [through polarity 0, +, –] in and through the body, they build it in a process similar to four threads in three shuttles. Then the twelve stations or centers are formed which constitute the individual zodiac of each person. This, then, is the miniature zodiac or microcosm by which man's finer forces are linked to the universe, and supply him with energy to attract the more solid forms of substance needed for his body ... These facts form the real field of Psychosomatics .. This

principle in Nature and in man is the basis of all
action as the finer energy operating in man and by
which he lives, breathes and functions ... Even as
the tiny atom is a universe in itself, so is man.
—Dr. Stone[16]

It is easy to hear the archetype that predominates in an individual's character structure. A person's voice usually expresses these archetypal qualities. Mercurial Air-predominant individuals are centered in the mind, and tend to talk rapidly (and some would say excessively). Martian Fire-predominant personalities are very action-oriented, with little patience for mere talk. Venusian Water-predominant individuals are centered in their feelings and can tend to be whiny. Earth-predominant individuals tend to be practical, slow moving, and down to earth.

In our model, the cells of the body resonate with archetypal forces. The astrological rising sign (the astrological archetype on the horizon at the time of birth) has a marked influence on the face and body type as does the ruling/predominant element/resonance in an astrological chart. Air predominance is a lengthening of the cells of the body. A thin build and a light wiry body are characteristic of the Air archetype of resonance. A long narrow face and thin frame of the classic ectomorph is Air-predominant. Fire is the archetype of radiance and directed force. Fire predominance is characterized by good muscle tone and "stereotyped beauty." The classic mesomorph is Fire-predominant. Water predominance is easily seen in the rounded moon face of Pisces, Scorpios, and Cancers. A Water-predominant body resonates with an archetypal attunement which is rounded and flowing. Water sustains a tendency to padding and excess weight. Earth predominance sustains a square, solid, stout, tree-trunk body with short wide neck, and thick arms and legs like the classic football player. Water and Earth parallel the endomorph of Sheldon's typology. While everyone is a blend of these archetypes, the elemental predominance reveals a great deal about body type and character.

> **Who's Talking?**
> People reflect their underlying predominant elemental/archetypal resonance.
>
> | | |
> |---|---|
> | **Ether** | Spaced out, expanded, cosmic |
> | **Air** | Lives in ideas, thoughts, talks fast |
> | **Fire** | Action oriented, no patience for talk |
> | **Water** | Lives in feelings |
> | **Earth** | Solid, practical, down to earth |

Archetypal Harmonic Integrity

Astrologically defined fields of resonance are a fundamental organizing principle of the body. The most fundamental features of the gross anatomy of the body offer a suggestive parallel to the astrological glyphs. These glyphs point to the fundamental types of arcs/archetypes, or harmonics of resonance, which manifest through the body. Each of the twelve glyphs, patterns, or archetypal energies sustains a body part/energy field with a unique predominant resonant frequency and archetypal/astrological attunement to the universe.

Differences in the physical qualities of the body are reflections of the differences in the energy fields. In each of the successive phases we picture energy moving downward and outward into manifestation, becoming more and more physical in our bodies as microcosms of the process in which energy crystallizes from Spirit into matter. Notice that each of the glyphs embodies a duality—two arcs in a dynamic, bilateral yet stable phase-relationship of polarity. This points to the fundamental resonant organizing processes of the universal holomovement of rajas and tamas. This gestalt of wholeness of the astrological "glyphs" can be called the "archetypal harmonic integrity" of the body.

Body cavities are thus identified as having unique predominant resonant frequencies, each of them in tune with universal, archetypal forces. For example there is a fundamental difference between the watery energy field of Cancer, the breast, and the Fiery energy field of Leo, the solar plexus. Each is a unique resonant environment. These astrologically defined fields of resonance are a fundamental organizing principle of the body. We now picture the twelve phases of "the four rivers of life" in their three polarities of emanation that sustain materialization.

The Astrological Glyphs

The twelve glyphs of the zodiac provide a suggestive paradigm in which the astrological archetypal resonances appear to be the fundamental organizing principles of the body, both as a body of consciousness and as a field of resonance. Each of the archetypal resonances is a unique resonant structure where a particular quality of consciousness predominates. The twelve glyphs point to twelve unique resonant processes that are sustained by the four material elements in their interaction with the tri-gunas of neutral, positive, and negative polarity. The body is a spectrum of living archetypes that connects the immortal Cosmic Person with its mortal expression over time.

The twelve phases of the zodiacal cycles differentiate the reflection of the spectrum of consciousness emanating from the nucleus of our solar cycle. All energy fields in our solar system are a microcosm of the solar cycle. The zodiac is the best tool for studying the archetypal spectrum of consciousness.

The all-pervasive holographic potency of Ved unfolds its "knowledge" as the astrological archetypes that underlie the order of creation. As above, so below—the body personifies the ubiquitous holomovement of the seed power of the Ved. Note that each archetype is essentially "psychological" and the process is a play of consciousness.

1. ♈ **Aries.** The essence of Aries, the first phase of manifestation, is the quality of existence. The glyph for Aries, the Fire-predominant, positive, first sign of the cycle, embodies a pattern of energy radiating outward into manifestation from a seed point, a bindu. The *birth* of every process of manifestation is this type of arc, the Aries-resonant frequency and pattern of energy emanating from a center.

   The glyph for Aries points to the archetype (type-of-arc) in tune with this harmonic of force. In the body this type of arc radiates upward and outward from the third eye and sustains the crown of the head as an energy field and as a body cavity resonating predominately with this keynote.

   The third eye resonates with the causal plane. The brain is the personification of the process of the mental plane. The two lobes of the brain, emanating from the causal plane resonance of the third eye, embody this archetypal energy pattern and personify the processes of the mental plane in the body. In the fields of force emanating from the third eye, the Aries archetype of resonance is embodied in the underlying sonic organization of the energy fields of the brain. The brain is a unique resonant field in the body.

2. ♉ **Taurus.** The Taurus type of arc is the keynote of the second stage in the manifestation cycle. Taurus adds the property of spatiality to existence and sustains the overall *space* for manifestation of the torso. The Taurus resonance sustains the body cavity of the neck—a nexus, a womb, where the mental plane images of Aries step down in vibration through the etheric plane into the physical plane. The top part of the glyph (◡) points to the receptive phase of the Tauran resonance, which receives impulses from the brain, the personification of the Mental plane, and then steps this energy down through Ether to physical expression sustaining the etheric body of the "tauruso." The fields of force sustained by the Tauran attunement receive the energies of the Aries archetype, concentrating and twisting them through the etheric into the physical vibrations of the torso. The Ether-predominant neck and etheric body of the "tauruso" resonate with this archetype.

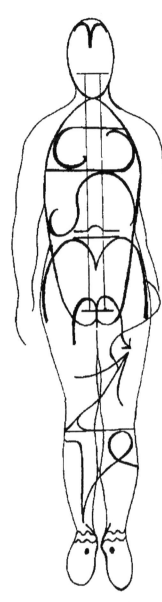

Form is an expression of energy. The astrological glyphs define a spectrum of resonance in the underlying sonic organization of the body. The archetypes Aries, Taurus, etc, manifest fields of resonance which are fundamental to structural and functional differentiation in the body.

The body cavity of the neck resonates with the harmonics of the archetype Taurus in the macrocosm. The sympathetic vibration between their energies is attuned to the same keynote, sustaining an isomorphism of energy and consciousness—as above, so below—in the microcosm of the individual.

The etheric body is the first physical field of resonance in the step-down from the unified field of the causal plane, through the mental plane. It is a field of resonance which is defined by the interaction of the emanation of the unified field with the emanation of the Earth. It sustains the pattern of resonance of an elliptical helix of the interaction of Heaven and Earth. Thus, the Taurus—the twisting of the interaction of the spirals of "above" and "below"—is the dinergy of the centrifugal and centrifugal resonance of "the two hands of God."

3. ♊ **Gemini.** The resonance of Gemini, the third moment in the cycle, predominates in the arms and spine. The spine is a unique resonant environment as the nucleus of the subtle body. Gemini unites the core and the periphery. The glyph for Gemini embodies the energy dynamic between the arms reaching out to touch the world and the spine as the core of our energy. Here we see the depth of being experiencing the world, and the knowledge of the world touching the center. It is an archetypal energy field found in the core of every unit of wholeness as Spirit animates nature. The radiation from the ultrasonic nucleus of the spine fills the space of the body with the rhythms of universal law. Its twin character is seen in the dynamic between the two phasing fields of energy, expressed in the bilateral symmetry of the body, and in the dynamic relationship between sacrum and cranium (the two horizontal lines of the glyph), expressed through the primary respiratory mechanism. In Gemini we have the lengthening of the energy field from the Etheric radiating into the spiral helix of Fire and Water and the "two hands of God" reaching out to play in the world.

4. ♋ **Cancer.** The next phase in this cycle of manifestation is Cancer, which rules the breast. The breasts are a unique energy field which resonates with the potency of divine love. Here again is a unique structure in the body, with a particular resonant frequency of vibration. It is the home of the neutral pole of the Watery Cancerian force, and the glyph represents the swirling spirals of energy emanating from the heart chakra representing the zenith of the sensitive Water element out of which flows the nurturing milk of divine love.

5. ♌Leo. So far we have archetypal energy patterns forming the head, neck, torso, spine, and breast. In Leo, we add the liver, diaphragm, stomach, pancreas, spleen, and the organs of the solar plexus, which resonate with the powerful radiance of the solar archetype (type-of-arc). The solar plexus is a unique field of resonance in the body.

6. ♍ Virgo. The shape of the ascending, transverse, and descending colon embodies the archetype of Virgo. The colon sustains the resonant frequency of Virgo and embodies the forces which archetypally characterize these energies. Thus, Virgo, representing the harvest, is a point in the cycle where discrimination is used to discern what is to be cast off as waste and what is to be saved to become a part of the body. The function of the colon embodies this process of discriminating wheat from chaff. People with Virgo strongly resonating in their astrological signatures express the same function in culture as scientists and artists.

7. ♎ Libra. The kidneys are attuned to Libra and embody this archetype in the correspondence of resonance that is personified in the body. Libra ruled by the goddess Venus rules beauty and midwives the birth of the wisdom cycle of the centripetal evolutionary force. In the cycles of the year, Libra mediates between spring/summer and fall/winter, the oscillation between centrifugal and centripetal predominance in the hemispheres. Similarly, in the body, Libra balances the forces of above and below, the upper arc of rajasic radiance and the lower line of tamasic crystallization. The Libra archetype mediates between love and fear, Fire and Water, expansion and contraction.

8. ♏Scorpio. Ruling the generative organs, Scorpio embodies the mystery of life emerging from the depths. The glyph depicts the hips, ovaries, and womb.

9. ♐ Sagittarius. Ruling the hara center and thighs, Sagittarius embodies the resonant pattern of the radiance of Fire (creativity) burning (manifesting) on the Earth (materiality).

10. ♑ Capricorn. The energy of Capricorn resonates with the knees and skeleton. Here you see a polarity in the symmetry between the two octaves of Ether and Earth, crystallized on the left of the glyph, and their phase dynamic embodied in the circles above and below on the right, embodied in the Ether-predominant inner marrow and Earth-predominant outer bone. The V-shape of the glyph embodies gravity lines radiating from the nucleus of the Earth, which predominate in the return currents that support structure—for example, the skeleton. The knees predominate as the

home of this archetype in the orchestration of the implicit order within the ultrasonic resonance of the body.

11. ♒ **Aquarius.** Near the end of this cycle stands Aquarius, which rules the ankles, a unique energy resonating in the body. The creature in the zodiac that is most characteristic of the Aquarian essence and epitomizes the end of the cycle of crystallization of Spirit into matter, of Heaven embodying itself on Earth, is the human. The ankles embody our fundamental standpoint in life, the values we pursue, and the direction of our personal and collective evolution. Aquarius represents the end point of evolution, theomorphic humanity, made in the image of the Creator.

12. ♓ **Pisces.** The feet resonate with the Piscean archetype, a Janus-faced field of force flowing between Father Sky and Mother Earth. The vertical curves represent rajas and tamas in their involutionary/evolutionary journey from Heaven to Earth. The horizontal line embodies the paradox of separateness within unity, the Janus face of being. The feet again are a unique energy field with a unique pattern of resonance, the home of a unique quality of consciousness in the body.

Masculine and Feminine Archetypes

In the ancient wisdom the key to understanding nature is through the "supreme ultimate principle" of the "two hands of God." Emanating from every center of creation is the spiral dance of centrifugal and centripetal polarity vortices. All creation is the play of light and shadow reflected through these mirrors of consciousness. Everything can be defined as patterns of resonance or "types of arcs." The most fundamental archetypes (types of arcs) are the gunas (universal modes of nature). The polarity of Heaven and Earth, Sun and Moon, east and west, sperm and ovum, manifests in every kingdom of nature. Male and female are the personification of this polarity of yang and yin.

The qualities of the rajas guna define the psychology of the masculine archetype. The Sanskrit root "rajas" means "to glow." Its essence is radiance. It is the light of the illimitable creative potency of consciousness reflected in creation. The passion of rajas guna is to project outward into manifestation. Rajas rules the conscious mind. The "male" archetype is characterized by the qualities of being: active, excited, directed, determined, competitive, goal-oriented, aggressive, persevering, constant, self-propelled. The Solar archetype is heroic: "Yang prefers and feels comfortable in a stimulating, complex environment, and acts quickly and impulsively as if endowed with an unlimited supply of stamina and an invincible immunity to harm."[17] Rajas guna is seeking Self-knowledge reflected through its manifestation. It is pushing, constant, logical, left brain, strong, passionate, physical. Rajas guna is personified in the young male and his blind passion for life and experience. In maturity the Father archetype of protector and ruler of his domain: "a man's home is his castle," also expresses the rajasic qualities. Its greatest need is to be recognized, to have its being reflected back through its manifestation, on its path to Self-knowledge.[18]

Tamas guna defines the psychology of the feminine archetype. Tamas guna is the receptacle of Spirit. Its nature is receptivity, emptiness, longing to provide a medium for essence. The "female" archetype longs for union and companionship, to hold the space for emotional intimacy and essence. Its dharma is to reflect the unity, wisdom, and unconditional love which are the quintessence of the mystery. Tamas is receptive, attractive, pulling, romantic, feeling, intuitive, emotional, cyclical. It wants to be filled—to be valued. "To be Yin is to enjoy quiet, calm, simple environments and poses a more limited capacity for food, work, and interaction. It is a preference of Yin to avoid prolonged stress and to desire regular, adequate periods of rest and rejuvenation."[19] The sweet fruit of the ever-ripening cycles of evolution, she embodies the Divine Mother through discrimination, wisdom, and unconditional love. The outer beauty of the feminine is a reflection of the wisdom of inner beauty. Tamas personifies the mystery of eternal life through birth and family as the chalice of the heart essence.

The Archetypes of Rajas and Tamas

Rajas/Yang/Sun	Tamas/Yin/Moon
Self-knowledge through the world's reflection	Medium for essence
To project into manifestation	Longing for union
Expanding	Containing
Push	Pull
Fullness	Emptiness
Strength	Endurance
Physical	Emotional
Passion	Romance
Physical Intimacy	Emotional Intimacy
Action	Communication
Logic	Feelings
Intellect/Left Brain	Intuition/Right Brain
Steady	Cyclic
Red	Blue
Fiery	Watery
Revealed	Hidden
Satiety	Yearning
Fear of criticism and limits	Fear of loneliness

The Gunas: Psychological Archetypes

In the Sanatana Dharma the process of creation is the evolution of consciousness. The *fundamental* nature of the evolutionary cycle is expressed in the gunas, the archetypal essences that underlie all creation. All of nature (prakriti) is woven of the three strands of sacred sound vibration: equilibrium, centrifugal and centripetal. The essence of these archetypes is to reveal, move, and restrain. In the individual manifestations of nature they are the psychological basis of all things.

In the psychological world the three gunas serve to illuminate, activate, and obscure. They manifest as the nature of pleasure, pain and indifference: sattva guna attaches to happiness, rajas to action, tamas to heedlessness. On the moral plane they represent affection, passion, and hate; in the psychic field they represent emancipation, affinity, and sin.

Sattva guna denotes "Ultimate Reality" or "Truth." It is the abstract principle of illumination. In the mental world it accounts for such qualities as joy, pleasure, enlightenment, faith, forbearance, forgiveness, courage, valor, concentration, humility, modesty, indifference, detachment, compassion, and pure action.

Rajas guna is energy, the abstract principle of activity. In the mental world it accounts for such qualities as argumentation, opinion, remorse, frenzy, wrath, attachment, jealousy, backbiting, egoism, selfishness, desire to afflict and kill, desire to buy and sell, the habit of evil thoughts, suspicion, insult, criticism, abusiveness, falsehood, deception, doubt, skepticism, animosity, envy, braggadocio, heedlessness, irregularity in conduct, treachery, disrespect, thievery, ostentation, lack of shame, gambling, scandal mongering, quarreling, drudgery, and all cravings of the senses. From the preponderance of rajas guna, all temptations and fancies arise.

Tamas guna is the abstract principle of restraint. It is that which veils consciousness and obstructs action. In the mental world it accounts for such qualities as indolence, carelessness, delusion, ignorance, indecision, sleepiness, laziness, fear, avarice, grief, lewdness, want of faith, pride, lassitude, and deluded conviction. It is the cause of want of discrimination, faith, knowledge, memory, and liberality. It is the cause of immorality. It is the power by which other things are measured.[20]

All phenomena take on their essential quality through resonance with these fundamental psychological forces of nature. These great archetypes of the evolutionary cycle are the basis of all manifestation.

Archetypes in Polarity Therapy

This paradigm offers a revealing way to view the body as a form of consciousness, and suggests a basis for a transpersonal and body/mind/Spirit psychology. It also points to the value of viewing astrology as an "alphabet" of cosmic creation and of studying the harmonics and cycles produced by the types of arcs emanating from the One.

Each body part, as illustrated above, is sustained by its capacity to resonate with an archetypal keynote. When a person tenses some body area in resistance to the experience or expression of a paralleling quality of consciousness, attunement with nature is blocked and the armored body part becomes more vulnerable to accident and disease. In our model, disease is the unconscious expression of issues that are not being dealt with consciously.

For example, a client may have problems with his knees. In this model we would work on foundation issues, such as early childhood, as a significant facet of the healing process. We would release the tension in the energy fields that creates resistance to attunement with the breath, the body's nuclear energy and Life Field. We would bring into consciousness issues of family and early childhood, supporting a safe space to experience and express feelings surfacing from the cellular memory as energy is released.

By understanding the consciousness underlying the archetypal integrity of the body we can break through tension and resistance and reestablish the sonic integrity of the body and its harmony with the Unitive Field. This is the fundamental aim of Polarity Therapy.

CHAPTER FIFTEEN

Conclusion: The Evolutionary Perspective

*The search goes on but Life is silence itself. We
are moved but know not the Mover. We think and
know not the Thinker.*
—Dr. Stone[1]

In the Sanatana Dharma the medium of creation is Intelligence. In an evolutionary approach we understand humanity as the microcosm and personification of an ever-evolving universe of consciousness. Each person is the personification of the leading edge of this Cosmic process. All creation is the universal form of the evolution of the universe. In the Sanatana Dharma, all energy in creation is the energy of an ever-expanding universe. The ubiquity of growth in our own subjective experience and in nature expresses this universal evolutionary quality of energy.

A basic premise is that Spirit has manifested the body as a vehicle for its evolutionary learning experience. The body is understood as an emanation of Spirit. The center of every atom, molecule, and cell resonates with a field of spiritual potency. This field of force provides the etheric pattern for the electromagnetic harmonics which determine and sustain the physical form. Through the evolutionary cycle of Spirit crystallizing into matter, each atom, molecule, and cell of our body is a miniature universe existing to manifest creativity. The body is itself an embodiment of the process of Spirit's evolutionary journey into matter.

Our Bodies Are Conscious Microcosms of the Universe

In Polarity Therapy the body is viewed as a spectrum of energies. This spectrum parallels a spectrum of universal consciousness, in which every element of the body expresses the evolutionary process. Problems in particular parts of the body reflect problems in experiencing and expressing particular parts of our evolving being. Polarity Therapy offers images for understanding and working with the person and the body as an evolving body of consciousness.

Each part of the body resonates with universal archetypes of conscious energy. If, in our evolution, we do not integrate the consciousness resonating in any part of the body, we tend to manifest tension in that area. This resistance further blocks our harmony with the sustaining forces of nature and makes us more vulnerable to accidents and disease. Each part of our body is the personification of an aspect of the spectrum of consciousness. Each area of the body resonates with a particular quality of our being and experience.

The cycle of Polarity between birth and death, the individual life, is also a cycle of growth. Life is understood as a sacred evolutionary journey, a learning experience for the Soul. This evolutionary perspective implies a deep respect for the person and the pathos of the human condition. Every moment of our subjectivity is rich with tenderness and pathos—to be human is a passionate condition. Our growth and the challenges we face in the process of maturation are a sacred journey in which we embody a moment of an ever-expanding, ever-evolving universe. Everything we face in life—every challenge to our growth, every element of resistance to our actualization, every realization—is grist for the evolutionary mill.

Anthropocosmic Mystic Doctrine

I died a mineral and became a plant,
I died a plant and rose an animal,
I died an animal and I was a man,
Why should I fear? When was I less by dying?
Yet once more I shall die as man, to soar
With the blessed angels; but even from angelhood
I must pass on. All except God perishes
When I have sacrificed my angel soul

I shall become that which no mind ever
conceived.
O, let me not exist! for Non-existence proclaims
To him we shall return.
—*Jelaluddin Rumi*[2]

The physical plane is the fruit of a polarity between two forces: Heaven and Earth, involution and evolution. Physical manifestation is always through a seed, a living crystal of the Earth's resonance, which embodies the return currents of the evolutionary force reaching back toward the heavens.

The anthropocosmic paradigm is fundamental to many esoteric and mystic traditions. This theory holds that man is both the final summarizing product of evolution and the original seed potential out of which the universe germinated. "We may use the analogy of the seed and the tree: the tree of the universe is the actualization of the seed potential which is Cosmic Man. I am using the word Man here in relation to the Sanskrit root *manas,* meaning 'mind', or consciousness, which can reflect upon itself."[3]

In our model we abandon the dominant paradigm, the conventional wisdom, in which the body is a vehicle for a mind and brain which are somehow "conscious." Consciousness is not epiphenomenal but is the innermost essence of all beings. In *advaita,* consciousness is omnipresent. Consciousness is the very stuff of all creation. A conscious being which we call the Universe emanates the body as a vehicle for its temporal expression in evolution.

Humanity may indeed be the vehicle of the "noosphere," the evolution of consciousness of the Mother Earth. Pierre Teilhard de Chardin, a Jesuit priest and an eminent scientist and philosopher, sought to integrate his insights as a scientist, theologian, and mystic.

> His key concept is what he called the "Law of Complexity-Consciousness," which states that evolution proceeds in the direction of increasing complexity, and that this increase in complexity is accompanied by a corresponding rise of consciousness, culminating in human spirituality. Teilhard uses the term "consciousness" in the sense of awareness, and defines it as "the specific effect of organized complexity" ... Teilhard also postulated the manifestation of mind in larger systems and wrote that in human evolution the planet is covered with a web of ideas for which he coined the term "mind-layer," or "noosphere." Finally, he saw God as the source of all being, and in particular as the source of the evolutionary force.[4]

In the Sanatana Dharma: "Cosmic evolution proceeds from the unconscious, unmoving, unknowable and unmanifest macrocosm to the conscious, moving, knowable and manifest microcosm. Human evolution is a return journey from the gross physical plane of the microcosm back to the unmanifest macrocosm."[5]

Cosmic Consciousness

All at once I found myself wrapped in a flame colored cloud ... there came upon me a sense of exultation, of immense joyousness accompanied or immediately followed by an intellectual illumination impossible to describe ... I saw that the universe is not composed of dead matter, but is on the contrary, a living Presence; I became conscious in myself of eternal life. It was not a conviction that I would have eternal life, but a consciousness that I possessed eternal life then; I saw that all men are immortal; that the cosmic order is such that without any preadventure all things work together for the good of each and all; that the foundation principle of the world, of all the worlds, is what we call love, and that the happiness of each and all is in the long run absolutely certain.

—R. M. Bucke, Cosmic Consciousness[6]

Our model is a theomorphic model. Each structure, process, being, kingdom of nature, and the cosmos embodies the great mystery of creation. In all dimensions, in every direction we look, we witness the synergy of the infinite creative potency of the Omnipresence that underlies all becoming.

From the mystery of the Absolute which is eminent as our consciousness, the Sacred Witness within each of us, to the manifestation of our daily drama as the crystallization of consciousness, our lives personify the infinite creative potency of the evolution of the universe. In the Sanatana Dharma we understand God as profoundly eminent: "God lives within you, as you." Each of us is an aspect of the omnipresence of the universal being. Our lives are the Sacred Mystery of Eternal Life, lived through a myriad of bodies and forms. Our model is an evolutionary model. Everything in our model is the play of consciousness of an ever-evolving universe. The only energy in our model on every level is the energy of the infinite creative potency of the evolution of consciousness. Through the Law of Three, the tri-gunas, and the Law of Seven, the Cosmic Octave, we point to the profound harmony and unity of creation. Our model is a paradigm of wholeness in which, through Consciousness, the ultimate medium, everything is contained within everything. Cosmic Consciousness, the knowledge of the omnipresence of the totality, is our innermost Self.

Everywhere and everything is united by a "Presence," an "Awareness," a "Consciousness." This eternal Presence is the life that manifests form and lives its cycles of evolution through each of us. This Presence is what is conscious and intelligent in each of us, and is the innermost Self of all that is. In the Sanatana Dharma there is no difference between Brahman, the Absolute (Being, Consciousness, Bliss), and Atman, the individualized presence of Purusha (Spirit, witness-soul, conscious-spiritual-person), as the heart of each person and of all of nature.

> Man as the knower, the Witness, the Atman, the Absolute Subjectivity, the Host, the Tathagatagarbha, the That in you which is reading this page, is the Godhead, Brahman, Dharmadhatu, Universal man of no rank, Mind, Reality itself; while man as an object of knowledge, as a perceived phenomenon, as Guest, as clothed in "the painted veil, the loathsome mask," is the ego, the individual person (from the Greek *persona,* "mask"), the separate alienated self.[7]

The present is the Eternal Present. This becoming is the Creative Intelligence whose evolution is creation. It is the oneness of this eternal presence as "the innermost Self of all that is," of a Universal Person whose center is everywhere and whose circumference is nowhere, which is the source of the theomorphic model we have illuminated.

> If Spirit has any meaning, it must be omnipresent, or all pervading and all-encompassing. There can't be a place Spirit is not, or it wouldn't be infinite. Therefore, Spirit has to be completely present, right here, right now, in your awareness. That is, your own present awareness, precisely as it is, without changing it or altering it in any way, is perfectly and completely permeated by Spirit.[8]

The true nature of everything as formless Consciousness or God cannot be grasped by the concepts of the mind or the data of the senses. It can only be grasped through the unique epistemology of personal evolution called *yoga,* which generates special categories of higher knowledge called "yogic perception" and "Self-realization."[9] Humanity is the only element of nature capable of consciously accelerating the process of evolution through a process of Self-cultivation.

The Life Field

"Lead us from the unreal to the real," begins the ancient prayer to the Spirit of Guidance. In the Sanatana Dharma the world is unreal, a fleeting appearance within consciousness. Reality is that which is unchanging—the Self. The ancient rishis realized the transitory nature of the world of appearances. In profound states of concentration they saw the world of appearances as an epiphenomenon (a secondary phenomenon accompanying another and caused by it) of consciousness. The yogis realized that underlying the unceasing change of the mind, emotions, intellect, and body there was an unchanging substratum of living consciousness which was the innermost essence of all beings, the Self. The ancient wisdom is centered in the real-ization of the transpersonal, the real-ization of the Self, the real-ization of the sacredness and unity of *all* life and the epiphenomenal nature of what we call "reality." Those who have awakened into a direct experience of the Self reveal that our reality is merely a dream, an appearance projected outward by consciousness. Indeed, the constant theme of this work has been that the world is a play of consciousness.

Throughout this explication we have again and again focused on the ancient axiom of the identity of all consciousness. Our focus has been the fruit of contemplation, the realization that all life and all consciousness are one. We have referred to this mystery as Higher Intelligence, the Unitive Field, the Present, the One Life, the Soul, Spirit, Brahman, God, Atman, the Self. It has become more and more clear that in the vision of the ancients, we exist in two dimensions, an unreal world of mere transitory appearances and a sacred transpersonal reality of immutable consciousness. The world around us is transitory and unceasingly changing. The Self, that which is actually alive and conscious in each of us, is eternal and divine.

Can it be that our ego's separateness is merely a painful illusion? Can it be that all the terrors of gaining and losing, having and changing, are an illusion, a mere play of consciousness? Can it be that we are as the sages tell us, untouched by the world? Can it be that as close as our life breath, there is an immortal realm of divine being? The Sanatana Dharma invites us to contemplate these questions and to consciously accelerate our evolution in a process of Self-realization.

Again and again throughout this journey our analysis of the ancient wisdom takes us to a Unitive Field of living consciousness, the Self. This Life Field is that which is real, as the unchanging awareness that manifests our consciousness of the world. Again and again we have understood that the world of appearances is sustained by the Soul, the animating force of the One Life. Once more we are drawn to the realization that the world is the self-projection of a Unitive Field of living consciousness which we call God. This Life Field is omnipresent as the One Life of a living universe and holographic as the omniscience of this Ultimate Intelligence. It is the innermost Self of all, and the basis of our realization of the sacredness and unity of all life.

Toward a Science of Harmony

This we know—the earth does not belong to man, man belongs to the earth. All things are connected like the blood that unites one family. Whatever befalls the earth befalls the sons of the earth. Man does not weave the web of life; he is merely a strand of it. What ever he does to the web, he does to himself.
—Chief Seattle (1854)[10]

The Perennial Philosophy offers a vision of the sacredness and unity of all life. It offers a model for the study of creation as *one conscious living being.* The Sanatana Dharma illuminates a model for an empirical science of conscious sound and the study of vibration, resonance, and harmony as levels of organization *sui generis* in the cosmos. A science based on the Perennial Philosophy could foster a healing art that "does no harm," and seeks to reestablish the harmony of our bodies' attunement with nature. The Sanatana Dharma unfolds a time-tested psychology that addresses the question of consciousness and honors the depth and sensitivity of human subjectivity. A theomorphic approach to psychology offers a model for understanding the body/mind as the personification of the process of creation. This model underscores the efficacy of somatic practices in psychotherapy and summons us to experience a profound psychology of liberation.

The Sanatana Dharma invites us to contemplate the identity of consciousness and to realize the divinity that is the essence of the human condition. The wisdom of the ancients calls us to topple the materialistic paradigm and to reestablish a sacred vision for living our life on this Earth. I trust that through this perspective we have come to a deeper appreciation of the Ultimate Intelligence, wisdom, and love that underlie and sustain this gift of life. We pray that this brief discussion will help to demystify the Hermetic science of the ancients and to encourage empirical research of the universe as one living conscious being.

Meditation/Contemplation

Enquiry into the Gross Body:
by Dr. Randolph Stone[11]

This gross body, I cannot be. Why? I am seeing it. It is an object for my sight.
I am the seer. It is separate and I am separate.

The Five Great Elements:
Space—Air—Fire—Water—Earth

This Gross body is not mine. Why? It belongs to the five great elements. It is a prod-
uct of the Pentamirus combination of the five Elements. It cannot be mine. Why?
Look. These represent the Five Elements.

Every Element is divided into two halves. One half remains unchanged. The other half
is further divided into four equal parts making each equal to ⅛ of the original. Thus
each is now found as five parts.

With the unchanged half of each Element, 1/8 part of each of the other four Elements
are combined, thus a size as whole as the original Element. But each now contains all
the Elements but one only predominates. Thus in the Pentamirus Combination,
twenty-five factors are manifested. How?

The Products of the Five Fold Combination of Sky[12]

Grief is the principle quality of Space (Ether), a feeling of nothingness. Desire is pro-
duced by the combination of Wind (Air) and Space. Anger is produced by the combi-
nation of Fire and Space. Attachment or love is produced by the combination of Earth
with Space. I am not these: Grief, Desire, Anger, Attachment, Fear. I am seeing them.
I am the Seer. They are not mine, they belong to the various Elements noted as above.
I should not claim these as mine.

The Products of the Five Fold Combination of Wind

Speed is the main quality of Wind. Lengthening is produced by the combination of
Space with Wind. Shaking is produced by the combination of Fire with Wind. Move-
ment is produced by the combination of Water with Wind. Contraction is produced
by the combination of Earth with Wind. Speed, Lengthening, Shaking, Moving, Con-
tracting, I am not. I am seeing these. They are objects for my observation. They are
not mine. They belong to the Elements noted above.

The Products of the Five Fold Combination of Fire

Hunger is the main quality of Fire. Sleep is produced by the combination of Space with Fire. Thirst is produced by the combination of Wind with Fire. Lustre is produced by the combination of Water with Fire. Laziness is produced by the combination of Earth with Fire. Hunger, sleep, thirst, lustre, and laziness, I am not. These are objects and I am able to see. They are not mine. They belong to the Elements noted above.

The Products of the Five Fold Combination of Water

Saliva is produced by the combination of Space with Water. Sweat is produced by the combination of Wind with Water. Urine is produced by the combination of Fire with Water. Semen is the main quality of water. Blood is produced by the combination of Earth with Water. Saliva, sweat, urine, semen, and blood, I am not. These are objects I am able to see. These do not belong to me. They are products of the Elements.

The Products of the Five Fold Combination of Earth.

Hair is produced by the combination of Space with Earth. Skin is produced by the combination of Wind with Earth. Blood vessels are produced by the combination of Fire with Earth. Flesh is produced by the combination of Water with Earth. Bones are the main products of Earth. Hair, skin, blood vessels, flesh and bones, I am not. I am able to see them. They are objects of my perception. They are not mine. They belong to the five Elements. They cannot be mine. I am not these. They are not mine. I am the Seer. Witness.

PART III

THE PRACTICE OF

SOMATIC PSYCHOLOGY

CHAPTER SIXTEEN

Somatic Psychology

In India we start everything with a prayer ...
Whether we are taking a bath, or naming a child,
whether we are taking our food or going to sleep,
whether we are building a house or observing the
Moon. Everything is an act of worship. Because
everything is imbued with The Divine Spirit.
Nothing is mundane. No act can be relegated to
the purely material. There is no division between
sacred and secular, for everything is Sacred.
Everything is Divine. Everything is permiated and
saturated with the Divine Spirit.
—Mathaji Vanamali[1]

You Are the Cutting Edge
of the Evolution of the Universe

Transpersonal Psychology is thousands of years old. But it is only in recent years that the transpersonal perspective has come to our attention in the West. Polarity Therapy offers a middle-range theory to apply these transpersonal insights in the research and practice of somatic psychology. Its resources include a revealing approach to character analysis through the elemental archetypes, an elemental model of pathology, and techniques for facilitating healing and consciously accelerating evolution through body-centered therapies.

Polarity Therapy understands the body as consciousness and works with the person as an evolving spiritual being. Each person represents a unique moment in creation, a unique facet of the Infinite Self in the process of the evolution of consciousness. Again quoting Paracelsus, "Gods are immortal men and men are mortal Gods." Our growth is a temporal microcosm of this timeless Being. The bodywork session serves as a space for working with issues in the individual's process of evolution and Self-actualization. The foundation of the conscious approach is a deep sense of respect for the person as an evolving spiritual being. Issues of health and disease are understood as components of the person's evolutionary/Self-actualization process.

Life is understood as a sacred evolutionary journey, a learning experience for the Soul. This evolutionary perspective implies a deep respect for the person and the depth of the human condition.[2] Our growth and the challenges we face in the process of maturation are a sacred journey in which we embody a moment of an ever-expanding, ever-evolving universe. Each of us is the cutting edge of the evolution of the universe. Everything we face in life—every challenge to our growth, every element of resistance to our actualization, every realization—is the turning of the cycles of the evolution of the cosmos.

An Esoteric Approach to Somatic Psychology

Working with an esoteric model of the body/mind can be a key to healing trauma. As previously discussed, in our theomorphic perspective the body is understood as a microcosm of creation and as the personification of the process of creation. The body is a materialization of Spirit, reflecting a process of steps in which Spirit crystallizes into matter. The body is a mind. The body is theomorphic, a crystallization of mental and emotional impulses. Each body cavity is a plane in our microcosm of the process of creation. The primary energy of the causal plane manifests into the body through the third eye, which steps energy down from the Over-Soul to the individual mind. The mind, which is a microcosm of the mental plane, is the seat of the images that we identify with and desire to manifest. The throat resonates with the Etheric plane of sound and vibration, which is the psychophysical nexus for giving voice to our attachments and for embodying our expression in the "tauruso." The thoracic cavity is the home of the Air element, the center of feelings and of human nature. We act on these feelings through the Fire-predominant solar plexus and hara, giving birth through the pelvis (Water) to completion on the Earth. When it is not safe to express one's being, elimination becomes a problem.

The unexpressed mental and emotional impulses become lodged as tension in the musculature of the chest, solar plexus, pelvis, and thighs.

The body is theomorphic. All energy vibrates through the universal step-down process. Form follows energy. The organization of the body reflects the theomorphic gestalt. The body cavities parallel a spectrum of resonance defined by archetypal forces that embody the stages of creation. The twelve astrological archetypes rule twelve stages in the process of manifestation. Universal creative energy (prakriti) is individualized through the third eye center (♈ Aries) which radiates a field of force resonating in the cranium that functions as the personification of the mental plane. Our mental plane images are the seeds of our manifestation. The resonant cavity of the neck area is the nexus, or etheric womb, which receives (♉ Taurus) these mental plane images and steps them down in vibration to be embodied through the etheric body of the "tauruso." The spine (♊ Gemini) is a unique environment, the ultrasonic core, for the step-down through the central nervous system. With the resonance of the breast (♋ Cancer) we have the heart of our attachments, the feelings that drive our manifestation. This passion becomes intention through the resonance of the solar plexus (♌ Leo). Intention is processed by resonating in the colon (♍ Virgo), which embodies science and discrimination, and by resonating in the kidneys (♎ Libra), which fosters beauty and balance through the womb of physical embodiment (♏ Scorpio), at the hara center. This resonance of the hara Fire is birthed through the muscles of the thighs into the physical-plane manifestation (♐ Sagittarius). The resonance of the skeleton (♑ Capricorn) offers the foundation of support for this vehicle, which is pursuing the standpoint, direction, and ideals of ♒ Aquarius, through the accumulation of cosmic experience (♓ Pisces), to Earth, where completion is manifested. Form crumbles and the evolutionary cycle of Heaven manifesting its infinite creative potency on Earth is completed to release Spirit for another round of evolution.

Every cell of the body is a conscious, sentient, universe of Cosmic Intelligence. The body is a crystallized expression of mental and emotional patterns which reflect the evolutionary process. As we move energy in the body, energy also moves on the subtler levels of emotion and mind to accelerate the process of evolution. Working with the body is an effective way of working with the emotions and mind toward liberation through Self-realization.

Mainstream science is intimidated by the vastness of human consciousness. The depth of intelligence that breathes through every atom of creation is relegated to the fringes of credibility. Materialistic science is boggled by the play of consciousness and the richness of human subjectivity.

In the evolutionary perspective *consciousness* is the medium of creation. God as absolute consciousness becomes the world. Our bodies and our lives are the veiled expression of this medium. The question for the somatic therapist that we have been exploring is, "What is the nature of these veils?"

Those of us who have focused on the body for a while may discover that the veils are responsive. The body is a body of intelligence. The body is a being. We find that the body will respond if we feel and listen care-fully. Every cell of its being is sentient, is feeling, is divine intelligence. "The Divine Mother . . . is in every molecule, in every atom, in all things which constitute the world."[3] The Intelligence in the body will respond to attention and intention, love and understanding. Consciousness is infinitely creative! Whether your technique is Rolfing or Shiatsu, Ortho-bionomy or Therapeutic Touch, Reiki or Reflexology, any therapy in the hands of an experienced practitioner can be effective. The omnipresence of God as absolute consciousness will relate to a caring practitioner, through any vehicle, any ritual, any technique, any etiquette that one uses with belief and intention.

Life Is Energy

Life is energy. The ways in which we integrate and express life are imbedded in the vital fields of our bodies and the energy patterns of our lifestyles. Our posture toward life is a statement in energy, and our bodies are a screen on which we project our emotional life. All of our emotions are registered through corresponding states of body tension. Each thought and emotion is a subjective state of openness or resistance that sends messages of relaxation or tension to our body.

Energy Blocks Defined

We understand energy blocks as excess, confused, and contradictory mental and emotional impulses to the musculature. This discord creates tension and resistance in the vibrating energy fields, which dam our attunement with the Soul. An energy block defines a place where it hasn't been safe to experience sentience, feeling, or aliveness.

When there is a perceived threat to the ego's fragile security, there is a fixation on the threat. The mind obsesses on the threat. The mind mulls over the complexity of the issues. The mind may spin for days, months, and years in its obsession with the threat. Often we are trapped in threatening relationships for years of our lives: dysfunctional families, oppressive schools, and repressive jobs. The mind obsesses on the threats to our emotional safety and self-esteem. Much emotional trauma reflects the wounds of this dilemma.

These fixations send impulses stimulating the fight-or-flight instincts of the sympathetic nervous system into the musculature. Day after day, month after month, each moment that the mind obsesses on these issues it is sending a myriad of confused, contradictory, and ineffectual impulses to the musculature. These impulses lodge in the nervous system and musculature as patterns of tension. This resistance drops the quality of vibration of the body from resonance with the finer forces of higher intelligence (sattva guna) and the conscious mind (rajas guna) to resonance with negative and unconscious forces (tamas guna). The body tension and armoring reduces the level of presence, aliveness, and the capacity to resonate with the sattvic quality of attunement to the Soul. Unconscious, negative, tamasic forces then predominate in the energy fields of the body/mind.

Mental and emotional impulses move through the central nervous system to physical expression through the musculature. The center of physical expression is in the area known in the Eastern arts as the *hara*. For our purposes, hara is the resonance of the Fire element (directed energy, action in the world) and the center of our *intention*. The resonance of the hara radiates up into the solar plexus (Leo) and down into our thighs (Sagittarius) from the Fire center near the navel. The fight-or-flight mechanism of the sympathetic nervous system is centered in the hara. Dr. Stone calls this area the "negative pole of the brain." It is where mental impulses are birthed into physical expression. The psoas muscle is a major avenue through which energy impulses move from the central nervous system to be birthed through the pelvis, through the thighs, and into physical expression out in the world.

When it is not safe to express our needs in the world, hara becomes tense and blocked. The myriad of confused and contradictory mental and emotional impulses foster armoring in the iliopsoas muscle to contract the pelvis, tense the abdomen and diaphragm, and restrict the breath. Tension in the hara limits the range of motion of the sacrum, the root of vitality, dropping the level of consciousness in the cerebrospinal system, from the finer forces of higher intelligence to negative and unconscious attunements. Restriction of the breath and of the cerebrospinal system lowers the level of consciousness in the body.

An energy block is a restriction in the body's capacity to resonate with the life breath. In body-centered therapy we use the cycles of breath to release tension and promote sound health and inner harmony. Each in-breath can focus a radiant, building presence at the point of contact. Each out-breath can collapse and release the energy field centripetally to promote elimination.

Healing is often a process of releasing the tension and resistance that inhibits the life-giving flow of the presence of the breath. Conscious breathing is a safe way to focus Divine Healing Presence on the energy block. To breathe fully is to resonate with the presence of innate intelligence of nature and the universe, to vibrate with the finer forces of higher intelligence which give life joy and inspiration. A willingness to breathe fully is a willingness to be fully sentient, fully alive.

Our natural state is to live an ecstatic life of surrender and inspiration attuned to the superconsciousness of sattva guna. Most people's desire-bound lives of alienation resonate with the conscious mind of rajas guna. In a troubled life the body/mind often resonates with the negativity and unconsciousness of the tamas guna archetype.

As discussed, blocked energy is an inability to be *present* in the body. An energy block is an inability to be alive, feeling, conscious. In an evolutionary perspective, an energy block is an unwillingness to be present within this theomorphic body: an unwillingness to feel, to be conscious. It is a place where the flow of creative evolution is stuck. An emotional complex is a fixation on the threats of the past.

In our evolutionary model, healing is releasing the resistance to being present, fostering attunement to the Self, aliveness, and health. Healing is the release of the fixation of the past and an opening to the gifts of the present.

Negativity

The psoas muscle is the major avenue through the pelvis for mental and emotional impulses moving from the central nervous system into physical expression.

It is common for clients to seek out therapy because they are stuck in negativity: depressed, uninspired, numb to life, etc. This is a situation where there is a fixation on the negative field. Excess, confused, and contradictory mental and emotional impulses create tension and resistance in the musculature which blocks elimination, keeping the physical from vibrating in harmony with the universal life force. Emotional trauma, a stagnant life-style, and toxic food tend to create resistance and negative attunement.

It is fundamental to understand that energy is always expressed as a cycle. Our task as Polarity Therapists is to support the integrity of these cycles, to reconnect energy fields to their source. Problems often arise in the elimination phase or negative field of the cycle. Energy tends to stagnate at the contracting, resistant, tamasic part of the cycle. By bringing the powerful force of sattvic attunement in the hands to areas of the body which are tense or blocked, we can release resistance and restore the harmonic integrity of the energy field. When we release resistance and reattune the energy fields, the cycle of Spirit 's reflection in matter can be expressed with fidelity. When we are attuned to Spirit,or the finer forces of higher intelligence, an intuitive sense of wholeness, meaning, and purpose is expressed as a joyful and inspired life.

Negativity is often the reflection of an experience that has not been adequately processed, expressed, and eliminated. Typically it is an experience which the ego (the executive of the personality) felt was too threatening to deal with consciously. So the area in the body where the feelings or expression was centered was tensed to resist consciously experiencing or expressing the feelings. This pattern of tension blocks the energy and armors the character from its feelings. The body then fixates on the incomplete elimination. For example, a client unable at the time of occurrence to consciously deal with incest might tense the pelvis and abdominal area to suppress the powerful mental impulses and confused emotions resulting from this experience. Years later the only conscious connection to the incidents might be an area of tension and a history of discomfort, disease, or injury in the abdomen.

When feelings have not been adequately processed, experienced, or expressed the ego tends to fixate on the threat and is resistant to elimination. Often the person becomes "negative—tense, anxious, defensive, closed, numb, uninspired, depressed, or generally caught in states of consciousness which can be seen as states of held or blocked energy patterns in the body as a whole.

Elimination

Restoring healthy elimination is often a key to healing trauma. Providing a safe and sacred space for clearing past trauma is where body-centered therapy excels. A goal of therapy is to release unconscious forces from the past that recapitulate past wounds. The threatened ego is often ruled by negative, unconscious, reactive, past images. Opening the pelvis, by releasing sympathetic nervous system impulses through perineal therapy, is a foundation of somatic emotional clearing. Releasing tension in the psoas and throughout the pelvis is often a key to releasing trauma from the cellular memory of the body/mind. Elimination becomes stuck when we fixate on a trauma. The pelvis is the home of elimination. It is the negative field of the brain where thoughts and emotions are birthed into expression in form. The perineum and thighs are the negative pole of the diaphragm. Releasing the perineal diaphragm is a key to opening the thoracic diaphragm. Restriction in the diaphragm inhibits our attunement to the fullness of the life-breath and the grace of inspiration. The sacrum, the root of vitality, is limited in its range of motion by perineal tension. Releasing pelvic tension restores the range of motion of the sacrum and cerebrospinal system for attunement to the healing presence of the Soul.

Polarity Therapy focuses on the release of tension in the body. Relaxed tissue can resonate with the potency of the Life Field to restore vitality to the body. Tension or resistance inhibits our ability to breathe fully and resonate with the finer forces of nature and higher intelligence that sustain health in the body. Body armoring represents an inability to be present in the body, to breathe, to be alive, to feel.

The classic example of the process that we use to release trauma in body-centered therapy is with mother and child: The child comes crying to the mother with the trauma of pain, hurt, or fear. The mother holds the child in safety and love, addressing the wound and reassuring the child that it is okay. The mother invites the child to be present in his or her body in the safety of her embrace. "What happened dear? ... Show me where it hurts ... " The child's experience is witnessed and validated in a safe and sacred space. The child is supported in being fully present in the experience, and thus clearing the trauma. Blocked energy is cleared when it is safe to be present in our experience. Safe to be present and feeling in our bodies. Safe to be sentient, breathing, and fully alive.

Compassionate touch through sattvic Polarity energy balancing facilitates a profound safety. Sattvic currents arise from the heart of being. Sattvic touch carries the compassionate intention and support of the practitioner to the client's center to facilitate a profound safety. In this sacred space clients are guided to experience that it is safe to be present in their bodies.

In the evolutionary perspective, the client need only experience that it is safe to be present in his or her body. To be present, sentient, and conscious—in the body—releases the evolutionary block. The client need only be present for a flash of emotion from the past. Once the client has been present for the emotion, immediately use Polarity Therapy techniques to prevent retraumatization. During emotional release the skilled practitioner can use perineal therapy to release the sympathetic nervous system and prevent retraumatization. The practitioner can also bring the client into a fetal position for nurturing shoulder hip rocking, which evokes the parasympathetic nervous system to release trauma. Sattvic contacts access the healing presence of the Soul. Profound realizations, and the personal experience of inner peace and well-being flow through sattvic touch. Compassionate bodywork with a trusted practitioner prevents retraumatization during emotional release.

The essence of body-centered therapy, then, is to support clients to be present in their bodies. Throughout the bodywork session we invite clients to bring their attention to their bodily sensations—to experience that it is safe to be present in their bodies and safe to be present with their feelings. In an energy block, we go unconscious because it is unsafe to experience our feelings, unsafe to be present, to be sentient. Facilitating a safe space where clients have support in being present for their feelings releases blocked energy from the body/mind. Presence releases blocked energy.

Our research indicates that clients do not need to reexperience trauma to release the past. It is only necessary for the client to experience that it is safe to be present in his or her body for trauma to be released and sentience to return.

The Parasympathetic, Sympathetic, and Cerebrospinal Nervous Systems

In Polarity Therapy, Dr. Stone focused on releasing trauma by working with the step-down process from within to without through the three nervous systems. His approach focused on working with the nervous system as a key to facilitating emotional clearing and reestablishing healthy elimination. In the Polarity model: "The parasympathetic system is the neutral conduit that channels thoughts from their pre-physical field of origin into the physical realm. The sympathetic system, in turn, becomes the positive channel to shuttle thoughts into action through the ... Cerebrospinal nervous system, [which is] the negative pole of the nervous system."[10]

The sympathetic nervous system facilitates the body's response to danger, adversity, stress, and anger by increasing heart rate, blood pressure, air exchange volume, blood flow to muscles, and the physiology necessary to spring into action. When a threat is perceived in the environment, the sympathetic nervous system is activated for fight or flight. Typically in our "civilized" way of life our "animal" has no mechanism for fight or flight. As discussed, excess, confused, and contradictory nerve impulses to "fight or flight" that cannot be safely expressed in the world become patterns of tension in the nervous system and musculature of the body/mind. Release of the sympathetic nervous system is a key to the release of trauma.

Dr. Stone writes: "parasympathetic emotional outlets can only be expressed when Nature, the sympathetic nervous system responds. This is the key to mental and emotional and functional treatment for frustration."[11] "Parasympathetic and sympathetic releases allow the cerebrospinal system to function harmoniously."[12]

When our well-being (parasympathetic) is threatened, the sympathetic nervous system is activated. Dr. Stone writes: "Parasympathetic impulses must flow into and work in conjunction with the Sympathetic, in order to work at all; for it is the Parasympathetic system that expresses balance and conscious mind impulses, and conveys them to the Sympathetic. What are mental frustrations? Parasympathetic or cranial impulses that could find no expression or response in the Sympathetic system ... Emotional frustrations are sympathetic and heart center impulses which are suppressed by the conscious mind impulses of the Parasympathetic system."[13] Mental and emotional impulses step-down from the mind into physical expression through the three nervous systems. The Air element predominates in the mind and parasympathetic nervous system. The Fire element rules energy directed into the world. Fire is expressed through the sympathetic nervous system.

Dr. Stone writes: "Sympathetic and parasympathetic nerve reflexes ... respond from below upward."[14] Thus, the key to their release is in the negative fields of the body that rule elimination. Perineal therapy releases the parasympathetic nervous system, which rules relaxation, and is the key to balancing the three nervous systems. It releases the sympathetic nervous system, which rules the autonomic functions of the fight-or-flight mechanism as well as body maintenance and building.

The perineum is the most negative reflex to the parasympathetic nervous system, which carries emotional impulses to the musculature (sympathetic) for expression. When it is not safe for the muscles to express our feelings in the world, both systems get blocked. "The perineal contact is the best treatment for nervousness and hysteria ..."[15] The perineum receives the final impulse of the caduceus which steps-down energy from Spirit into the body. Dr. Stone believed that he had identified the key to reestablishing healthy elimination and restoring the integrity of the nervous system in perineal therapy. The perineum is the most negative pole of the parasympathetic system, the sacrum is neutral, and the shoulders near the neck are positive.[16]

The ganglion of impar, located behind the coccyx, is the negative pole of the sympathetic nervous system. To balance the sympathetic nervous system, Dr. Stone recommends contacts at the sphenoid sutures or occiput as the positive field, sacrum and ganglion of impar as the neutral field, and Achilles tendon and heel as the negative field.[17] Perineal therapy releases trauma from the sympathetic and parasympathetic nervous systems to release fixations from the past and establish healthy elimination.

In a theomorphic model it is at the pelvis that we birth our expression into the world. Air, desire (thoracic cavity), stimulates Fire, intention (solar plexus), to bring into form through Water (pelvis) expression on the Earth. The Water element which predominates at the pelvis rules the way we connect with others and nourish ourselves. When we feel unsafe to connect we tense the pelvis. Pelvic tension closes off both our receptivity (left hip) and our ability to express our needs (right hip). Polarity focuses on releasing trauma from the pelvis to reestablish nourishing receptivity, and to restore safety for expression and healthy elimination.

Compassionate Presence

As healers, our most potent tool is to be deeply present for our clients. Being present is understood to be the very essence of the healing relationship and the key to effective therapy. As body therapists, our healing focus is to be present in our hands. As therapists we hold a safe and sacred presence of support, respect, and faith in the wisdom of the growth process.

Sattvic contacts facilitate a receptivity to the healing presence of the Soul—the profound intelligence that animates the body. The transverse currents arise from the core of the body—the heart of being. Sattvic contacts resonate with the nucleus of every atom of the body. Tender contacts with a compassionate intention touch the Soul, the life essence. Sattvic contacts facilitate a superconscious state of lucidity on the cellular level, which invites healing. Being present in our hands, fully concentrated in our caring, and in an exalted state of consciousness supports a sacred space where the client has permission to release the past and open to the gifts of the Present.

Sattvic touch is a gentle touch of stillness and presence. Sattvic currents arise from the core of the body, the heart of the being. In sattvic contacts the practitioner's consciousness touches the intelligence which is the medium of creation. Sattvic contacts promote a receptivity to the Soul.

Counseling Protocols: Defining a Safe Space

It is fundamental to "define the situation" for clients, letting them know that this is a "safe" and appropriate place for them to process and clear their emotional issues. By defining the situation you give the client permission to process and let the client know that the bodywork session is a *safe* and *appropriate* place to clear emotions. Say something like this during the first sattvic contacts as you begin the session:

"Take a few deep breaths, deep into your umbilical center ... feeling yourself relaxing more and more with each breath. Deep breaths, deep into

your umbilical center . . . relaxing breaths, feeling yourself relaxing more and more with each breath . . . This is a time for you to relax and allow yourself to be nurtured and taken care of . . . A safe space where you don't have to be anyone or do anything . . . A safe space in which you can just relax and let go.

"During this session we will be balancing your energy . . . Energy balancing facilitates natural processes of emotional release and personal integration. Feelings and images from your past may come to the surface of your awareness as a part of a healthy process of release and clearing. I invite you to be as open as possible to fully experiencing or fully expressing whatever feelings or images may surface for you. Appreciate that this a safe and appropriate place for you to express the truths of your own feelings. I want to encourage you to experience or express as fully as possible whatever feelings may surface for you.

"Understand that this is a safe space and you are in control in this situation. If at any time you feel that anything going on is too intense or in any way inappropriate for you, simply say: 'please stop.' During the first part of the session we may be talking about issues in your life, later we will be working in silence."

Restructuring the Personal Unconscious: Rescuing the Inner Child[18]

Nobody deals with reality. We all relate to images in our minds that are our representations of people and situations from our lives. These images can be called your personal unconscious. Your biological mother is not the mother you carry around in your head. The mother in your personal unconscious is a very selective representation of your experience of her. She is your personal creation.

We have tremendous power to restructure our personal unconscious. Researchers who have worked with the personal unconscious tell us that it has no concept of time. Thus we can rerun past events and change them from trauma and loss to positive, successful outcomes. The following process is profoundly effective in clearing the deepest life issues and facilitating the transformation of our personal unconscious.

The facilitator begins: *"I'd like to invite you to open to an image from your past. It may be a photograph from your childhood that's easy to remember, a picture from a dream, or an image in your mind. The picture that is the key to your growth will spontaneously come into your awareness right now. If more than one picture comes up, pick the earliest one."*

"Are you in touch with the picture?" If the client says no, repeat the preceding paragraph. If repeating the invitation to the past brings no image to the surface, you assume that the client has present issues which need to be dealt with. You simply ask: *"What is one of the issues you're working on in your life?"* You continue with the work supporting the client in experiencing his or her feelings, in experiencing that it is safe to be present in his or her body, and to be present somatically for his or her feelings.

If the client indicates that he or she is in touch with a picture, continue: *"Focus on this image in your mind's eye. Tune in to the parts of your being and time in your life that resonate with this image. Open to the parts of your being that this image represents. Allow this image access to your body, to your heart and feelings, to your mind, and to all your adult resources for understanding and expression. Allow yourself to become the part of yourself that's represented in the picture, to be the you that's in this picture."*

"Speak to me now, in the first person. How old are you?"

"What name are you called?" (Continue to refer to the client by that name, for example, "Joey" or "Patsy Anne." And continue to speak to the client as though the client were a child of the age given. Soften your voice and use the kind of language you would with a vulnerable little child, for example,*"Yes, little one," "Yes, dear."*)

"What are you wearing? What do you look like?" Let the client describe his or her dress, facial expression, and emotional space in the picture. Continue: *"Take a deep breath, and as you release the breath, allow your body to relax and open and feel what it feels like to be you."* [Use active listening to feed back the client's communication, for example, *"Eight year old Tammy, in your white pajamas with the little flowers, curly golden hair, big eyes, and puzzled face."*] Repeat: *"Take a deep breath, relax, soften, open, and feel what it feels like to be you.* [*"Eight year old Tammy, in your white pajamas with the little flowers, curly golden hair, big eyes, and puzzled face."*] Give the client space to begin to get in touch with his or her body. *"Take a deep breath. What are your eyes and face saying? What does it feel like to be you?"*

The picture from the past allows the higher mind to guide us to a key place where there are unresolved issues or resources for healing and growth. Often clients will begin to emote immediately in the sanctuary of the sattvic cradle as they experience that they have the resources to safely get in touch with unresolved feelings. Support the client in being present for his or her feelings.

Feeling Tone

We create our world through mental/emotional images that can be called the feeling tone. A goal of therapy is to support clients in experiencing that it is safe to be present in their bodies and to fully experience their feeling tones. Often clients will begin the above process with a picture that represents a profound resource of well-being. The photo from early childhood will put them in touch with the feeling tone of their original nature. Innocent, joyous, present, energetic, spacious—this can be a profound resource in therapy. Homework may include time spent nightly, tuning into this resource as the client drifts off to sleep, cultivating the feeling tone of well-being.

Active Listening

Active listening is a key technique to support clients in experiencing that it is safe to be present for their feelings in the body. The practitioner seeks to be fully present in the active listening process. In active listening we witness the client's experience by repeating back to him or her their communication. The client says: "My chest feels tight, my throat constricted." You say: *"Your chest feels tight, your throat constricted."* If emotions surface you receive them with a tender: *"Yes, yes, good, breathe, feel,"* etc.

"What are you feeling in your body?" Start to explore the client's feelings about family members. Identify who represents a threat to the child, or a trauma in the child's growth process. Support the client through active listening to get in touch with his or her feelings and to feel safe to be present and express those feelings. Typically the client has not felt safe to experience or express feelings.

Experience And Express the "Truth of One's Feelings"

We validate the client's feelings and support the client in experiencing the "truth of his or her feelings." "Truth" in this context, is that which can not be argued about, for example, *your* feelings. You can affirm to the client the truth of his or her experience; you can affirm that it is valid. Affirm that everyone is entitled to their feelings. The feelings in one's body are the truth of one's experience. You can validate the client's experience through active listening. You can receive his or her communication and witness experience through active listening. Tears need only to be witnessed in a safe place to be released.

EPC: Etheric Plane Communication

Facilitate etheric plane communication (EPC). EPC really works. All consciousness is One. Through this knowledge, our healing intention can communicate without boundaries, and across time, space, and personae.

Telling The "Truth of Your Feelings"[19]

"Truth" defined: That which cannot be argued about: *your* feelings, *your* experience.

"Truth"	Personal experience
Core Feelings:	I'm scared, I'm sad, I'm angry, I'm excited.
Sensations:	My stomach is tight. I feel a numbness in my shoulder. My eye is twitching. My palms are sweaty.
Specific thoughts:	I just heard a voice in my head telling me to shut up!
Familiar feelings:	This tight feeling in my belly.
Contrasts and qualities:	It feels like my heart is stone. I feel pressure on in my head and my legs feel very heavy.

As discussed above, often clients are holding on to unresolved issues from the past. When there is a treat to our security the mind fixates on the threat. The obsession sends mental and emotional impulses into the musculature. We use EPC to safely communicate feelings to release and clear the past. Use EPC to facilitate a safe space where the client has support for expressing feelings and getting things off his or her chest

Invite the client to picture, in his or her mind's eye, the individual with whom communication is needed. Ask the client to look into that person's eyes, speaking out loud to that person as though he or she were really present. Encourage the client to experience and express his or her feelings as fully as possible, venting anger, vulnerability, sadness, needs, love, etc. Remind the client that this is a safe space, that he or she can't possibly be hurt or hurt the other person's feelings. If the client feels threatened in speaking to the person have him or her address a photo or television image of the person. Let the client know that the EPC will change the psychic space and transform the actual relationship. EPC moves energy—it releases mental and emotional impulses from the musculature and nervous system as the client is supported in dramatizing issues. Feelings that have been buried for decades easily surface and are released in the safety and sacredness of the Polarity Therapy session.

Support Cathartic Emotional Release

As part of EPC, you may invite the client to yell or scream while on the table. You may invite the client to get off the table to beat a pillow, strangle a towel, do woodchoppers, or do something else to express his or her frustration or rage and to move energy through the nervous system musculature. The expression can be gentle. The musculature simply needs to be moving for this to work. It doesn't have to be intense. Keep bringing the client back to his or her somatic experience, to the experience that it is safe to be present with feelings in his or her body. Making it safe for clients to experience and express their feelings helps them to process, resolve, and clear past issues. Expressing feelings helps to fulfill the cycle of energy moving from Spirit to matter and to release energy blocks.

Tamasic Contacts

Tamas, the principle of resistance, can be stimulated to flush resistance out of the body, in a process which releases blocked energy and restores healthy elimination. Always ask the client's permission before doing tamasic work. Only do tamasic work for a moment—only long enough to evoke a deep cellular response of resistance from the client. Painful tamasic contacts should only be held for a moment, one or two seconds, and should always end in a sattvic contact to ameliorate the pain and support a safe space for emotional release. Tamasic contacts are most effective in the areas of the body that resonate predominantly with the tamasic harmonics of force at the negative fields—toes, feet, soleus/Achilles.

Tamasic contacts are meant to evoke the client's resistance through an instant of pain. Tamasic contacts bring resistance to the surface and often facilitate cathartic emotional release. After a tamasic contact invite the client to get in touch with his or her feelings. Extreme care should be taken not to retraumatize the client.

When emotions surface, support the client in being present with his or her feelings in a safe space for a moment, then immediately move the client into a fetal position for perineal therapy to release trauma from the sympathetic nervous system or use shoulder hip rocking to foster a parasympathetic space of profound well-being.

The breath is the life essence. States of consciousness are breath patterns. To prevent retraumatization bring the client's attention back to the feeling tone of the life breath. Have the client breathe safety, peace, and love into the body. Support the client in being present somatically in the experience of safety, peace, and well-being.

Somatic Emotional Clearing

In threatening or traumatic situations the mind fixates on the threat sending a myriad of contradictory and confused mental and emotional impulses into the musculature. Confused and contradictory mental and emotional impulses to the musculature sustain tension and resistance in the energy fields. This resistance is emotional armoring. This hardening drops our level of aliveness and feeling; health is a capacity to be alive and feeling. Moving energy in the negative field eliminates armoring from the cellular memory to release fixation and reestablish healthy elimination. Tamasic contacts may provoke a process of elimination which can be dramatic and cathartic. It can be frightening to the client, and the practitioner, as the worst emotional moments of a person's life are released from the cellular memory, surfacing to be experienced consciously and expressed in the safety of the bodywork session.

Do not identify with the elimination. It is only energy being released. Hold a safe and sacred space for the client. If you, as the practitioner, get scared, pray for the client's well-being.

As suppressed pain is released, suggest the client breathe peace and safety into the space. Dr. Stone writes: "Comforting suggestions and inspirational uplift will help the mind to concentrate its scattered energy and to repolarize its various centers."[20] As resistance is released, the energy fields of the body can come into harmony with the peace and guidance of the finer forces of higher intelligence of the subtle body. The client may have released emotional armoring during the session and may be more feeling, sensitive, vulnerable, and open. Suggest that clients avoid anything intense after the session, take time to integrate, and make space to nurture themselves.

Archetypal Attunement and Inner Guide Work

When you feel it's appropriate, you can move the client into a fetal position and hold the perineal contact. Let the fetal position serve as an "anchor," which the client can use to rerun the visualization in order to cultivate the healing feeling tone.

Invite clients to get in touch with an image that can nurture or support them (for example, his or her inner child). The image could be the adult self, a spirit guide, a God or Goddess, his or her essence, a healed parent or grandparent, Mother Earth, a nature spirit, a dear friend, or an idealized parent. Guide the client through a process in which he or she invokes the guide and the guide offers idealized, sublime, and perfect comfort and nurturance, in a way that he or she can really experience with the five senses: feeling held, feeling the heartbeat of the guide, feeling his or her head against the guide's breast, being held tightly, feeling the quality of the angel's feathers against the skin, the guide's eyes feasting upon his or her with love, speaking words of tenderness, of unconditional love, etc.

Facilitate a feast of support, picturing the guide as the most trusting, nurturing, supportive being imaginable. Ask clients if they would like the guide to visit in dreams, teaching them, helping them to grow, bringing them guidance and inspiration in dealing with life's problems, giving them quality time and attention, and promising to always be with them, never abandon them.

Invite the client to move into the fetal position "anchor" at night as he or she is going to sleep or whenever the need is felt, and to invoke the guide, basking in the feeling tone of the healing experience. Invite the client to go to sleep through this process as a way of cultivating the feeling tone, restructuring the personal unconscious and clearing the past. Through these simple techniques, in a sacred space, guided by higher intelligence, healthy individuals can actualize tools for emotional clearing and personal growth.

Reparenting the Inner Child

Often clients will choose to have their adult self parent their child. This offers a powerful vehicle for the cultivation of a nurturing relationship with a part of his or her being. A goal of therapy is to establish conscious resources to care for vulnerable parts of the being. Often the client has been moved by unconscious forces to create safety in his or her life. We can invite the vulnerable part of the being to realize and experience that it is safe in the adult's contemporary life and to realize that the adult can parent it with unconditional love, understanding, tenderness, and support. We invite the vulnerable part to experience that the adult has the resources in his or her ego structure and the capacity to maintain safe boundaries to care for his or her vulnerable aspect. We can support the vulnerable part in experiencing that the adult's connection to higher guidance is adequate for its protection and care. We can develop resources so that

> **Healing Affirmations for the Inner Child (Newborn)**
> I'm so glad you're here.
> I'm so glad you're a boy (girl).
> Welcome.
> I've prepared a special place for you.
> I'm here to give you
> everything you need.
> I'm so glad you're who you are.
> You're such a beautiful baby.
> You're beautiful, you're perfect.
> I love you so much.

when the client feels somatic signs of vulnerability his or her can consciously respond to the inner child and care for a vulnerable part of his or her being with these resources.

The Five States of Energetic Pathophysiology

The theomorphic dynamic of the ubiquitous evolutionary cycle has been called the "supreme ultimate principle." The Polarity cycle of the elements underlies all movement in creation. The five elemental phases of resonance undulate through every spiral of creation. As Spirit becomes blocked in its life expression, the body experiences varying states of ill health or disease. These states, varying from mental and emotional disturbance to acute inflammation, chronic illness, and finally to full degeneration, are characterized by energetic states corresponding to the step-down of the five elements.

The Five States of Energetic Pathophysiology	
Ether:	Free flow of energy
Air:	Mental and emotional energy blocks
Fire:	Acute inflammation
Water:	Chronic fixation
Earth:	Tissue death, autonomy from source

The essence of Ether is freedom. We experience the Ether element in balance as psychic space, emptiness, peace. The quality of Ether predominance is nonobstructive motion radiating lines of force in all directions and sustaining the space in which the other forces operate. A threat to our space, to the freedom and integrity of our Being, disturbs the Ether element.

If the threat to our space and the security systems that support our space does not abate, the dis-ease moves to the Air-predominant level of mind. The energetic disturbance has now stepped down from Ether, or space, to Air, or movement. The mind, disturbed by the threat, searches for solutions.

The mind obsesses on

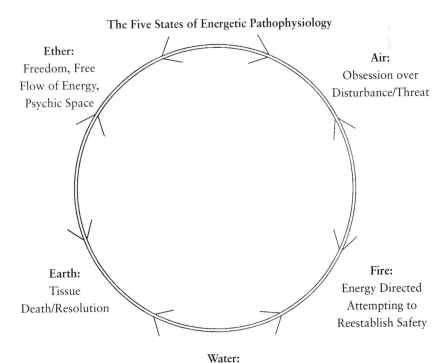

The Five States of Energetic Pathophysiology

Ether: Freedom, Free Flow of Energy, Psychic Space

Air: Obsession over Disturbance/Threat

Fire: Energy Directed Attempting to Reestablish Safety

Water: Fixation, Emotional Armoring/Catharsis

Earth: Tissue Death/Resolution

Focus on the Out-Breath

It is very important to have the client focus on breathing during release work, especially on the out-breath. Breath is the omnipresence of universal law. A fundamental organizing force, breath moves through the body much like waves move through the ocean. Breath rules the cycles of respiration, which are reflected in the peristalsis, oscillation, and vibration of every body cavity, organ, cell, molecule, and atom of the body. The out-breath resonates with elimination. A wave of centripetal, contracting, releasing, eliminating force moves through every atom, molecule, cell, organ, and cavity of the body with each out-breath. This wave releases tension and resistance. The radiant in-breath "respires" the vibrations of energy in centrifugal attunement to the core.

the threat, mulling over possibilities for the next phase, Fire, or action in reality, to resolve the disturbance. If the threat to the integrity of our being is not resolved, there may be a mental or emotional fixation on the threat. We obsess, seeking a plan of action. Again, obsession sends a myriad of confused and contradictory mental and emotional impulses to the nervous system, musculature, and organs. Energy is blocked by this armoring process. Unresolved mental and emotional disturbance may manifest on the level of the Fire element as an acute illness or an accident. The illness may embody the person's exhaustion in the face of the continuing threat to his or her perceived security and the accident may express the need for space to marshal strength and mobilize inner resources.

The disturbance may become Water-predominant as a chronic health problem. As discussed, in our model disease is often the unconscious acting out of feelings that were too threatening or debilitating to deal with consciously. A chronic health problem often is metaphor, the personification of the problem.

An Earth-predominant block in the cycle is one of complete breakdown, crumbling, falling apart, losing consciousness and autonomy from the Source, and tissue-death.

Pathology is complex, the product of many levels of being. Yet the practitioner may find this simple model based on disturbance in the harmony of the elements useful. If a client comes to us with a perceived threat to their security or freedom (Ether), we support them in thinking clearly about the situation (Air). If the disturbance is mental or emotional (Air), we facilitate a plan of action (Fire). If the client presents a Fiery rage we support him or her in experiencing the feelings (Water), or in thinking about the issues (Air). In the case of acute illness (Fire), it may be useful to facilitate a process by which you support the client in becoming more conscious of mental (Air) and emotional (Water) issues and to articulate a plan of action (Fire). If they are embodying the problem (Water), we give support in clearing emotions, or making a plan of action (Fire), and thinking and becoming more conscious of the issues (Air). Problems on the level of Earth need to be lifted to Water, a cleansing of

Emotional Postural Statements

Collapsed Posture: *Consciousness of defeat, "I am overwhelmed."*
Head forward, sunken chest and shoulders, rib cage fallen, belly tight. Tissue quality: no resiliency.

Rigid Posture: *Consciousness of defiance and control, "You can't get at me!"*
Chest thrust up and forward, pelvis tilted back, tight spine, forced straightness.

Tissue quality: *rigid, ungiving, tight, and hard.*

Inflated Posture: *Consciousness: "I need to be big!"*
Body cavities are big—belly, chest, pelvis. Often the body is split where the upper body, which is visible and expressive, is "big," and the foundation of the lower body is smaller. Sometimes "loud" as a person, jolly, boisterous, etc., yet often there is a denial of sensitivity and vulnerability.
Tissue quality: expanded but tight.

Swollen Body type: *Consciousness: "I feel so much."*
Tends toward fatness, water retention. Everything is spilling out.
Tissue quality: mushy on surface, often rigid on deeper levels.

Tight Body type: *Consciousness: "I need to be tense to hold it all together."*
Thin, tense, chronic anxiety, nervous.
Tissue quality: tight, stringy, wiry.

Head: *Left tilt: feeling-oriented, yin, watery, feminine.*
Right tilt: mental, yang, fiery, airy, action-oriented.
Down tilt: defeat, heaviness, shame, guilt, punished.

Neck: *Neck down, head forward: defeat, yet denial, can-do attitude on top of fear and feeling overwhelmed. Duty over fear.*
Rigid: intellect overriding the emotions.
Rigid straight: "Soldier's neck," dedication to duty, overriding personal feelings.
Pulled back: fear response, startled, retreating.

Shoulders: *Raised: fear.*
Retracted, pulled back: anger, wanting to strike out but holding it back.
Rounded forward: protecting the heart, vulnerability.
Left leading: more feeling, receptive.
Right leading: more action, thinking, yang.

Chest: *Collapsed: depression, grief.*
Inflated: boisterous, yang, forceful.

Pelvis: *tipped forward: overemphasis on emotions, sexuality, often with swayback.*
Tipped back: emotions repressed, holding back sexuality.
Tight Buttocks: controlling anger, repressed anger, goes with contracted shoulders.
Tight perineal floor: fear response, chronic anxiety, control.
Rotation: left, more receptive.
Rotation: right, more active.

Legs: rotated outward: fear, falling back.
Bowlegged: defiance, forcing forward.
Thin, underdeveloped: weak foundation, lack of personal power, lack of ability to move in the world.
Fat, weak: Stagnant, slow, lethargic, lacks direction, focus, low fire.
Strong, overmuscled: forcefulness, determination.

the past, and the restoration of effective elimination. Pathophysiology describes a cycle. One can facilitate movement at any point of the cycle. All energy on every level moves through the elemental phases. This understanding can be a key to effective body-centered therapy.

Health: The Ability to Take Responsibility

Polarity Therapy deals with the whole person, not merely the body. The body is understood as a crystallization of the mind and emotions. Moving energy in the body as the "negative pole" of the mind facilitates mental and emotional integration. The body is seen as a process in which mental impulses move into physical expression. Our work is to facilitate the release and integration of mental and emotional contents that have not been adequately processed through the body into Self-expression.

In Polarity Therapy we are working with fundamentally healthy individuals who are willing and able to take responsibility for actualizing a more integrated state of being. Health is defined as *the ability to take responsibility*. It is important that somatic therapists assess the apparent emotional health of the client. In your client intake raise questions like the following: Is the client currently under a doctor's care or has he or she ever been under a doctors care for emotional issues? Is the client currently or has the client recently been on medication? Does the client appear to be anxious, stressed, and distraught? Is the client depressed or hyperactive? Does the client appear to be disassociated, confused, unrealistic, or overwhelmed? Assess the client's resources. Does the client have healthy supportive relationships with friends and family? Is the client having a positive experience at work or school? Does the client have a spiritual practice? Is the client dealing with physical health issues? What mind-altering substances does the client use? Is the client taking a high degree of responsibility for his or her life by making healthy lifestyle choices? Have the name and number of a counselor or psychotherapist in whom you have confidence readily available so that you can refer individuals whose problems lie outside of your scope of practice (health).

Health is a balance of positive, negative, and neutral forces. In therapy it may be appropriate to facilitate catharsis, emotional clearing, and elimination in the negative phase of the healing process. But more important are the health-building lifestyle resources of the positive phase, in which we educate the client and support lifestyle changes that support balance, health, and well-being. The neutral phase, of connection to source through time in nature, unselfish service, or spirituality, is the fulcrum of healing. Health is an organic, day-to-day, thought-by-thought process. Polarity health educators approach healing holistically with resources for the three phases of healing. It is important to work with clients holistically and educationally in a balance of the three polarities.

We are not psychotherapists. We are educators sharing techniques that get people in touch with their own inner resources for guidance, healing, and growth. Moving blocked energy in the body facilitates a receptivity to higher guidance. Polarity is "self-directed therapy," guided by the client's innate, divine intelligence. We do not provide theories or answers—we encourage clients to listen to their own inner guidance and to find personal insight and realizations within.

In the wisdom of the Sanatana Dharma all energy is the energy of an ever-evolving universe of Creative Intelligence. In this model we are made of growth. This higher guidance has the capacity to illuminate every aspect of our lives. Inner guidance inspires the healing process. As we release resistance from the negative fields of the body/mind, insights and realizations guide the healing and actualization process. Through Polarity energy balancing we facilitate a profound personal experience of Essence, Presence . . . the Self. This profound attunement to the healing presence of the Soul is the capstone of a somatic, transpersonal approach to healing.

Healing is Creative

Healing is a creative process! We create our healing from moment to moment. The point of power in the healing process is *now*. Yes, the past contributes to the matrix of the healing process, but healing is not a mechanistic process determined by the past. Healing is a process guided by creative evolution in the present. The point of power is the present. Personal evolution and healing are rooted in realizations within consciousness. This natural growth process is guided by the evolutionary intelligence. Healing and personal growth are often the expression of a leap in consciousness. This process is creative. Moving crystallized energy in the negative fields of the body/mind facilitates the creative process. Releasing blocked energy promotes evolution through personal realizations and an expansion of consciousness. In an evolutionary perspective personal realization and the expansion of consciousness is the foundation of the healing process.

In every moment the body is sustained by its attunement with Spirit. In Polarity we understand the theomorphic organization of the body/mind and use these principles to promote somatic clearing.

The work of the somatic practitioner is to facilitate the release of resistance in the body, to cleanse the body's cellular memory of trauma so that its energy fields can resonate with the finer forces of spiritual attunement. When the resistance is dissolved and armoring is released, the person is in tune with the source of joy, and feels alive and in tune with higher intelligence. Being in harmony with the finer forces of Creative Intelligence brings inspiration, guidance, clarity, connectedness, purpose, and meaning into our moment-to-moment living. In the peace of this attunement the individual experiences a sacred communion with Life. In our natural state of Soul communion we feel *Wonder-full*. There is a *Great-fulness* and experience of the natural spirituality inherent in aliveness and good health. This attunement to the *Well-of-Being* is the fruit of Polarity energy balancing.

CHAPTER SEVENTEEN

Elemental Imbalance and Character Disorders[1]

Balancing the Ether Element

The hands-on principles which underlie the practice of somatic psychology focus on releasing tension, resistance, and character-armoring from the body in order to bring it back into harmony with "the finer forces of higher intelligence." The cerebrospinal system is the ultrasonic core of the body and resonates with the Source—The Life Field. The etheric body radiates from this core. All of the energies of the body are a step-down from this nuclear field. The etheric body holds the neutral space in which the energies of the body emanate. Dr. Stone, using a biblical metaphor, referred to it as "the one river from which the other four rivers arise." Ether is the bridge between the mental and the physical plane.

Ether rules the overall *space* of the body, and provides the sonic container defining the organism's field. Ether, also called Akasa, Space, or Wood, is the ultrasonic medium of creation. Ether resonates with the emptiness of the Unitive Field. It is the stillness at the heart of movement. It holds the neutral space in which the other elements move. It is the *potent* neutral element out of which the other elements emerge. From a theomorphic perspective, the neck and throat chakra, as the center of the Ether element, is the point at which the subtle inner elements combine into the pentamirus (fivefold) combinations of gross elements for physical manifestation.

Ether rules communication and the emotions in general, and combines with Air, Fire, Water, and Earth to manifest the range of emotions. Specifically it governs the emotion of grief. It is the home of the powerful neutronic force of sattva guna, the ground state, the force of equilibrium, stillness, essence. Ether rules the throat, predominating in the neck, which is the nexus or womb through which mental impulses are given birth into emotional and physical expression. The plasma generated in the bone marrow is Ether-predominant.[2] Ether predominates in the ultrasonic core, through the midline from the crown chakra through the caduceus of the spine and through the reproductive organs to the big toe.

Ether holds the *space* in which the physical body manifests. It is associated with the overall well-being of a person and his or her sense of self-esteem. Thus, when Ether is in balance, the person experiences a great deal of "psychic space" in life and a sense of safety and freedom. Balanced Ether manifests a "mature" persona that is accountable and makes conscious choices in a process of self-actualization. In contrast, when a person is centered in his or her ego structure, the mind is consumed by the ego's agenda of dealing with perceived threats or with ploys to pump itself up, all of which consume psychic space. Overly contracted (tamasic) Ether manifests as grief and as poor self-esteem, a sense of worthlessness, shame, and victimization. Overly expanded (rajasic) Ether manifests as arrogance, tyranny, or a false sense of superiority in its yang mode of expression. Overly expanded Ether may also manifest as a "spaced out," ungrounded yin mode. In balance (sattvic) Ether manifests qualities of self-esteem, self-love, and humility. Ether comes into balance through the deepest kind of Self-knowledge, which lifts the being out of the ego's drama into the transpersonal realms of Self-realization.

The quintessence of Ether is Liberation (*moksha)*, freedom from the ego's constant drama and liberation into the security of Higher Knowledge. To balance Ether involves Self-knowledge, detachment, self-acceptance and a humble, genuine self-esteem. Balanced Ether flows from a sattvic lifestyle and a life focused on Soul-communion. The Ether-predominant ultrasonic core is the center of the finer forces of vibration of the transpersonal realms of consciousness in the body. It resonates with a higher potency of energy and consciousness. Polarity energy balancing of this harmonic of the resonance often results in

Ether predominates in the throat, cerebrospinal system, and plasma. Ether steps down to Air through the joints.

Techniques to Balance the Ether Element	
Balance Ether element	Joint session
Cranial-sacral balancing	Core session
Spinal energy balancing	Spine session
Parasympathetic release	Perineal therapy session
Squat	Six-pointed star
Structural work with the sacrum	Pelvic and sacrum sessions
Perineal therapy	Perineal therapy session
Toning: used throughout this work	Use *beja* seed mantras
Sattvic diet	Polarity cleansing diet

a turning inward, toward the profound stillness and harmony of our essence in peaceful communion (yoga) with the Self.

Polarity Therapy offers a broad range of techniques which focus on balancing the Ether element. These include the simple, easy-to-follow protocols listed in the table, Techniques to Balance the Ether Element.

Balancing the Air Element

Air rules *movement* in the body. The rhythms of the breath and heartbeat, the peristalsis of the organs, and oscillation of the atoms are all ruled by the Air element. The Air harmonic of vibration lengthens the energy field from the unitive resonance of the Ether phase of the cycle to the polarity of centrifugal/centripetal resonance that rules physical manifestation. Air is the first material phase in the quantum step-down in vibration from above to below, within to without. As the waves of vibration move through the field, Air is a lengthening of the vibrations emanating from within to without.

In Chinese medicine, Air (Metal) is said to be the relationship between Fire and Water. Air is a servant of wholeness and maintains a profound balance between yin and yang. All movement in nature takes place as an expression of the cycle of the Tao from Fire to Water. Thus Air is the neutral transverse force that is the fulcrum for the manifestation of centrifugal Fire and centripetal Water. Energy emanates from the source though the bindu of the fulcrum which is the center of the energy field. The Sun (Fire) and the Moon (Water) rise and set, but the air (Air) is a constant in nature.

Air sustains the fundamental rhythms that fill the body with attunement to the forces of the cosmos. The beat of the heart is the neutral source; the cycle of in-breath and out-breath of the lungs is the positive field; and the peristalsis of the colon is the negative field of the Air-predominant torso. In the anterior body the positive field of Air is the thoracic cavity; the neutral field is the colon; and the negative field is the calves. From a posterior perspective Gemini, which rules the thoracic cavity, is positive; Libra, the kidneys/adrenals, is neutral; and Aquarius, the ankles, is negative. These relationships are called "triads" in Polarity Therapy. Bipolar contacts work with the underlying triad fields of force, which resonate in the body and are deeply therapeutic.

Air rules the mind, whose mercurial nature is incessant movement as well as attention and thought. An Air imbalance may manifest as "analysis leads to paralysis," where there is excessive mental activity, worry and anxiety, but no action. An Air imbalance may be characterized by a person who is caught up in thoughts but out of touch with their feelings and not present in the body.

It is said in the Sanatana Dharma that: "All our problems are caused by not knowing who we are." We are so identified with our ego-mind and its drama that we forget the Sacred Self. From a therapeutic perspective the essence of the ancient wisdom is: "Identification is ignorance." Self-knowledge and a sattvic diet and healthy lifestyle are the keys to quieting the mind.

Air rules the attachments, desires, and aversions that fundamentally motivate our behavior. There are two forces in desire: attraction, pulling toward, and aversion, pushing away. In the East it is said that the "mind" is the surface of the "heart." When the heart desires, the mind seeks. The mind is a slave to our aversions and attachments. If the heart is broken or threatened the mind is incessantly active, searching for healing. When the ego is threatened the mind is incessantly fixated on the threat—first in denial, then in a complex of defenses. To heal the mind is to foster physical safety and emotional security. This process can take decades, if done unconsciously. It can be dramatically accelerated through the conscious use of these Polarity principles, which tune us into a deeper level of innate intelligence and guidance within our own being. Air rules desire. The key to quieting the mind and maturing toward inner peace is to cultivate moderation and self-regulation and to let go of the insatiable wanting of the self-centered life.

Air rules the nervous system. Excess mental activity, worry, and anxiety create excess, confused, and contradictory mental and emotional impulses, which lead to imbalances of the nervous system. When it is not "safe" for us to experience or express our feelings, we tense the areas of the body where these feeling are centered and we create character armor. Air rules speed and motion and predominates in the centers of movement of the joints in the body. The excess mental activity of an Air imbalance may foster tension, rigidity, and character armoring, which limits the range of motion in the joints and the range of emotion in our character structure.

When the freedom or security of the Spirit (Ether) is perceived to be threatened, the mind (Air), is obsessed attempting to deal with the threat. If the threat is overwhelming to the ego or unceasing over a period of time, we tense the muscles of the thoracic cavity, diaphragm and heart to cut down on our level of aliveness and feeling, and we suppress our expressive energy. In our model every atom and cell of the body is conscious and sentient. Cancer, the breast, is the center

Techniques to Balance the Air Element

Balance sattva guna	Sattvic session
Open diaphragm and heart	Diaphragm and heart session
Mobilize joints	Joint session
Quiet mind	Perineal therapy
Parasympathetic balancing	Core session
Sattvic diet	Polarity cleansing diet

of the consciousness "I feel"; Leo the diaphragm and heart is the center of the consciousness "I am." We tense and armor areas of the body to suppress our feeling, our consciousness, our pain. An Air imbalance often manifests structurally as shoulder tension, a contracted diaphragm, immobile rib cage, and collapsed chest. Energetically, an Air imbalance is often a pattern of tension that contracts the "heart" to protect it from feelings that were perceived as threatening. Giving and receiving love is defensive and inhibited when the heart center is armored.

Air rules the incessant attachment of desire, greed, and aversion that characterize the involutionary cycle. In the evolutionary cycle Air manifests yearning for liberation, desirelessness, charity, universal love, and compassion.

Sattva guna rules the Air principle (made up of the Ether and Air elements). When in balance, sattva guna fosters integrity, honesty, contentment, and moderation. Balancing the Air element fosters a receptivity to sattva guna and the stillness, essence, and peace of attunement to the Unitive Field. Polarity Therapy offers a broad range of techniques that focus on balancing the Air element. These include the simple, easy to follow protocols listed in the table, Techniques to Balance the Air Element.

Balancing the Fire Element

The Fire element rules directed force in the body. Fire expresses purpose, the intelligent willpower, motivation, desire, and excitement of creation. Fire rules the head (Aries) as vision, intention, focus, and concentration. Fire rules the solar plexus and heart (Leo) as warmth, drive, vitality, will power, and enthusiasm. Fire rules the thighs (Sagittarius) as purpose, the power to move in the world, to care, and to take responsibility.

Fire rules personal power (umbilical center) and purpose in the world (thighs). Fire seeks self-knowledge, to know itself through its reflection in creation. Overly expanded, Fire is exploding, callous, and insensitive, and acts in ways that are oblivious to others' realities. Fire governs the emotions of anger and resentment. It can be controlling, blaming, and shaming. Suppressing Fire leads to seething resentment and inner-directed rage. Too little Fire is imploding and manifests as insecurity, self-blame, and powerlessness; in this state a person is easily controlled by others,

Techniques to Balance the Fire Element	
Bloodless surgery	Rajasic session
Balance Fire triad	Rajasic session
Stimulate Fire principle	Rajasic session
Sympathetic balance	Core session
Sympathetic balance	Spine session
Balance Fire of digestion	Polarity cleansing diet
Easy stretching postures	Woodchoppers, ha breaths

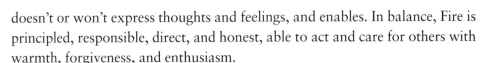

doesn't or won't express thoughts and feelings, and enables. In balance, Fire is principled, responsible, direct, and honest, able to act and care for others with warmth, forgiveness, and enthusiasm.

Fire rules direction in the body. The eyes, sense of sight, and thighs are Fire-predominant. The solar plexus, muscles, and all the red tissue and organs of assimilation are ruled by Fire. Fire is the driving force of vitality in the body. The umbilicus is the radiant center of vital energy and radiates warmth and vitality throughout the body.

Polarity Therapy offers a broad range of techniques that focus on balancing the Fire element. These include the simple, easy-to-follow protocols in the table, Techniques to Balance the Fire Element.

Balancing the Water Element

Water is the phase in which consciousness precipitates into substantial solidity. Water is the medium of life. Water refers to the centripetal crystallization of resonance that we experience as "form." Water predominates in the pelvic basin, the center of the organs of cleansing, renewal, and healing. Water resonates with the tamasic, contractive quality which is always flowing downward and which rules elimination. The pelvis is the center of the reproductive organs, which reenact the mystery of creation to birth new life.

Water is the realm of feeling. Dr. Stone called the Watery pelvis the "irrational pole" because Water predominates in unconscious emotions and deep feelings. (Air is the "rational" pole.) Water governs attachment, the strength of emotion that binds us to things, ideas, and people. The pelvis can be understood as the negative pole of the brain, where the images from the brain are embodied into emotions and birthed into the world as action. When it is not safe for our Watery feelings to flow outward into expression, we have a crisis in emotional elimination. Chronic fears and emotional tension lodge in the pelvis.

Attachments and love resonate with Water. Its focus is "I need," and its primary quality is emotional self-care. In balance (sattvic), Water is flowing, growing, receptive, sensitive, intuitive, compassionate, nurturing, moderate. Expanded (rajasic) water is isolated, anti-dependent, unfeeling. Contracted (tamasic) water is overly sensitive, stuck, dependent, needy, despairing, paranoid, possessive, lust- and

Techniques to Balance the Water Element

Open perineum	Perineal therapy
Release pelvic tension	Pelvic session
Release hips and buttocks	Sacrum session
Balance Water triad	Sacrum session
Release physical toxicity	Polarity cleansing diet
Release negative field	Polarity foot reflexology
Easy stretching postures	Squat, ha squat, pyramid

love-addicted, ruled by irrational and unconscious forces.

The Water element relates to the ways in which we nourish ourselves. Water rules the lips where we take in nourishment and express our needs. Water also rules the hips and pelvis where gender issues center in the body. Pelvic tension may indicate issues relating to giving and receiving sexual and emotional nourishment. Tension in the buttocks, hips, and pelvis may indicate isolation and a lack of nourishing connections.

Water sinks into the ground cleansing the field of impurities. The Water triad of Cancer the breast, Scorpio the pelvis, and Pisces the feet is a key to restoring healthy elimination. The feet are the most negative field of the body. They are where we ground to the Earth, and are a key to releasing chronic tension and chronic health problems. In Dr. Stone's work there is a major focus on the negative field. Energy always moves in cycles. It is in the negative field of the cycle that problems of resistance, tension, and elimination dam the flow emanating from the center. A key to Polarity Therapy is working with physical and emotional elimination in the body/mind. A major focus is working with the negative field of the perineum and pelvis to release chronic tension and unexpressed emotions in order to restore healthy elimination.

Polarity Therapy offers a broad range of techniques that focus on balancing the Water element. These include the simple, easy-to-follow protocols in the table, Techniques to Balance the Water Element.

Balancing the Earth Element

Earth rules completion, boundaries, and form in the body. The positive pole of the Earth triad is Taurus, the neck, which rules the etheric body and its precipitation into the overall boundary of the torso. Virgo, the colon, rules discrimination and the boundary between body nourishment and waste, taking in and letting go. Capricorn, the knees, rules the form of the skeleton on which everything else hangs, as well as issues of emotional foundations.

Earth relates to boundary issues, self-protection, and survival. Excess Earth manifests as limited imagination and intuition, resistance and defensiveness, inertia, fear,invulnerability, paranoia, being closed-down, untrusting, and gripping the Earth. Contracted Earth

Techniques to Balance the Earth Element	
Earth triad balance	Deep perineal therapy
Release coccyx	Perineal therapy
Balance sacrum	Pelvic session
Balance sacrum	Sacrum session
Neck rigidity	Spine session, core session
Emotional release	Perineal therapy, tamasic contacts
Cleanse colon	Polarity cleansing diet
Easy stretching postures	Squat

manifests as boundary problems: over-sensitivity, self-doubt, or being too trusting, too vulnerable, defensive, or anxious. Balanced Earth is grounded, supporting, stable, down-to-earth, and practical, with clear boundaries, patience, perseverance, and courage. Earth-predominant people are often masters of the physical plane. The inertia of Earth may lead to laziness, an addiction to routine, and a lack of imagination.

Tamas guna, the force of contraction, rules the Earth element. Any form of contraction—physical, mental, or emotional—is Earth-predominant. Inertia, rigidity, withdrawal, and narrow-mindedness are all symptoms of an Earth imbalance. The emotions of fear and courage are ruled by Earth. When boundaries are threatened, fear manifests and first chakra survival responses are stimulated. Chronic fear may keep us out of the body and unable to be present and grounded. "The Earth element governs fear responses and a common fear response is shaky knees, uncontrollable bowels and a stiff rigid neck. If you have ever experienced a major earthquake, where the ground is literally taken out from under you, you may have experienced some of these Earth reactions."[3] In daily life, fear may manifest bodily symptoms through a spastic colon, colitis, constipation, diarrhea, rigidity in the neck, and weak, injury-prone knees.

Tamas guna rules the return currents—moving from below to above and without to within—which rule structure in the body and the muscular-skeletal system. These patterns of resonance have their foundation in the heels, sacrum, and occiput. Profound responses often result from energy balancing that works with these principles.

Polarity Therapy offers a broad range of techniques which focus on balancing the Earth element. These include the simple, easy-to-follow protocols in the table, Techniques to Balance the Earth Element.

Character and the Play of the Elements

Rajasic *Overly* *Expanded*	Sattvic *Self* *Actualizing*	Tamasic *Overly* *Contracted*
ETHER		
Egoism	Identity	Grief
False sense of superiority	Self-esteem	Poor self-esteem
Arrogance	Humility	Shame
Tyranny	Psychic space	Victimization
AIR		
Lust	Contentment	Discontented
Greed	Equanimity	Aversion
Pumped up	Centered	Deflated
Unrealistically optimistic	Honesty	Unrealistically pessimistic
Creating illusions	Realistic	Hopeless
Speedy	Moderate	Depressed/immobile
Impatient	Charitability	Jealous
FIRE		
Raging at others	Respectful	Raging at self
Blaming/shaming others	Forgiving	Self-blaming
Judging	Insightful	Enabling
WATER		
Isolation	Flowing	Enmeshed
Anti-dependent	Caring	Dependent
Compulsive	Receptivity	Needy
Passionate	Compassionate	Possessive
Charming	Nurturing	Despairing
Sex addicted	Chaste	Love addicted
EARTH		
Walls up	Protective	No boundaries
Defensive	Supporting	Too vulnerable
Paranoid	Grounded	Frozen
Self-concealing	Accepting limits	Over sensitivity
Overly vigilant	Stable/courageous	Anxious
Invulnerability	Self-respecting	Self-doubting

CHAPTER EIGHTEEN

The Ultimate Medicine

On earth there is no purifier as great
as this knowledge ...
—Sri Krishna[1]

In the Sanatana Dharma reality is defined as that which is unchanging. For the rishis "reality" is that which is birthless, deathless—immutable. The sages ask, how can something that is merely a transitory appearance be considered real? Our world, our science, and our minds are a phantasmagoria of change. The sages tell us to let go of this transitory world of appearances and seek the Real.

In the Sanatana Dharma reality is always present. Indeed, in the Sanatana Dharma reality is the *present itself*. The past is now a memory, the future a mere dream. The moment is the only "reality." Everything comes and goes, but the present has always been here. In the Sanatana Dharma the present is the Eternal Present.

Who is present? Sat Purusha is universally present as the one consciousness, the innermost Self of all beings. That which is actually present in each being, which is actually conscious in everyone, the unqualified subject, is the One Self that is Real.

Ken Wilber, writing about Dzogchen, the Tibetan practice of instant enlightenment, explains:

> If Spirit has any meaning, it must be omnipresent, or all pervading and
> all encompassing. There can be no place where Spirit is not, or it wouldn't

be infinite. Therefore, Spirit has to be completely present, right here, right now, in your own awareness. That is, your present awareness, precisely as it is, without changing it or altering it in any way, is perfectly and completely permeated by Spirit . . . according to Dzogchen, you are already one with Spirit, and that awareness is fully present, right now . . . Spirit is already complete and fully present in the state of awareness that you have now . . . The one thing we are always aware of is . . . awareness itself.[2]

The Unitive Field is the consciousness itself of all beings. Each life is a leaf on the tree of eternal life. Each one of us is a flowering of the beauty and creativity of Consciousness into a moment of time and space. Though the leaves come and go with the seasons, the Tree of Consciousness remains as the unchanging medium of creation.

Sri Krishna reveals in *The Bhagavad Gita:* "There was never a time when I did not exist, nor you . . . Nor is there any future in which we shall cease to be." Consciousness itself, that which is actually present in each of us—not our bodies, not our minds, not our emotions, but our Self—is eternal. "That which is the principle of consciousness in all beings, and the source of life in all."[3] is Real. What it is that is actually conscious in each of us is the One Self. "Innermost element, Everywhere, always, Being of Beings, Changeless, Eternal, For Ever and Ever."[4] Each of us is the eminence of The Omnipresence of Consciousness and the personification of the Mystery of Eternal Life.

We again quote the wisdom of *The Ultimate Medicine.* The great twentieth-century sage Sri Nisargadatta (Mr. Natural State) explains:

> . . . treat the life force as God itself, there cannot be consciousness without life force. So consciousness and life force are two components, inextricably woven together, of one principle . . . Once you consider that the life force is God itself, and that no other God exists, then you raise the life force to a status enabling it, together working with consciousness, to give you an understanding of the working of the whole . . .[5]

Each of us is the personification of the leading edge of an ever-evolving universe. Our life is the art and poetry of the Cosmic Drama. We are nothing less than the atoms in the body of an omnipresent consciousness that we call God. The pathos of our condition is unfathomably deep. All of life, literature, and art is a testimony to that.

We are tender, delicate, sentient beings. Each of us is very sensitive, a universe of feeling and passion. Each of us is unfathomably creative in the pathos of our wounds and the psychodynamics of our inner lives. The qualities we bring to life and the challenges we face in life are grist for the mill of cosmic evolution. Our lives are the leading edge of the creative potency of the universe. The Over-Soul, Sat Purusha, is the constant witness as the subjectivity of each individual.

This Mind is present within every cell of our bodies. As previously discussed, when life is physically or emotionally threatening, the cells of our bodies contract in defense of their boundaries. Tension and resistance dam the flow of life force in its step-down from within to without. This energy block is a fixation upon trauma that has us hooked into the past. In an energy block we lose our ability to resonate with the finer forces of higher intelligence.

The body is a mind. Each of the elemental archetypes is an aspect of Immortal Being personified as an aspect of the body. Each archetype is conscious, sentient, present as the omnipresence of the Intelligence which *is* the universe. The Cosmic Person (Sat Purusha) is present as every atom of creation. In the Sanatana Dharma, God as illimitable consciousness *becomes the world*. Spirit is present as the illimitable depth of sentience within each being, indeed, within each atom of being. In the wisdom of the Sanatana Dharma, God lives within you, as you.

Self-Directed Therapy

Transpersonal therapy is self-directed. We respect the profound intelligence that is both the source and constant witness of life. We support individuals in getting in touch with their own inner guidance. We validate and support their capacity to find their own solutions to the challenges to their growth. In a transpersonal perspective we hold a sacred space with a profound respect for clients as sacred beings. We respect the depth of their intelligence as Cosmic Intelligence, the omnipresence of Purusha, the Cosmic Presence. We respect the depth of their ability to direct their own processes of healing. Transpersonal therapy is self-directed therapy. We constantly turn clients toward their own guidance in the therapeutic process. The therapeutic process is sacred; it is the evolution of the Soul and a journey to an inevitable Self-realization.

The concepts we have in our minds, the paradigms we have in books, are merely maps and not reality. In the Sanatana Dharma Consciousness/God/Reality is the "Present" itself. But *Freedom* is veiled by the noise of our minds, a slave to our unceasing desire and the disturbances of our trauma and insatiable insecurity. The sages tell us to be humble, not to try so hard to be somebody, and to give up the self-cherishing which is the root of our ignorance and suffering. They tell us to let go of desires. If we give our lives to this prescription and let go, what is left after all the busy-ness of the mind and insatiability of the sensual gratification has been dropped is the Eternal Present.

The mystics tell us that in every moment we are sustained not by our ego, but by Totality, Dharma, the Eternal Life. They invite us to surrender to a higher Intelligence within our being and to lose our "self" (ego) to find our "Self" (Mystery). The task of the transpersonal facilitator is to support this process of Self-inquiry and Self-realization.

In the Sanatana Dharma all life is a process of becoming, and a microcosm of cosmic evolution. Growth is an organic process, thought-by-thought, breath-by-breath, gesture-by-gesture, throughout a lifetime. Each season offers its challenges. Healing is learning how to take care of this temple of the Spirit. Traditionally this life came with an owner's manual called "culture." With the breakdown of family and society, and the overthrow of tradition by godless scientism, people are left to their own wits in a chaos of possibilities. The Sanatana Dharma offers a manual for life, a way of life that, when practiced, tunes us into a higher level of presence, intelligence, and guidance in our lives.

The Healing Power of Self-Knowledge: Suffering and Ignorance

India offers a culture whose constant focus is a concern with the deepest questions of our human ignorance and the path to salvation. The most conspicuous and remarkable quality of its way of life is the centrality of its focus on liberation. As making money is the central focus of American culture, *liberation* is the constant theme of the traditional culture of India. Professor Coomaraswamy asserts: "The heart and essence of the Indian experience is to be found in a constant intuition of the unity of all life, and the instinctive and ineradicable conviction that the recognition of this unity is the highest good and uttermost freedom."[6] As Atman is the source of life and consciousness within each being, Atman is at the heart of the Sanatana Dharma as well.

In the West, before industrialization, God was at the hub of traditional culture. It is valuable to understand that in traditional India it is not *Brahman* (God) that is the focal point of the world view but *Atman*. Atman, the portion of the divine supporting individual existence, is one with Brahman. It is the unity of all life and consciousness and the possibility of liberation that is the constant cultural theme. The individual's Self is one with the Great Mystery of life and consciousness. S. Radhakrishnan, describing the "Life Eternal," writes: "The fact that the individual consciousness has for its essential reality the Universal Self implies the possibility that every human being can rend the veil of separateness and gain recognition of his true nature and oneness with all beings."[7] In the West, philosophy is seen as academic and irrelevant to the life of the common man. India is a nation of philosophers. People live and breathe a cosmology whose constant theme offers a pathway to the realization of the meaning and ultimate purpose of life. "In India ... philosophy is not regarded as mental gymnastics, but rather with deep religious conviction, as our salvation (*moksha*) from the ignorance *(avidya)* which forever hides from our eyes the vision of reality."[8]

Bharat is the name Indians give to their own heritage. The heroes of the culture of Bharat are not merely kings and warriors, as in the West. When kings are remembered, it is because they have attained the highest good and have served a realization of the unity of all life. When warriors are remembered, it is because they have conquered illusion. The history of Bharat is a history of sages and saints, those who have conquered ignorance and realized *Reality.*

Indian history is not linear and serial as in the West. Names, dates, and dynasties are left behind in the rubble of history. What is cleaved to as history is a remembrance of the epic and inspiring teachings which offer the possibility of liberation. Typically more important than the names of the sages is the wisdom that they realized and shared.

The Sanatana Dharma, the revelation of natural law which is at the heart of the culture of Bharat, offers a constant vision of the sacredness and unity of all life. The wisdom of the Sanatana Dharma represents the largest and most coherent corpus of the ancient wisdom available to us today. These teachings—the *Vedas, Upanishads,* and *Bhagavad Gita*—have been chanted daily in an unbroken succession of generations for millennia.

In the Sanatana Dharma, knowledge is not pursued as "philosophy," the love of learning, but to alleviate suffering. The purpose of Brahmavidya, ultimate truth, is to address the root cause of suffering ... ignorance.[9]

In the Sanatana Dharma, Purusha (Spirit) manifests the body and personality as part of cosmic evolution, expressing the infinite play of the creativity of consciousness. Eternity masks itself in the persona of time and Spirit is imprisoned in the body, bound by ignorance and a slave to insatiable desire. In the cosmology of the Sanatana Dharma, *desire* is the cause of the world. God, as illimitable consciousness, manifests creation, nature, and life itself as a stage to fulfill our desires. Our world comes into existence through the strength and avarice of our desires. Our little worlds that make up this universe are ultimately not made up of substance, but of desires.[10]

Our attachment to our desires and our identification with our persona blind us to our true Self. *All of our problems are caused by not knowing who we are.*

The insecurity of our persona, the hell of our fears, the pain of our wounds, the confusion of our minds, the insatiability of our appetites, the greed of our egos, our emptiness and longing all are rooted in *ignorance* of our true nature.

Identification Is the Essence of Ignorance

All of our problems are rooted in *ignorance*. All of our problems flow from not knowing who we are. Twenty-six hundred years ago, the scion of the Sanatana Dharma, Siddhartha the Buddha, said: "I have come to teach one thing ... the alleviation of suffering." The first truth that the Buddha taught is that *all life is suffering*. The Sanskrit term *duhkha* (*dur*, "bad," and *kha*, "state"), which is usually translated as suffering, does not merely mean the suffering of pain, sorrow, disease, poverty, and war. It is translated variously across the Buddhist literature as unhappiness, alienation, disenchantment, and incompleteness. Five thousand years ago people were suffering with the same existential problems of alienation and ontological insecurity that plague us today. The greed, fear, and insecurity of the ego are the timeless subjects of Buddhist psychology. The Sanatana Dharma offers us a psychology whose goal is *nirvana*—liberation, the extinguishing of the painful illusion of individuality into the mystery of *sunyata*, the unbounded.

In Buddhist psychology all life is suffering, essentially because of our attachments. Attachment in life is *the cause* of suffering. Attachment has to do with both desires and aversions. The deepest attachment is to what is called "self-cherishing," our identification with the "false self" of our individual persona. This is the deepest attachment and the root of suffering. Sogyal Rinpoche, writing in *The Tibetan Book of Living and Dying*, explains:

> Self-grasping and self-cherishing are seen, when you really look at them, to be the root of all harm to others, and also of all harm to ourselves. Every single negative thing we have ever thought or done has ultimately arisen from our grasping at a false self, and our cherishing of that false self, making it the dearest most important element in our lives. All those negative thoughts, emotions, desires, and actions that are the cause of our negative karma are engendered by self-grasping and self-cherishing.[11]

Ontological Insecurity Is the Cause of Suffering

The all-consuming insecurity of the ego—the obsession of self-importance, the unending personal drama of the inner life, the incessant noise of the mind—is the fundamental suffering which the Buddha addresses. The malaise of the human condition is ontological insecurity, where we live in a subtle hell of constant insecurity. Like a house built on a poor foundation, the ego, built on the foundation of ignorance, is fundamentally insecure, always threatened, always shaky. The Sanatana Dharma asserts that ignorance is the cause of this suffering.

Identification is the essence of ignorance. Most people are totally identified with their every thought and every emotion: they think that they are each thought and feeling. Nowhere in Western psychology have we been taught to realize that we are the pure subjectivity that is the witness to the thought process, or that a thought or a feeling is merely input. Nowhere in traditional psychology have we been taught not to identify with our thoughts. Indeed our psychotherapists makes a fetish of every thought and feeling, ignorant of that peace which is beyond the capricious mind and impetuous senses.

Most individuals identify with every thought and feeling and suffer in a blind attachment to this identification. Ken Wilber explains: "Anytime you identify with a problem, an anxiety, a mental state, a memory, a desire, a bodily sensation, or emotion, you are throwing yourself into bondage, limitation, fear, constriction, and ultimately death."[12] In incessant obsession, our ego-minds whine, "I am lonely," "I am angry," "I am fat," "I am a victim of—" Our emotions blindly follow our thoughts. Each of us is deeply identified with the story that we tell ourselves. Our minds are fixated upon our personal dramas. Incessantly our minds chant the liturgy of our desires and fears. Our ego-minds are consumed by our childhood wounds, insatiably searching for a balm for our traumas and insecurities. We are possessed by the demon of fear, and obsessed with our vulnerability and insecurity. We are totally identified with the persona's drama, with little personal knowledge of our deeper natures.

We are slaves to our desires. Seeking union outside ourselves, we long for another who will love us enough to heal the wounds of our insecurity, shame, and self-hatred. We yearn for a relationship that will be a balm to the emptiness of our lives. We distract ourselves from our emptiness and self-hatred, our longing and ontological insecurity, with food, drugs, sex, fantasy, sports, business, and consumerism—all in a myriad of unending desires.

Shame as Ontological Insecurity

In contemporary America we translate our experience of *"duhkha"* as *shame,* the inability to feel good about oneself. For many, this shame manifests as a frightening emptiness. We have created a culture whose religion is distraction from this emptiness. The conventional wisdom, politics, art, entertainment, and science all take place within a superficial arena of mere information that is irrelevant to the deeper questions of life. In the traditional society of the Sanatana Dharma, the deepest meaning was woven into the very fabric of life. For example, the greeting of Hinduism, *namaskar* or *namaste*, "Honor unto thee," is a salutation to the divine within, a constant reminder of sacred Self remembering and the divinity within all beings. The prayer over food has a similar message of ultimate truth. "The act of offering is Brahman. The offering is Brahman. Into the fire (of digestion) which is Brahman offering is made by him who is Brahman. By him alone who is absorbed in the offering to Brahman is unity with Brahman attained."[13]

I can remember being terrified, in graduate school, hearing that when you peeled off the layers of the onion of personality *there was nothing there.* Though I had an insatiable appetite for the wisdom of the day in psychology and sociology and was an avid student of the literature on alienation, nothing in my experience in Western culture had prepared me to confront the fear and loathing of my own emptiness.

The sages of the East peel the same onion of the persona, and find the same emptiness. But the mystics explain that Emptiness is the container for everything. What else is vast enough? Emptiness is Freedom itself. Emptiness is the quality-less absolute of the present, which is the unchanging background, the witness to the play of consciousness. Emptiness, unqualified consciousness, is the omnipresence of the Creator. Rather than tremble at the vastness and its challenge to the deepest doubts of our ontological insecurity, the sages invite us to leap into totality, to lose ourselves/find Ourselves in Everything.

Everything and Nothing

In the *dharmakaya,* a holographic universe whose center is everywhere, it is our ontological task to individuate, to define our being, to find our center. In the deepest sense, it is our profound challenge to individuate when we are at the same time everything and nothing.

Our ego-mind emerges from the void. It volunteers to be the center of our personality. It assures us that it can function as the center and executive of the personality. Facile, it attempts to define our relationship to ourself and the world, yet at every moment it trembles with insecurity at its task: to maintain the illusion of separateness.

Who am I? is the question that you have been asked to contemplate, again and again through this essay. *Am I this body,* a body which in one season is an embryo, then an infant, toddler, youth, adult, elder? Which one of these bodies am I? How can I be this body which is always changing? *Am I this Mind?* The mind is changing faster than the body. Which thought, which part of this capricious mind am I? Am I these feelings, which I am so deeply attached to? My insecurity, fears, hurt, loneliness, confusion, ecstasy? *Who am I?* Am I my beauty, my strength, my knowledge, my car, my job, my family, my name?

Ultimately we all must face our ignorance. Nothing our parents taught us, nothing we learned in school, nothing we saw on television, nothing in the phantasmagoria of consumer society has prepared us to address this question. *This is ignorance.*

The sages tell us that when we *earnestly* contemplate this question, with a burning desire to know ... *Who am I?* ... the ignorance will fall away and we will directly experience our essential nature.

In the Sanatana Dharma an omnipresent God becomes the world. The world is not merely the creation of Divinity but is the Divine Itself. Purusha (Spirit) is profoundly eminent. In the Sanatana Dharma, *Vasudeva*, the all-pervasive Spirit, is the innermost Self of all that is. Man, nature, and the cosmos are emanations of this omnipresent being. An eternal portion of Divinity becomes the Life of the living world. The Life in each being is the Living Mystery of Eternal Life.

The challenge of individuation, the process of defining our individuality, is to integrate the transpersonal into the personal. Self-actualization takes on a deeper perspective in a transpersonal approach when the self we are seeking to actualize is the Self of all being. Ken Wilber explains:

> The discovery, in one form or another, of this transcendent self is the major aim of all transpersonal band therapies and disciplines ... it is a center and expanse of awareness which is creatively detached from one's personal mind, body, emotions, thoughts and feelings. I am what remains, a pure center of awareness, an unmoved witness of all these thoughts, emotions, feelings, and desires.[14]

The sage S. Radhakrishnan asserts: "When the Supreme is seen, the knots of the heart are cut asunder, the doubts of the intellect are dispelled and the effects of our actions are destroyed. There can be no sorrow or pain or fear when there is no other. The freed soul is like a blind man who has gained his sight, a sick man made whole. He cannot have any doubt for he is full and abiding knowledge."[15]

Self-realization has profound implications for psychotherapy, as Wilber explains:

> To the extent that you realize that you are not, for example, your anx-
> ieties, then your anxieties no longer threaten you. Even if anxiety is pre-
> sent, it no longer overwhelms you because you are no longer exclusively
> tied to it. You are no longer courting it, fighting it, resisting it, or running
> from it. In the most radical fashion, anxiety is throughly accepted as it is
> and allowed to move as it will. You have nothing to lose, nothing to gain
> by its presence or absence, for you are simply watching it pass by. Thus,
> any emotion, sensation, thought, memory, or experience that disturbs you
> is simply one with which you have exclusively identified yourself, and the
> ultimate resolution of the disturbance is simply to disidentify with it.[16]

As you realize your true Self, you are liberated from the obsessive suffering of the persona's drama. The incessant insecurity drops away and a
knowing strength emerges from Self-knowledge.

Brahmavidya: Ultimate Truth

Brahmavidya, ultimate truth, is the healing balm of a transpersonal perspective for the suffering of our ignorance. If attachment is the cause of suffering,
detachment is the stuff of liberation. Mathaji Vanamali explains:

> The highest form of detachment comes from knowledge, knowledge of
> the oneness of reality, that exists equally between subject and object ...
> detachment ... flowers from the recognition of the omnipresence of the
> supreme person ... which at once, deals a death blow to all desires. Just
> as, when we wake up from the dream, the desires of the dream fall away
> by themselves. So also, the problems of life melt away in the presence of
> God, who is the reality of all existence. There is an automatic rising of the
> Soul, to an awareness where desires lose their significance altogether.[17]

Sri. Radhakrishnan concurs: "He who knows himself to be all can have
no desire."[18]

Sri Krishna: The Father of Transpersonal Psychology

The Bhagavad Gita is a treatise in transpersonal psychology. It offers a step-
by-step methodology for healing painful ignorance and identification with the
limited ego-mind and emotions.

The teaching of *The Bhagavad Gita* is given to the warrior Arjuna in
response to an emotional breakdown from an ethical crisis as he is going into

battle. Arjuna cries out: "My limbs are weakened, my mouth is parching, my body trembles, my hair stands upright, my skin seems burning, the bow ... slips from my hand, my brain is whirling round and round, I can stand no longer .." Torn by a crisis in Dharma, Arjuna, leader of the forces of right-eousness, continues: "What can we hope from this killing of kinsman? ... Evil they may be, worst of the wicked, yet if we kill then our sin is greater. What is the crime I am planning O Krishna? Murder of brothers! Am I indeed so greedy for greatness? Rather than this, let the evil ... come with their weapons against me in battle: I shall not struggle, I shall not strike them. Now let them kill me, that will be better."[19]

Lord Krishna offers the teaching of *The Bhagavad Gita* as a transpersonal psychologist. He offers this profound knowledge as a healing balm for the ter-ror and insecurity of the ego. Mahatma Gandhi writes: "When I first became acquainted with the Gita, I felt that it was not a historical work, but that, under the guise of physical warfare, it described the duel that perpetually went on in the hearts of mankind .."[20] This is no other-worldly teaching but is for every one of us who cringes facing the day-to-day crises on the battlefield of life. Arjuna's despair as an actor on the stage of a drama with no winners parallels the despondency we face today, some five thousand years later.

The transpersonal psychology of *The Bhagavad Gita* offers nothing less than nirvana, the spiritual disillusion of the separate individual self. Realiza-tion of the Eternal Truth of the Sanatana Dharma is *Brahma Nirvana,* a state of extinction of all ego limitations in a state of at-one-ness with Totality. This understanding is Self-realization. Sri Krishna reveals: "If a yogi has perfect control over his mind, and struggles continually in this way to unite himself with Brahman, he will come at last to the crowning peace of Nirvana, the peace that is in me."[21]

Brahma Nirvana as explained in *The Bhagavad Gita,* is a state of Self-realization that can be practiced in any station in life. In the West, nirvana is usually understood as an egoless, other-worldly state that can be practiced only in a Himalayan cave. Sri Krishna explains that it is the rare soul who can truly renounce the world. He asserts that to think you have renounced the world when you sit in a cave dreaming of past hurts or future glories is mere hypocrisy. Again and again, throughout the eighteen chapters of this great treatise on yoga, beloved Krishna contrasts other-worldly renunciation with life in the world. Krishna's injunction is not to an other-worldly nirvana. Lord Krishna's teaching is that through dedicated, unselfish service, every action in life can be yoga. The consciousness of the omnipresence of the One dissolves the painful illusion of separateness: all action becomes divine action.

Indeed, he councils Arjuna to do his duty as a warrior and hurl himself into the battle as yoga and his path to liberation.

Krishna clarifies again and again throughout the gospel that in karma—the law of cause and effect—the cause is our consciousness. Whether an action binds us in suffering or is a step to liberation has not to do with the outward effect of action but the cause in our intention. Whether an action leads to bondage or freedom has to do with the level of knowledge or ignorance that lies behind it. The cause of action (karma) is in the consciousness that motivates that action. The karma from the action of a knife cutting into flesh is different for a life-saving surgeon than for an angry murderer. The karma of an action is a product of the knowledge motivating the action.

The sage Mathaji Vanamali explains:

> The force that causes the emanation of things is said to be Karma. It is a power of the Lord, which ejects all particulars. All the little individual karmas, or actions that we perform here, are a reverberation of this cosmic impulse, for the sake of the great universal purpose. This is the secret of karma, all action is, in the end, a universal action. It is not my action or your action—every ripple on the crest of the wave is nothing but the work of the ocean, though the wave may claim it to be its own particular handiwork. Every breath that we breathe is nothing but the cosmic breath. Our intelligence is a faint reflection of the cosmic Intelligence. Our very existence is part and parcel of the universal existence. This is the great secret behind the Law of Karma and only by thoroughly assimilating this knowledge can we become karma yogis. Only when we consciously and willingly accept the fact that we are but mere instruments in the movement of the great universal Karma can we give up our egos and our personal desires and thus become fit to be called karmayogis.[22]

Sri Krishna explains that the profound despondency of our worldly dilemma is a painful fiction rooted in our ignorance and identification with this separate ego. To liberate us from the fear and insecurity of the tiny ego he offers nothing less than Self-knowledge, the teaching of the illusion of our separateness and the immortality of the soul. Mathaji Vanamali continues:

> To move towards God and to feel an aspiration for God is the birth of wisdom. Everything else is ignorance ... *Jnana* [knowledge] is knowledge of the Self ... the nature of which ... Brahman is that which is both transcendent and universal. In its supercosmic state, it is transcendent ... bodiless, yet a million-bodied Spirit, with hands of strength and feet of swiftness, whose heads and eyes and faces we see wherever we turn, whose ear is listening to the silence of eternity, as well as to the music of the world.

This is the universal being in whose embrace we all live. It is the perceiver of all sense objects, but has no sense organ, without attachments and without attributes, yet the enjoyer and sustainer of all attributes. Attached to nothing, it supports in immortal freedom all the action and movement of its universal prakriti [nature]. He is at once the inner and the outer, the far and the near, the moving and the unmoving. He is the subtlest of all subtleties which is beyond our grasp, as well as the density of force and substance which we can see. Though indivisible he seems to divide himself in innumerable forms. All is eternally born from him as Brahma, sustained in his eternity as Vishnu, and taken eternally back into him as Shiva. He is the light of all lights, the luminous beyond all the darkness of our ignorance. He is both knowledge, as well as the object of knowledge. One does not have to go to universities to seek this knowledge, for He is seated in the hearts of all.[23]

Action based on the selfish attachments of the separate ego binds us in ignorance. So long as this deluded separate ego thinks it is the doer, it is bound in ignorance. Unselfish deeds offered in a spirit of loving service are liberating. To the extent that the knowledge of the One Life permeates the action, one is free from the illusion of doership. "When your intellect has cleared itself of delusions, you will become indifferent to the results of all action . . . To unite the heart with Brahman and then to act: that is the secret of non-attached work."[24] To realize the unity and sacredness of all life is to be released from the ontological insecurity, the shame, suffering, and selfish grasping of the separate ego. Sri Krishna explains that unselfish service based on unitive consciousness, without attachment to the outcome, is the key to freedom.

Throughout the *Gita* there is a constant emphasis on cultivating equanimity as the foundation of yoga. It recommends the practice of discrimination (*viveka*), a constant vigilance to ascertain the Real from the unreal, as a key to liberation. Yoga is nothing less than union with the Self. Sri Krishna explains: "In this yoga, the will is directed singly toward one ideal . . . [it requires] concentration of the will which leads to absorption in God . . . Perform every action with your heart fixed on The Supreme Lord. Renounce [your] attachment to the fruits . . . Poise your mind in tranquility . . . for it is this evenness of temper which is meant by yoga."[25]

In the *Gita* there is a constant emphasis on cultivating detachment. Sri Krishna tells us to cultivate dispassion as we break through the fetters of desire, aversion, and identification with the ego, which bind us to the endless suffering of our lives. The Preceptor of The Universe teaches that to serve in the world without attachment to the outcome of our actions is the key to the breaking through the attachments which bind us. The *Gita* counsels us to give up doership—the identification with the limited ego—and to leap into an awareness that our separateness is a painful illusion. Its constant teaching is transpersonal, the Supreme Knowledge that you are not the body, mind, or emotions but the Atman, unborn, undying, never ceasing, never beginning, deathless, birthless, unchanging for ever, innermost element, everywhere, always, Being of beings, changeless, eternal, for ever and ever.[26]

The Dharma of Loving Service

Dharma in the Indian conception is not merely
the good, the right, morality and justice, ethics;
it is the whole government of all the relations
of man with other beings, with Nature, with God.
Dharma is both that which we hold to and that
which holds together our inner and outer
activities, and in this its primary sense it means
a fundamental law of our nature which
secretly conditions all of our activities . . .
Dharma is all that helps us to grow into divine
purity, largeness, light, freedom, strength,
joy, love, good, unity and beauty.
—*Sri Aurobindo*[27]

Sanatana Dharma is often translated as "Eternal Truth," or "Natural Law." It prescribes a way of living that if upheld by humanity, will uphold humanity. Unlike Western science, which claims to stand outside of questions of values and ethics, the Sanatana Dharma prescribes a way of life. The Sanatana Dharma invites us to live our lives as though the God within each being mattered. As a timeless statement of the unity and sacredness of all being the Sanatana Dharma invites us to realize the sacredness of our own life, the life of all humanity, and the life of all creation. In the Sanatana Dharma, "Truth" is not a concept, but a way of life. Dharma can only be practiced, it can only be lived.

A transpersonal approach to psychology is holistic. We are made out of wholeness; all our boundaries are Janus-faced. We are a part of everything, and everything is a part of us. Indeed we are wholeness itself. In Ayurveda everything that touches us is food. It all becomes a part of us. Our attitude, our lifestyle, our environment, our friends, our ancestors, our stars—all contribute to who we are.

The Sanatana Dharma invites us to practice Dharma and to live our lives as though the God within each being mattered. It is naive to think that we can ignore universal law and be happy or healthy. Health and happiness flower from Dharma, the practice of upholding natural law. The more holistic our approach, the better its chance for promoting physical and emotional health. The fruit of Dharma is yoga, union with the Self. (See Appendix D for a discussion of Polarity Therapy as right livelihood.)

Yoga

In a transpersonal perspective we support the client in a process of *Yoga*. The word *yoga* comes from roots which mean to "yoke" or "unite." Yoga can also be translated as "integration." Yoga is the timeless science of integrating the persona and transpersonal Self. Yoga is a science because its paradigm has been clearly thought out, systematically presented, and continuously replicated for thousands of years. Yoga *is* transpersonal/somatic psychology.

There are many schools of yoga. They all focus on a common theme: "From the unreal, lead me to the real. From shadows lead me to Light. From death to immortality ... "[28] The purpose of yoga is the integration of the personal self into the infinite Self. Since divine Intelligence is omnipresent, this union can be approached from *any* direction.

Hatha yoga facilitates health practices that promote resonance with the finer forces of higher intelligence to unite the individual life with the Eternal Life. *Jnana yoga* is the way of knowledge, a pathway of the most earnest searching and contemplation to elevate one's personal consciousness to a transpersonal state of Self-realization. *Raja yoga* offers a system for stilling the workings of the mind to unite with that which is beyond the mind. *Bhakti Yoga* offers a devotional psychology and practices to facilitate integration of the personal with the Mystery through rapture of ecstatic Soul communion. *Karma yoga* is a pathway to the transpersonal, where the sacredness of life is experienced through dedicated unselfish service.

The truth of yoga can only be lived. The laboratory for the science of yoga is life itself. Its truth cannot be found in a book or learned in universities. It is an inner experiment seeking to replicate a timeless truth. What our heart and mind dwells upon, we become. Yoga must be lived in the most earnest and devoted way—breath-by-breath, thought-by-thought, day-by-day, for decades—as a way of life called Dharma. The fruit of practicing Dharma is the moment-to-moment experience of Union.

Truth Is God

Mahatma Gandhi was a devoted student of the Sanatana Dharma, and the *Bhagavad Gita* was his gospel. His life exemplifies the power and possibilities of practicing dharma. Experiencing the injustice of racial prejudice, Gandhi saw that moral and political injustice were based on ignorance. He became convinced that it is not enough for the individual merely to do good, but that we had a duty to actively face evil.

Gandhi hurled himself into the battlefield of life. He taught *satyagraha*, "clinging to Truth." Truth was Ghandi's method and Truth his goal. For Ghandi, Truth was not a concept, but was a way of life. For the *Mahatma* (one in touch with the vastness of Being), the means were as important as the end. His every action was based on *Brahmavidya*, the knowledge of the oneness of all life. In a state of consciousness of *Brahma Nirvana*, a realization of the unity of all life, Gandhi embraced each being as his own God Self.

Gandhi taught by example. He dealt with all beings in a spirit of humility with the deepest respect for their Divinity. He insisted on seeing the divine in his most ignorant adversaries. The Truth of the *knowledge* behind his action was a profound experience for everyone whose life was touched by this Self-realized being.

He described his life as "my experiments with truth." As he matured in this experiment, he came to the realization that Truth is God. In the practice of his life he experienced that as he cared for the life around him in a "truthful," unselfish way, he experienced the sacredness of all life. As he cared for people and nature in a nonviolent *(Ahimsa)*, unexploitive way that honored the divinity of all life, he experienced a life where every relationship was service to God. Through practicing Dharma, his every action was an experience of the sacredness and unity of all life.

Namaskar

Honor to the Divine within

PART IV

RESOURCES FOR THE

PRACTICE OF

POLARITY THERAPY

APPENDIX A

Conscious Sound Vibration as a Healing Technique

In the ancient wisdom of the Sanatana Dharma, the medium of creation is conscious sound vibration. According to the *Vedas,* the universe was sung into being. The Logos or the Divine Word became all the worlds. The power of Parabrahman the Supreme Being is Sabda-Brahman the Divine as primal sound energy.

Sanskrit, which means "perfected, well-made, polished," is an artificial language, "patiently refined sound by sound, bearing in all its details the imprint of conscious work, constructed on the very principles of thought, of creation, in a fashion very similar to mathematics but more flexible and wide ranging in its applications."[1] The Sanskrit language "is scientifically formed and co-ordinates with universal basic principles that build and unfold all manifested things. Its very alphabet is a mantra ... revealing the song that was sounded in space when the world sprang into being. The Sanskrit language is constructed in harmonious relation with the very truths of existence ... "[2]

Mantra, which means "mind instrument," is psychoactive. Mantra is the scientific use of sound to manifest states of consciousness. The rules of Sanskrit parallel the rules of manifestation. In Sanskrit, "the word is not just a sound arbitrarily connected with an object or event, but is essentially a voice, a force producing an effect directly on the substance of being. It possesses to the utmost the power of any true and genuine poetry or music, to create a resonance in the subtle substance of being and to bring about in the listener a fine attunement to the experience of the seer, poet or composer."[3]

"The essence of a mantra is to concentrate and vitalize will power."[4] We can consciously use the psychically potent sound syllables of mantra to influence the human system. Tantric Guru Harish Johari writes: "According to all spiritual sciences, first God created sound, and from these sound frequencies came the phenomenal world. Our total existence is constituted by this sound, which becomes mantra when organized by a desire to communicate, manifest, invoke or materialize, excite or energize manifested or unmanifested energy."[5] The British jurist, Tantric scholar, and yogi Sir John Woodroffe explains: "The essence of all this is ... to concentrate and vitalize thought and will power ... Mantra, in short is a power (Sakti); power in the form of Sound. The root man means to think. Like the Greek word Logos, mantra means thought and word combined as an expression of creative will."[6]

Woodroffe explains the dynamics of the continuity of thought and sound in the Tantric model. Thought and sound manifest in four states with sound at one end and thought at the other. *Para* is the highest state. Here sound is transcendental. It has no particular wavelength and is above name and form. It is the unchanging primeval stratum of all language. It is prime energy or shakti. In its undifferentiated stage it corresponds to Shabdabrahma, the divine vibration that unites all. The second stage is *pasyanti* or visible sound. It is the telepathic state in which one can almost see the form of the thought. It is the universal level on which all thought takes place. An Indian, a German, and an Eskimo can look at a flower and experience the same thought on a nonverbal level. The third is *madhyama* in which the speaker selects his words through a prism of mental conceptions. The fourth state is *Vaikhari,* the spoken word, the most concrete state of thought. This is thought translated into the coded state called language. Here the thought implies both name and form.[7]

Seal parallels Woodroffe's analysis in his discussion of the model of sound offered in the Mimamsakas paradigm, defining Sphota or "significant sound" as "transcendental" or "intelligible" sound representing the Platonic ideas or logoi, which are eternal, ubiquitous, and noumenal.[8]

Woodroffe goes to great length to emphasize that the bija mantras of the sanskrit alphabet are the archetypal seeds, the actual mechanisms in the One Mind that underlie manifestation. "It is true and certain that supreme Kulakundalini Herself, who is the fifty letters, from A to Ksa, has given birth to this entire universe, consisting of moving and nonmoving things ... in the eternal body of the Devi. Mantras also, in the form of letters, are eternal Brahman, full of energy, and aspects of herself. Mantras which are seeds, out of which grow the fruit of the universe are eternally present in Her body. For this reason they are called Bijamantras [seed-mantras]."[9]

Each chakra is the center of an archetypal seed or bija. Bija mantras are said to be more powerful than other mantras. They parallel the tattwas or essential forces, which are the archetypal principles in consciousness, the seeds of manifestation.

Toning

Dr. Stone always urged us to work with subtle energy principles emphasizing that the without is an emanation from within. He emphasized that the real cause of disease was imbalances in the tattwas, the inner essences of subtle energy. In toning we use the mantric principles of concentration and visualization outlined above to materialize our healing intention. Sound is the motor activity of Ether. We can consciously use sound which resonates with the etheric body to influence the subtle energy fields of the client. It is at the throat chakra that intention steps down into the etheric body for coordinated expression in the physical body. It is at the throat that thought and intention are expressed into physicality. Our healing intention resonates with the manomayakosa, "the sheath built of mind," to influence the client's healing.

Speaking is a sacred microcosm of the primordial act of creation. In the perspective of the ancients, sound is the medium of creation. Through sound we focus intention into manifestation. Breath is the psychophysical nexus. Breath is the vehicle through which thought and intention are expressed as sound. Breath as the carrier of prana, the intelligent life force, is the link between mind and body. The focused breath resonates with the pranamayakosa, the vital body of the client, to influence the nervous system and vitality. Our lips are Shiva and Shakti. Where Shakti centrifugal and Shiva centripetal radiate, a vortex

Mantras to Balance Tattwas[10]

	Tantric[11]	Ayurveda[12]
Unitive Field	Aum (pronounced "om")	Aum (pronounced "om")
Sun	Aung	Sum ("soom")
Moon	Mang	Som (rhymes with "om")
Ether	Hang	Ham (short mantra, *a* sounds like vowel in "but")
Air	Yang	Yam (short mantra, *a* sounds like vowel in "but")
Fire	Rang	Ram (short mantra, *a* sounds like vowel in "but")
Water	Vang	Vam (short mantra, *a* sounds like vowel in "but")
Earth	Lung	Lam (short mantra, *a* sounds like vowel in "but")
Mars		Am (pronounced "um")
Mercury		Bum (*u* as in "put")
Jupiter		Gum (*u* as in "put")
Venus		Shum (*u* as in "put")
Saturn		Sham (pronounced "shum")

is sustained. This vortex of life breath is the medium of creation. As we speak, tone, and breathe, centrifugal and centripetal vortices of prana carry our thoughts and intention into form. Our utterance is a microcosm of the sacred word, the primordial act of creation.

A simple method of toning is to focus the tone "Ah" and diamond white light into the energy block as harmoniously as possible. Blend client and practitioner's tones, climbing the inner octaves of vibration (Jacob's Ladder), toward the Unitive Field. You will be amazed at how tension melts, energy blocks dissolve, and pulses balance using these subtle energy techniques.

We can use golden or diamond white light in the visualization or we can use color breathing. Color is a pure form of vibration. Color breathing and visualization is a way of focusing consciousness and a healing intention toward the client. In the Tantric tradition, the color yellow resonates with the Earth, white with Water, red with Fire, green with Air, violet with Ether. Because the lower chakras are blessed by the cosmic guru Brihaspati (Jupiter), his fire of wisdom predominates in red, orange, and gold, radiating from the first three chakras, which are driven by the evolutionary return currents.

For example, if there is a Water imbalance or if cleansing is needed, we would tone using the bija "vang" or "vam," picturing diamond white light flooding the pelvis. For an Earth imbalance, we would visualize red, yellow, or golden light focused on the colon or knees as we tone "lung "or "lam." For a Fire imbalance or to stimulate vitality, we would tone using the sound "rang" or "ram" and visualize red or golden light flooding the solar plexus. For an Air imbalance, we would tone to "yang" or "yam," focusing on the thoracic cavity and visualizing the color green. For an Ether imbalance, we would focus on the throat with the sound "hang" or "ham," visualizing the color violet.

Bija Mantras for Health[13]

Aim (pronounced "aym")	Best for mind. Improves concentration, thinking, rational powers and speech.
Shrim (pronounced "shreem")	Promotes general health, beauty, creativity, and prosperity.
Ram	Promotes strength, calm, rest, and peace Good for excess Air and mental disorders.
Hum (*u* as in put)	Wards off negative influences. Promotes digestive Fire.
Krim (pronounced "kreem")	Improves the capacity for work and gives power and efficacy to action.
Klim (pronounced "kleem")	Gives strength, control of emotions, and sexual vitality.
Sham (pronounced "shum")	Promotes peace, detachment, and contentment Good for mental problems.
Hrim (pronounced "hreem")	After toning, promotes cleansing, purification, energy, joy, and ecstasy. Aid to detoxification.

It is valuable to be mindful that the elements are aspects of Cosmic Intelligence. We can reverently invoke (invite) these transcendental beings to bring their blessings to the healing process as devas (lights of Cosmic Intelligence) and angels or grandmothers and grandfathers. It is powerful to consciously invite the Angel of Earth or Grandmother Water or Father Sky (Air) to bless the healing process through color toning.

Many practitioners work with overtones and are guided to make remarkable sounds. There is no musical instrument more perfect than the human voice, no tool more fascinating than our Creative Intelligence. Experiment and be creative. This work represents a leading edge in the exploration of the power of healing intention.

Toning, color breathing, and visualization may simply be metaphor. Metaphor is a finger pointing to the mystery. Metaphor is very powerful in the healing process. Metaphor goes beyond our simplistic images of the real world. Metaphor invokes healing mechanisms beyond the linear mind. It speaks to the limitless Creative Intelligence of being. Each of us is a study in creativity and mystery. Metaphor empowers the mystery in the healing process. Any model or paradigm of healing ultimately is metaphor. Consciousness is illimitable Creative Intelligence and will play by any set of rules you generate. The universe will respond to any etiquette that you evolve in an attentive, compassionate, and caring relationship.

Toe Reading: Every Wrinkle Tells a Story

Stress lines (wrinkles), shape, color, texture, size. Everything about the toes is conscious energy.

Deep "valley" line = blocked energy. Line no "valley" = healing.

Right Foot Relates to Self Expression

Left Foot Relates to Environment

Gap between toes = lack of integration between areas of life expressed in toes. Toes gripping (hammer toe)= Fear, shy, very sensitive. Toe off ground = ungrounded.

A. Accumulation of wealth/possessions
E. Eliminationn
G. Grounding

V. Vision
F. Frustration/ Liver
W. Worry
P. Processing
A. Anger, rage
R. Results

Me. Mental activity
Bs. Broken Spirit/Depression
D. Diaphragm
K. Kidneys/adrenals
Bv. Basic values
Ba. Balance
A. Ambivalence

H. Heartfelt attention from envir...
Bh. Broken heart
Ae. Anger in envir...Sp. Passion
W. Worry   Sf. Sexual frustration
St. Stomach   E. Emotionality in envir...
V. Vision   P. Psoas
A. Ambivalence

N. Nurturing   A. Accumulation
M. Madness in environment C. Concentration
E. Elimination
G. Grounding

N. Nurturing
M. Madness in environment
SP. Passion
SF. Sexual Frustration
E. Emotionality
P. Psoas

Physical Plane — ♉ G, ♍ E, A ♉

Emotions nurturance passion — Pelvis P, ♅ E, Sf, Sp, ♏ M, N ♋

Work and creativity — Solar Plexus ↗, Ae, ♌, W, D F, V ♈

Passion for life — ♒ A, M, Bv K, ≈, Bs H ♊

Higher nature
Responsibility Shoulder tension
Neck tension Stubborn
Spirituality highest ideals

Higher nature
Responsibility Shoulder tension Gap =
Neck tension Stubborn
Spirituality highest ideals

Passion for life — ♒ A, M, Bv, k, Bs Ma ♊

Work and creativity — Solar Plexus R, ↗, Ae, ♌, P W F D, V ♈

Emotions nurturance passion — Pelvis P, ♅ E, Sf, Sp, ♏ M, D N ♋

Physical world — ♉ G, ♍ E, A ♉

Head   O   =

■ **Earth**
♋ Cancer   ♉ Taurus
▲ **Water**   ♍ Virgo
♏ Scorpio   ♑ Capricorn
♓ Pices

▲ **Fire**
♈ Aires
♌ Leo
♐ Sagittarius

= **Air**
♊ Gemini   ♎ Libra
♒ Aquarius

Body Reading: Every Wrinkle Tells A Story

Today there is growing body of research exploring the holographic paradigm.[1] In the wisdom of ancient India we are offered an understanding of creation, nature, and the body as an emanation from a Unitive Field of Ultimate Intelligence. Here we review a timeless paradigm of a holographic universe and—as above, so below—holographic body. We apply these insights in practice as we explore the art and science of toe reading and learn that every wrinkle tells a story.

Toe Reading

One of the most telling systems of body reading is the reading of the toes. The body exists through its attunement with nature and the cosmic forces. The toes are the most negative field of the body, the most negative and thus the most crystallized expression of the body/mind. Patterns of resistance and blocked energy and the fundamental mental and emotional energy patterns that define character are most easily viewed at the feet. The body is a fabric of vibration, and how open or closed a person is to the vibrations that predominate in the immediate environment, nature, and the cosmos is reflected in the feet.

Chronic health conditions first manifest as crystallizations of the more material negative energy fields. Crystallization in the negative fields inhibits the ability of the body to vibrate with the finer forces of nature and the cosmos. Sensitivity and pain at the feet are reflections of problems in corresponding areas of the body. Acute symptoms in body functioning may manifest as pain in the analogous areas of the hands as well. The body is the microcosm of a process where Spirit is expressed in matter, and the feet are the most material expression of spiritual, mental, and emotional patterns.

In toe reading, "every wrinkle tells a story." Every feature of the toes reflects differences of energy in the individual. Every difference of color, size, shape, flow, space, and every pattern of lines, wrinkles, calluses, etc., is a reflection of states of energy in the body as a whole.

Every feature of the toes is energy. The size, shape, and color of each toe and each of the phalanges are a reflection of the flows of energy. The space between the toes, whether a toe points upward or downward, and the way the toes merge, are reflections of energy patterns.

The left foot reflects the left side of the body and our receptivity to our environment in general. The right foot reflects the right side of the body and issues that focus on our self-expression.

The five toes of each foot resonate with the five fields of energy radiating outward from the core of the body: Ether, Air, Fire, Water, and Earth. The big toe resonates with the Ether elemental harmonic and with the ultrasonic core; it reflexes to the head and neck. The second toe resonates with the Air element and predominates in reflexes to the thoracic cavity. The third toe resonates with the Fire element and solar plexus. The fourth toe resonates with the Water element and the pelvis. The little toe resonates with the Earth element and the colon. The toes thus reflect the five elemental archetypes that govern all vibration.

The phalanges of the toes also reflect the pathway of Spirit to matter, of energy moving from above to below. On the big, or Ether, toe the first phalange resonates with the head; the second phalange with the neck; and the first metatarsal resonates with the sternum and chest. On the second, or Air, toe the first phalange of resonates with Gemini; the second phalange with Libra and the kidneys; and the third phalange with Aquarius. On the third, or Fire, toe the first phalange resonates with Aries; the second phalange with Leo and the solar plexus, and the third with Sagittarius and the hara. On the fourth, or Water, toe the first phalange resonates with the nurturing breast of Cancer; the second phalange with Scorpio and the pelvis, psoas, and hips; and the third phalange relates to Pisces. The little or Earth toe reflects grounding: the first phalange resonates with Taurus, the neck, and the upper torso; the second phalange with Virgo and the pelvis; and the third with Capricorn and the knees.

The Ether, or big, toe resonates with the head, neck, spiritual core, and higher nature. Lines in the distal areas of the second phalange correspond to neck tension and more proximally to shoulder tension. If the first phalange is turned upward, this marks the "mystic toe" of an individual who has an affinity for spirituality and has a natural mystical attunement. If the big toe on the right foot is turned toward the smaller toes, the individual has a dedication to service and to bring higher ideals into everyday life. If there is a gap between the big toe and the second (Air) toe, the person is an idealistic seeker with a gap between his or her dreams and life.

The Air, or second, toe reflects matters of the heart, passion for life, and mental activity. A large first phalange on the Air toe indicates ambition, desire, or a very mercurial person with an agile mind. If the Air toe is the longest toe, it reflects an individual who is very "spirited," radiant, and positive, who projects a passion for life at the vanguard of their personality. On the left foot, a line between the first and second phalanges indicates complexity and defensiveness in matters of the heart. A deep line, wrinkles, depression in the skin, or discoloration may reflect a broken heart. On the right foot a deep line can mean a period of depression earlier in life with a spirit which has been broken or a loss of one's passion for life.

The quality and coloring of the lines can be an indication of the process of healing. Lines which are still discolored but otherwise not deep reflect a healing process and energy flowing again. Grays and browns reflect armoring and crystallized emotions that remain to be cleared. Red and pink spots reflect current issues in the client's life.

An hourglass shape centered at the first phalange indicates a diaphragm block, as does a line between the first and second phalange of the Air or Fire toes. Stress lines between the second phalange and the base reflect tension around the kidneys related to fear, ambivalence, and uncertainty about self-expression when found on the right Air toe; and being thrown off center by the environment when they are on the left Air toe.

The Fire, or third, toe resonates with the organs of the solar plexus, and with work and self-expression. A pronounced first phalange indicates an individual with powerful mental images (the vision of Aries). Lines between the first and second phalange indicate frustration and a solar plexus block. Discoloring and stress lines between the second phalange and the base of the toe indicate issues with expression of anger (on the right) or resisting anger in the environment (on the left). Stress lines in this area, on either foot, indicate tension and resistance in the solar plexus, spleen, and umbilical center. A "waist" or narrowing of the toe on the right foot toward the base can indicate problems with grounding one's vision, work, and creativity. Gray and brown discoloration, depression, and wrinkles in the skin of the left Fire toe often relate to an abusive childhood environment. Wrinkles and discoloration on the middle phalanges reflect anger and on the lower phalanges reflect rage in the environment.

The Water, or fourth, toe reflects emotional expression, bonding, and the capacity to share warmth and nurturing. A large or wide Water toe reflects expanded attachments and a warm, nurturing person. A Water toe turning into the Fire toe reflects a person who puts a lot of warmth and feeling into work and self-expression. Darkness and stress lines between the second phalange and the base of the Water toe may indicate rage and issues with emotional abuse in the environment (on the left toe) or in the individual's character (on the right toe). A narrowing of the toe in this area indicates psoas muscle tension and an energy block in the pelvis. Excessive wrinkles usually point to emotional disturbances (except in Virgo Sun or Virgo rising).

An Earth, or little, toe held off the ground indicates an individual who is not grounded in life or money or other matters. A twisted Earth toe can reflect issues with elimination.

Toes that bend under and grip the earth reflect fear. Toes that bend up indicate a lack of grounding in that area. Toes turning upward can also indicate chronic illness. All the toes turning upward may indicate a process of leaving the body due to chronic illness. Space between toes indicates a lack of integration between the levels of consciousness and areas of personality represented by those toes. Short toes indicate directness and spontaneity, knobby phalanges indicate the opposite. Large gaps between all the toes, and toes that go off in different direction, indicate ungroundedness or emotional disorder. Toes that are straight, full, and even indicate clarity, health, and emotional integration.

The toes resonate with the astrological signatures. Indications of stress in the toes can be related to the corresponding astrological signs and elements of character. Great insight into character and stress can be initiated through the astrological correspondences.

The toes of individuals with prominent Water-harmonic energy—Cancers, Scorpios, and Pisces—who appear to "flow" on the surface while reacting on deeper levels, may lack surface stress lines and be difficult to read. The toes of individuals with the Sun in Virgo, or Virgo rising, may show so many lines that they may also be difficult to read.

As a negative field of the body the toes are relatively stable and usually change slowly. The patterns that you read in the toes are chronic. Be aware that your toe reading may reflect the past rather than the present and that the client may have healed the wounds and outgrown patterns revealed in the reading. Only read toes in a well-lighted area and make sure that the client's feet are clean and that you are not reading a tuft of hair or shadow as a life tragedy.

It is illegal to offer any form of diagnosis of physical or psychological illness or disease. I believe that it is a form of violence to tell someone that something is wrong with them. Clients will really believe what you say to them. Your comment about "a broken heart" may become part of their identity. You have a responsibility to phrase comments positively and to offer your reading in terms that will empower clients and support them in taking more responsibility for themselves. Instead of telling the client that their kidneys are shot (as my first Polarity Therapist told me, I later found out that he told that to all his clients) you could indicate that regular exercise or less sensory indulgence would be health building.

Toe reading is a valuable assessment technique for the somatic therapist. At the toes we view the most crystallized expression of a person's mental and emotional patterns. Patterns of resistance inhibit our resonance with the tattvas, the universal creative essences. Toe reading is an excellent tool for identifying neck, shoulder, diaphragm, and pelvic energy blocks. With toe reading we can easily view areas of stress, tension, and body armoring through their reflex points in the corresponding areas of the holomorphic patterns in the feet.[2]

The Polarity Cleansing Diet

A Fast for the Ego
Is a Feast for the Spirit[1]
Chew, Chew, Chew, Your Food
(To the tune of "Row, row, row, your boat")

Chew, chew, chew, your food,
Gently through the meal
The more you chew,
The less you eat,
The better you will feel!

The Polarity cleansing diet is an opportunity to experience the cleansing and health-building effects of a safe and simple alkaline diet. Based on the teachings of Dr. Randolph Stone, this diet is the fruit of over seventy years of experience in the study and practice of natural therapeutics.

The Polarity cleansing diet is a safe cleansing and health-building regime. It is a diet that you and your clients can use without experienced professional supervision. It can be used for health building under almost any circumstances, by any constitutional type, and at any time of the year. The diet offers an opportunity to break dietary patterns and examine emotional patterns associated with eating. It also offers understanding and support for making a commitment to a healthier lifestyle.

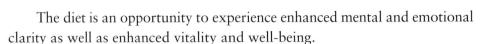

The diet is an opportunity to experience enhanced mental and emotional clarity as well as enhanced vitality and well-being.

One should stay with this diet a minimum of seven days or until an enhanced state of health has been achieved. This is a cleansing diet, not a maintenance diet, and should not be used indefinitely.

It is important to pay attention to your elimination in a cleansing diet. Use herbal laxatives or enemas if necessary to promote daily elimination.

Polarity Diet Self Test[2]

Who can benefit from the Polarity diet? Answer the following questions to determine whether the Polarity diet would be appropriate for you.

Do you experience any of the following?

pain anywhere in your body	wake up feeling groggy
tension, especially in the neck and shoulders	bowel irregularities
	often feel tired after eating
general body aches	excess mucus
aching, tight muscles	frequent colds and flu
stiff joints and/or decreased range of motion of joints	frequent craving for sugar, salt, and bread

Modifications

You may need to modify the basic diet if you fit in the following groups:

You may need to add sprouted millet and quinoa, eat more cooked food, avoid fruit, and drink only a small amount of liver flush if you are:

underweight

weak, spacy, and ungrounded

experiencing dizziness

You may need to eliminate or use very small amounts of ginger and garlic if you experience:

hot flashes	night sweats
heat in palms or feet	red face and eyes
dislike of heat	frequent thirst
insomnia	red tongue
dizziness	dry mouth
nervousness or irritability	

You may need to use larger amounts of ginger and garlic, and eat less oil, nuts and seeds, if you experience:

feeling sluggish after eating

dislike of cold damp weather

Polarity Liver-Flush Cocktail

Take 4 to 8 ounces as your morning meal. *More is not better!* In a blender add:

3 to 4 tablespoons of cold pressed olive or almond oil

the juice of one peeled organic lemon

1 or 2 medium size cloves of garlic

an equal amount of fresh ginger (a piece about 1 inch long)

the juice of one organic grapefruit

Polarity Cleansing Tea

To boiling water add a blend of equal amounts of:

ginger root,

anise or fennel seed

flax seed

fenugreek seed

twice as much licorice root,

After simmering the above for 20 minutes take it off of the stove and add *peppermint*. Serve with *lemon* and *fresh ginger juice* to taste. Start the tea first and let it brew while you are getting the liver-flush together.

Drink four to eight ounces of tea immediately after liver-flush and throughout the day as you like. If you are constipated, eat lots of roughage, add pears to your salads, drink lots of water, use more licorice root and fresh garlic, and take the Polarity cleansing tea and liver-flush twice daily. If constipation continues, take two tablespoons of flax seeds with warm water or eight ounces of orange juice and olive oil.

No other food should be eaten for two hours after the liver-flush and tea.

If you have diarrhea, use no licorice, liver-flush, or ginger, but substitute cinnamon bark in the Polarity cleansing tea and eat ground cinnamon with baked apples and/or dates and raisins, chewed thoroughly.

Fresh vegetable juices may be taken between meals as a snack, as well as a little fresh fruit. "Chew" your juice and do not take more than eight ounces at a time.

Meals

Eat less and chew more. Only alkaline-forming foods are allowed on this diet.

Take a raw salad made of finely chopped and grated fresh vegetables. Choose from the following: carrots; cabbage; radishes; turnips; boston, red leaf, or romaine lettuce; cucumbers; green onions; or other greens; in season, blended to taste. Your salad should include sprouts of alfalfa, mung, aduki, sunflower, or fenugreek seeds. Four soaked almonds may be taken twice a day.

The meal begins with how much food you put on your plate. Eat less and chew more. Allow yourself to eat consciously and to experience the taste and energy of the food you are eating.

Enjoy the blessings of this diet. Be creative in making your salads. Play with salad dressings of olive oil, ginger, lemon juice, and herbs. For constipation add pears to your salad. Eat less and chew consciously.

Vegetable soups as well as steamed and baked vegetables may be eaten. Soups that have a distinctive character, for example, carrot/dill, celery/onion or beet/lemon, offer more of a sense of variety, making the diet more enjoyable. Grains, beans, and starchy vegetables such as potatoes and winter squash are not allowed. Eat less and experience your food.

Only alkaline-forming foods are allowed on this diet: No starches or protein are to be eaten. No meat, dairy, grains, potatoes, rice, bread, coffee, tea, sugar, eggs, yeast, or tamari.

Keep things simple. Do not try to get around the diet. If you need to ask the question: "Is this on the diet?" The answer is, "No . . . " Eat less and chew more.

Diet has its foundation in your attitude. An important part of the diet takes place in your mind. Get clear about your willingness to totally embrace this health-building regime. Take the time to clarify your goals in doing this diet. Experience how good you feel on the diet.

Take space for yourself while dieting. Understand that you are doing a great deal of inner work. Try to get away from stressful outer commitments.

Pay attention to your breathing. Eat breath, prana, universal life force. Understand that breath is the fundamental source of sustenance. When you find yourself thinking about food take a few deep conscious breaths to realign your energy.

The Polarity Purifying Diet will often facilitate a "healing crisis." This is a health-building process of elimination. You may experience negativity and resistance as emotional states and physical conditions from your past come to the surface of your consciousness in the elimination process. Try to keep a positive attitude and not identify with the symptoms and negativity.

This diet offers you an opportunity to become conscious of your ego's food habits and attachments and to break out of patterns that are not healthy.

It is very important to get very clear about getting off the diet. Be vigilant at the end of the diet. Do not let an ego impulse re-establish disharmonious food habits.

You should end this diet consciously, gradually adding one new food at each meal. Express your intention by writing down a meal-by-meal, food-by-food plan for consciously moving from the cleansing to your maintenance diet. See the section, "Breaking the Polarity Diet," which follows.

Align your will and focus your intention on getting very clear about the eating habits and foods you are giving up. Clarify your relationship to food and plan your new foods. Identify your power foods.

Before you put anything in your mouth, ask, What will this do for my spiritual life?

The Polarity Cleansing Diet is a fast for the ego and a feast for the Spirit. The diet can be used "spiritually," as a quest for vision and a means to contact a higher intelligence within, for guidance and transformation in your life. This regime will purify your body—harmonizing and concentrating your body's vibration, uplifting your being into attunement with the healing and revitalizing spiritual potency. The experience of renewal, of dramatically enhanced emotional clarity, physical well-being, and spiritual presence, is meant to be educational. Your experience of wholeness is meant to encourage you to cultivate an everyday lifestyle that promotes this quality of spiritual attunement and personal well-being.

Understanding Toxicity and Energy Blocks: Acid and Alkaline

The Ph balance in the body chemistry is a foundation of health. Dr. Stone explains:

> A concentration of electromagnetic light wave energy with an increase of hydrogen ions deposited in the cells makes the cell over-acid in its chemistry and produces acute symptoms of redness, heat, swelling, and inflammation. These symptoms are an excess of the positive charge and the action of the fire principle, which produces acute disease . . .[3]

A deconcentration of electromagnetic light waves in the fields of the cell with an increase of hydroxyl ions makes the cells excessively alkaline, cold, constricted and dehydrated. This is the basis for all chronic disease and limitation of motion. Here the water element is functioning in excess, as the primary polarity factor in the energy fields' imbalance, as excess negative charge with its constrictive over-alkalinity, limiting function and motion, and its crippling effect like in arthritis, rheumatism, etc.[4]

Breaking the Polarity Diet

*Chewing your food and systematic under-eating
are the keys to health and longevity.*

Continue with a simplified diet for a week. Allergy test by adding one new food at each meal or one new food a day. First, add corn and see how you react. Does your energy drop dramatically after the meal? Does your pulse go up measurably? Do you feel pain in your back or internal organs? Are you getting a headache? Do you feel vulnerable, or are you falling apart emotionally a few hours after the meal, or the next morning? These are all symptoms of food allergy. Next try wheat. Test for dairy, soy, yeast, sugar. Keep your meals simple; experiment by adding one new food at each meal. Pay careful attention to the relationship between what you put in your body and how you feel.

Days one and two, eat moderate-sized meals of simple steamed vegetables, salads, and whole grains with fresh or soaked fruit for treats (if necessary). Days three and four, add legumes and seaweed or low fat yogurt, whole grain breads, crackers, tortillas, pasta, dry cereals, nutritional supplements. Days five and six, add raw or low-fat dairy products (if you wish to resume them), organic eggs, fish, and other flesh foods. Begin with half-portions. Day seven, eat a light dinner, but skip breakfast and lunch. Drink only vegetable juices and water until dinner. Fasting for two meals once a week is an excellent health-building practice for most people. You can do the re-entry diet outlined above for seven or fourteen days. Take as long as possible to add new foods. Simplify as much as possible. To get the most out of your digestive power, follow the rules of simple food combining. You can get a postcard-sized food-combining chart in any health food store.

To maintain the level of clarity and well-being that you have experienced on the Polarity Diet:

1. Start your own sprout garden. Sunflower seeds sprout overnight and alfalfa sprouts take only a few days. Any health food store will have everything that you need to get started. Sprouts are a convenient, inexpensive way to keep the highest quality "live food" in your diet.

2. Eat whole grains and avoid refined and chemically-treated flour, such as bleached white flour. Avoid refined sugar—use honey, maple syrup, and fruit-juice sweetened treats.

3. Eat a low-protein diet. Minimize or eliminate eggs and flesh foods. Minimize dairy or eliminate commercial dairy products. Goat's milk and raw milk products are better.

4. Eat higher quality snack food. Eliminate soda. Cut down on fruit juice. Use quality herb teas. Buy a juicer. Use carrot and vegetable juices, apples and apple juice, fresh or soaked fruit, popcorn, or almonds for snacks. Check out the Life Stream carrot/raisin/date sprouted bread.

To retoxify, add stimulants (caffeine and sugar), fried foods (chips) and rancid cooking oils, refined and chemically-treated foods, and eat unconsciously. If you do choose to abuse your body through junk food, you will be on the path to better health if you eat these foods in moderation and eat eighty to ninety percent healthy foods with only ten to twenty percent toxic foods.

Your diet is a reflection of your consciousness. It can be valuable to keep a diet diary, to become more conscious of eating patterns.

If you are serious about wanting to be healthy and at peace you will need to educate yourself and become quite conscious and disciplined about your nutrition. For recommended reading, see the Polarity diet section of the bibliography.

Alkaline and Acid Foods[5]

When foods oxidize in your body they leave a residue or ash. In this residue, if the minerals sodium, potassium, calcium, and magnesium predominate, they are designated as alkaline foods. The converse of this is true such that if sulfur, phosphorus, chlorine, and incombustible organic acid radicals predominate, the food is designated as acidic.

With long, painstaking analytical work and by comparing the excess of one group of minerals over the other, the numerical values of acidity and alkalinity can be determined. In the human body, these pH values can be validated or verified by observing increased or decreased pain free range of motion.

Foods for the Polarity Diet

Beans:
Sprouted Only
Aduki
Mung

Nuts and Seeds
Sprouted Only
Alfalfa
Almonds
Fenugreek
Radish
Red Clover
Sunflower

Grains
Sprouted Only
Amaranth
Buckwheat
Millet
Quinoa

Oil
Flax oil
Olive oil

Young Greens
Buckwheat
Barley
Sunflower
Wheat

Vegetables
Beets and greens
Broccoli

Burdock
Cabbage, all
Carrot
Cauliflower
Celery
Chives
Collards
Cucumber
Eggplant
Endive
Escarole
Fennel
Green Beans
Kale
Kelp
Kohlrabi
Leeks
Lettuce
Mustard greens
Okra
Onion
Parsley
Peas, fresh
Peppers
Rutabega
Spinach
Summer squash
Swish chard
Tomato

Turnip and greens
Watercress
Zucchini

Fruits
Apple
Apricot
Blackberry
Cantalope
Cherry
Fig
Grape
Grapefruit
Huckleberry
Kumquat
Lemon
Lime
Mango
Nectarine
Orange
Papaya
Pear
Peach
Pineapple
Pomegranate
Raspberry
Strawberry
Tangerine
Watermelon

Transitional Simple Diet and Allergy Test

	Day 1	Day 2	Day 3	Day 4
	Warm lemon water, first thing in the morning.	Warm lemon water	Warm lemon water	Warm lemon water
Breakfast	Millet	Corn meal	Porridge	Oats
	Miso vegie soup	Miso vegie soup or yogurt	Miso vegie soup or yogurt	Miso vegie soup
Lunch	Millet	Brown rice	Simple grain	Cheese
	Miso vegie soup	Miso vegie soup	Miso vegie soup	Simple grain
	Simple vegie	Simple vegie	Simple vegie	Simple vegie
	Salad	Salad	Salad	Salad
Dinner	Quinoa	Basmati rice	Tofu Bread	Simple soup
	Simple vegie soup	Chapati	Simple soup	Simple vegie
	Simple vegie	Simple vegie	Simple vegie	
		Simple dhal		
	Test for yeast			
	Test for honey			
	Test for eggs			

Allergy test by adding one new food to a simple diet at each meal. Pay careful attention to the relationship between what you eat and your energy level, emotions, and well-being.

Signs of stress or allergy: pulse speeds up, energy drops, pain in the back or internal organs, headache, anxiety, or falling apart emotionally. See this chapter section "Breaking the Polarity Diet" for more information.

Right Livelihood:
Your Career as a Polarity Therapist

Polarity Therapy as a means of right livelihood offers a deeply fulfilling experience of giving and receiving love and nurturance. It is a career that offers the opportunity to actualize your compassionate nature and to be of genuine service to others. It is a means for you to take personal responsibility for healing humanity and serving our planet. It offers an opportunity for you to help make a better world through a profession in which you have permission to be a caring and responsible human being.

The Calling of Polarity Health Education

Polarity health education offers a way of life that upholds Dharma. Polarity Therapy is self-actualizing work, through which you can bring your wisdom and love into a life of dedicated, unselfish service. Polarity offers an opportunity to serve your most cherished truths of the sacredness and unity of life. It offers the individual an opportunity to take personal responsibility for the healing of humanity. Polarity Therapists work with the deepest insights about consciousness and the human condition to accelerate the evolution of the soul. Polarity offers right livelihood and is nonviolent to man and nature. Through nurturing touch the Polarity health educator ameliorates suffering and brings balance to life. As a Polarity Practitioner you can respond to an inner calling to compassionate service and integrate your higher nature into your livelihood.

Loving Hands Are Healing Hands

Caring touch is the embodiment of the compassion of a conscious universe. Compassionate touch has a profound capacity to heal and to make whole. Touch has a very special power. It resonates with the keynote of the heart. Sattvic touch on the surface of the body aligns with the yin sensory currents, which, in the balance of yin and yang, penetrate to the deepest yang core of the body.

Sattvic touch facilitates a receptivity to sattva guna, the Unitive Field. Sattvic touch transcends the loneliness, separation, and alienation of the ego. Sattva guna arises from the heart. Caring intention, nurturing feelings, and healing love are carried by the sensory currents to the heart of the being. Nurturing touch brings us into the healing power of the present. Through loving touch, the somatic therapist holds a safe and sacred space where the client has permission and support to process and release unresolved emotional experiences from the cellular memory of the body.

Service is love in action. To touch in a caring way is to actualize our higher nature. Compassionate service puts us in touch with our essence. It integrates our unselfish and wise higher nature into the practice of our daily life. Touch is a deeply healing human experience. Working with compassionate touch is working with the essence of the healing force, the healing presence of the Soul. Our touch resonates with divine love and is the living presence of Spirit in the world. Respect your clients as sacred beings. Cherish your clients; touch them with warmth and safety. Touch with tenderness, as you have always wanted to be touched. We are instruments through which love is expressed in the world. *Loving hands are healing hands.*

For information about studying Bruce Burger's unique approach to Energy Healing, visit us at:

Web Page: http://www.weare1.us
E-mail: bruce@weare1.us
(707) 923-3387

American Polarity Therapy Association Standards for Practice and Code of Ethics[1]

Theory and Basic Principles of Polarity Therapy

Associate Polarity Practitioner

The beginning student is first introduced to the notion that all life is an expression of energy in motion, and that energy emerges from and returns to a central unified source of life energy. Energy in the body is a manifestation of the cyclic journey of spirit: from its cosmic unified source, into the duality of the physical realm, and back to its source.

This theme is further developed with the concept of unity and neutrality and the creation of the polarized opposite forces of attraction and repulsion. These two forces, in turn, follow the blueprint inherent within their source as they mutually interact to weave the fabric of manifest life. Consequently, all forms and processes are generated from within the universal source of life energy, and are expressions of that life energy in motion.

The theory is further delineated with the formation of the three primary principles of motion as Sattva, the neutral Airy Principle, Rajas, the positive Fiery Principle, and Tamas, the negative Watery Principle. These three principles set the stage for energy to manifest as five discreet qualities or steps, characterized by the five elements of Ether, Air, Fire, Water, and Earth.

By virtue of these elements, the physical human body is created. The body is a composite of the interaction of the five elements (which are electromagnetic in nature) in the physical domain. The body, then, is an electromagnetic energy system expressing the dynamic interplay of the five elements from their most subtle to their most dense forms. Each form manifests a unique anatomy, with the most subtle form expressed as a "wireless" circuitry directing the flow of energy in its etheric state. Thus are created not only all aspects of the body, but also the totality of human expression, described by Dr. Stone as the Pentamirus combination of the elements.

The student is also introduced to the idea that healing and disease are processes that can be described in energetic terms. Generally, healing and health are attributes of energy flowing in its natural and unobstructed state, while disease is a reflection of energy in an obstructed condition in one or more of its several levels of manifestation.

In addition the life process can be described as states of energy stepping down, as in the case of disease, and stepping up, as in the case of healing, through the five elemental states of Ether, Air, Fire, Water, and Earth. The five states of energy are defined by the quality of energy motion through the "wireless" circuits of the body.

Ether is seen as the manifestation of a free, uninterrupted energy flow, experienced as a state of freedom of expression, creativity, movement and health. As energy becomes blocked in its pathway of motion back to its energy center in the body, the body experiences varying states of ill health, or disease. These states, varying from mental and emotional disturbance, to acute physical inflammation, to chronic illness, and finally to full degeneration, are characterized by energetic states corresponding to the four step down elements of Air, Fire, Water, and Earth, respectively.

As disease is regarded as the effect of energy being disrupted in its flow, healing is seen as the process of releasing the obstructions to the free flow of energy so that it may fully reconnect with its source. Consequently, a return to physical health is the reconnection of energies in the body with their respective centers or sources.

This relationship between health and energy connecting with its source is expanded to include any system. That is, the health of any system is dependent upon its full, undisrupted connection with its source. With this understanding, the underlying principle of Polarity Therapy emerges as the idea of "reconnection to a source" and pertains to each facet of the work. This includes establishing reconnections to sources of energy in the body, in nature, in interpersonal and societal relationships, and within oneself, to the source of life energy itself. Health is considered in its larger context beyond just the body, recognizing the interconnectedness of all things to each other and to a unified central source of life energy.

Registered Polarity Practitioner

Once an understanding has been derived as to how energy relates to form and process, the study delves deeper into the dynamics of how the body is created and functions as an energy system. The body is seen as a system of energy fields and centers, "wireless" energy pathways, lines of force, and geometric and harmonic relationships. The energies of the body are derived from and organized around a central ultrasonic core. The ultrasonic core is an expression of primary energy as the neutral essence. This neutral essence is stored in and conveyed by cerebrospinal fluid throughout the whole of the human system. The neutral essence is considered to be the basic ordering and healing principle in the human body.

The ultrasonic core is the central pathway of motion along which the five primary centers are created, located anatomically along the central canal of the spinal cord. Two intertwining currents of *pranic* energy, each the polarized opposite of the other, pursue a double helical pathway along this core, creating the five chakras or whirling energy centers of Ether, Air, Fire, Water, and Earth, located in the center of the throat, behind the heart, in the solar plexus behind the umbilicus, at the lumbosacral junction and at the sacrococcygeal junction, respectively. In this way the five sensory pranas are formed in order to allow the perception of our physical experience.

Similarly, five motor pranas are formed from two currents also emergent from the entry point of the ultrasonic core, at the eye center. These two currents form the five primary oval fields of the head, neck, chest, abdomen, and pelvis, representing the five elements of Fire, Ether, Air, Earth and Water, respectively. These five fields enable the body to interact with its environment. They also contain the five primary chakras of perception and establish an energetic physiology as the motor and sensory pranas commingle.

From the ultrasonic core and the five primary chakras emerge currents of subtle electro-magnetic energy that fill the space of the etheric body with etheric energy waves. The pathways of these etheric energy waves create a "wireless" circuitry that interconnect every point of this subtle form with every other point within the energetic system. The pathways follow three distinct directions as vertical North-South Watery currents on the right and left sides of the body, coronally spiraling Fiery currents on the anterior and posterior aspects of the body, and transverse East-West Airy currents intertwining horizontally from the feet to the head. These three primary currents create an interwoven field which expresses the three principles of energy movement (also known as the Gunas).

The energy body created by these three etheric energy wave circuits describes a spatially oriented triaxial set of polarity relationships as to top (+) and bottom (–), anterior (+) and posterior (–), and right (+) and left (–). The center of the body lying in between any of the three axes is neutral.

The concept of "Three Geometries" is a synthesis of Dr. Stone's work, developed by subsequent Polarity teachers, which is helpful in organizing the interrelated energetic anatomies described by Dr. Stone. This concept is included in the Standards for Practice to help RPP students understand energetic relationships in the body.

Three primary sets of geometric relationships emerge in the formation of the etheric energy body. The first (Geometries of Projection and Reflection) interconnects the five elements and three principles with the various fields and centers of the body. The second (Geometries of Involution and Evolution) interconnects the generating lines of force with internal and external gravity. The third (Geometries of Symmetry and Balance) interconnects the five elements and three principles with spatially and functionally related harmonics and specific and formative energy-focusing body postures.

Geometries of Projection and Reflection

The geometric relationships that interconnect the five elements and three principles with the various fields and centers can be called the Geometries of Projection and Reflection. These describe geometric harmonic relationships, created as the three principles of positive, neutral and negative forces, which induce the creation and sustain the motion of the five elements.

Dr. Stone demonstrates how the three principles are expressed in the body in Vol. I, Book 2, chart 4, p. 11, showing the Anterior Structural Relationships, and in Vol II, Book 5, charts 2 (p. 85) and 19 (p. 163), showing the Posterior Structural Relationships. . . .

The three principles of energy movement are the primary forces that are expressed as fundamental relationships in the centers and fields of the body. These relationships allow the creation and interweaving of the five elements and are first expressed in the creation of the eye center and head.

From this level, energy is projected into the physical domain and the physical body via the five elements. The projected energy creates the arena in which life is experienced. These same energies are then reflected back as sensory feedback. This reflection of sensory information modulates the continued projection which, in turn, modifies additional reflected feedback. In this way the projection-reflection loop tempers our perception and experience of life.

Dr. Stone depicts the fundamental field and center relationship in his chart called the "Primordial Mind Pattern in the Head," found in Vol. II, Book V, chart 7, p. 118. From this representation are derived other reflex patterns created as the five elements express the dual relationship of projection and reflection. These ... describe such specific harmonic reflexes as those linking the cranial structures to the pelvic girdle, the tongue, thumb webs, and ankles to the abdomen and pelvis, and the ear pinna, hands, and feet to the overall body.

Geometries of Symmetry and Balance

A second set of geometric relationships is created by interconnecting the lines of force that direct energies as they initially form the body. This set of relationships can be referred to as the Geometries of Symmetry and Balance. These are represented by the Interlaced Triangles (in Vol. I Book II, chart 11, p. 18), the Five Pointed Star (in Vol. I, Book II, charts 9 and 10, pp. 16–17), and the Geometric Gravity Lines (in Vol. I, Book I, charts 6 and 7, pp. 76–78). Together these charts depict an overall relationship between internal lines of force as they reflect our net response to gravity, internally and externally. When the body's energies are flowing freely and concentrically with respect to the ultrasonic core, then the lines of force are aligned such that the body experiences structural symmetry and physiological balance.

Geometries of Involution and Evolution

The third set of geometric patterns are depicted by the Geometries of Involution and Evolution. The Involution Pattern is represented in Vol. I, Book I, pp. 48 and 49, and the Evolution Pattern is depicted in the composite of charts 3 and 6 of Vol. II, Book V, pp. 102, 155.

The two postures of involution (known as the fetal position) and evolution (known as the "bow" posture in Hatha Yoga) are polarized energy focusing body positions. The fetal posture intensifies the focus of energy within the body, and also promotes the downward and outward flow of "Apana" forces for expulsion and elimination. The "bow" posture represents the pattern of selfless openness and reconnection with the source of life energy. The geometry depicts an orientation of the torso and all the primary body centers toward its energy source, while the feet and lower extremities stay grounded and in an opposite orientation.

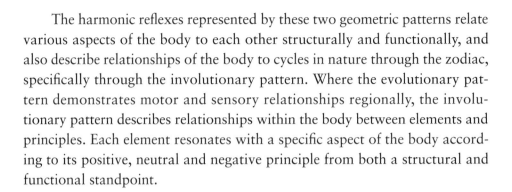

The harmonic reflexes represented by these two geometric patterns relate various aspects of the body to each other structurally and functionally, and also describe relationships of the body to cycles in nature through the zodiac, specifically through the involutionary pattern. Where the evolutionary pattern demonstrates motor and sensory relationships regionally, the involutionary pattern describes relationships within the body between elements and principles. Each element resonates with a specific aspect of the body according to its positive, neutral and negative principle from both a structural and functional standpoint.

Expressions of the Elements

The body is seen as an intricate and dynamic interplay of the Three Principles and Five Elements which creates its subtle anatomy and physiology. Furthermore, the Three Principles and Five Elements are understood to make up more than the body alone. They intercombine in their many states to create all aspects of human expression, including body, emotions, mind, and spirit.

The five chakras, which first define the elements in the physical realm, are expressed negatively as the emotions of pride (Ether), greed (Air), anger (Fire), attachment (Water), and fear (Earth). This is the experience of humankind when the attention is directed entirely downward and outward into the world for the purpose of identifying the ego with the worldly creation, thinking itself the prime mover of life. In contrast, when the attention is directed inward and upward, acknowledging the source of life energy as the doer behind our experience and understanding of the world and ourselves, the passions transform into the five virtues of courage (Earth), detachment (Water), forgiveness (Fire), contentment (Air), and humility (Ether).

Beyond the passions and virtues, the elements express our experience of emotion, gross and subtle body movement, and lifestyle habits as well as the fluids and tissues of the coarse metabolic body. Dr. Stone describes these element intercombinations through the Pentamirus Combinations. Each element predominantly influences the other four elements in some aspect of our experience.

Dr. Stone points out that the representation of the elements as the various qualities of our personal experience of worldly life are separate from the observer of life. In short, we are not the elements. We watch them weave their play. We can watch with respectful distance and appreciate the tapestry, or get caught up in the weave and lose sight of who we are, where we are, and where we are going.

Dr. Stone strongly acknowledges the role of consciousness as it affects the energies of the body. He describes a hierarchy of consciousness from the most sublime states of being to the coarsest levels of life. In the realms of mind, consciousness may be experienced in many ways and forms, all of which embody particular frequencies or vibrations of energy. Dr. Stone's statement "Energy is the real substance behind the appearance of matter and forms" therefore applies to consciousness as well as physical phenomena.

Thoughts impact form according to the energetic charge and wave form of the thought; these are significant forces or stresses to be reckoned with in the physical body. The energies of the body change and transform in accordance with the form of the impacting thought, thereby affecting the integrity and health of the system.

The Registered Polarity Practitioner develops an understanding of how the mind affects the energetic status of the principles and elements, and, therefore, all aspects of human experience. From emotions, to physical states, to the overall quality of life, the interplay of the principles and elements expresses our state of mind and health. As the energies are disrupted in their motion away from and back toward their respective centers both in the body and in nature, we experience the stepped-down states of pathology, both organically and psychologically.

The more disturbed the flow of energy in the body, the more inefficiently the body functions. The body's energies shift from their concentric flow around the ultrasonic core toward an increasingly eccentric motion. The intent of the Polarity practitioner is to assist the body's re-establishment of concentric energy movement by releasing blocked energy in the body and guiding the client to do the same in all aspects of his or her life. Once blocks to this free flow are released, energy naturally returns to its source, whether in the body, in nature, or throughout the domain of consciousness.

American Polarity Therapy Association
122 N. Elm Street, Suite 512
Greenboro, NC 27401
(phone) (336) 574-1121
APTAoffices@polaritytherapy.org

International Society of Polarity Therapists
Dian Hunt, Secretary,
54 Ashford Road
Topsham, Exeter, EX3 OLA
England
(phone) 011-44-1392-877-015

Glossary

Advaita: Nondualism, monism. The sanskrit term *advaita* means "not two." The doctrine of advaita asserts an identity of all consciousness.

Air Element: The pattern of vibration/resonance that underlies all movement in nature.

Ahamkara: I-maker, ego, self-conscious, individuating principle.

Ajna chakra: The center of will, inner mind, occult vision, third eye.

Akasa, Ether, space element.

Amrita: Immortality, nectar of immortality.

Anahata chakra: Heart center, emotional center.

Ananda: Bliss, the experience of the Self. "The Self knows and loves the Self." It is this love which is Bliss or "resting in the Self," for as it is elsewhere said, "Supreme love is bliss."[1]

Anandamayakosa: Finest and most subtle aspect of mind is the balanced, blissful resonance.

Annamayakosa: The physical body, "sheath built of food."

Antaratma, Inner Self, embodied Soul.

Apana, Downward, eliminating force.

Apas, Ap: Water element, quality of liquidity, function of contraction.

Atman: A portion of the divine supporting individual existence. "the *Atman* or Self is all-pervading, immense, the ground, and agent of knowledge. It is thus the seat of consciousness."[2]

Atmos, Greek, "breath."

Aum, Om, ॐ: The original Unity. The *Mandukya Upanishad* teaches: "The syllable ॐ (Aum), which is the imperishable Brahman, is the universe. Whatsoever has existed, whatsoever exists, whatsoever shall exist hereafter is ॐ." All creation is but an echo of ॐ. the *Taittiriya Upanishad* teaches: "Thou art Brahman, one with the syllable ॐ . . . the mother of all sounds."

Axiom, Widely accepted maxim, self-evident truth.

Ayurveda, The ancient East Indian science of life.

Bhagavad Gita: (The Song Celestial), an ancient Gospel cherished by many as the most lucid formulation of the Sanatana Dharma. "The Bhagavad Gita . . . is . . . one of the clearest and most comprehensive summaries of the Perennial Philosophy ever to have been made."[3]

Bharata, devoted to light, The name Indians give to their own heritage, India.

Bhurloka: Earth, material plane.

Bija: Bija are the archetypal seeds, the actual mechanisms in the One Mind, that underlie manifestation.

Bindu: Point, dot, source, seed, basis for the emanation of the first principle, Mahatattva.

Brahma, The Creator in the trinity of the Sanatana Dharma, with Vishnu (Preserver) and Shiva (Destroyer).

Brahman, The Supreme Being, that which has reached its ultimate evolution, the Ultimate Principle.

Brahmadanda: Stick of Brahma, spine.

Buddhi: Intellect, understanding, reasoning, faculty of comprehension.

Centrifugal: moving away from a center.

Centripetal: moving toward a center.

Cit: To perceive, to know. Universal consciousness. The static aspect of unchanging universal consciousness.

Chakras: Whirl, the step-down mechanism from the ultrasonic core into the lower vibratory fields of the body.

Chit: Consciousness, a self-aware force of existence (from the verb root *chit*, "to think," "to be aware"); "the divine counterpart to lower mind."

Cymatics: Process of sound generating form.

Devi: Goddess.

Dharma: "Dharma in the Indian conception is not merely the good, the right, morality and justice, ethics; it is the whole government of all the relations of man with other beings, with

Nature, with God. Dharma is both that which we hold to and that which holds together our inner and outer activities, and in this its primary sense it means a fundamental law of our nature which secretly conditions all of our activities, . . . Dharma is all that helps us to grow into divine purity, largeness, light, freedom, strength, joy, love, good, unity and beauty."[4]

Dharmadhatu: Boundless, fundamental ground of all being.

Dharmakaya: The Absolute, Reality, Transcendental Self, the Void.

Dhyanas: Celestial guides.

Dinergy: Nature is woven of a fabric of two spirals—centrifugal and centripetal—whose polarity universally forms the underlying structure and process of everything in nature. This polarity relationship has been labeled "dinergy" by Gyorgy Doczi. The spiraling double helix is the foundation of all organization in nature.

Dosas: The three humors of the body—*vayu, pitta, kapha* of Ayurveda.

Duhkha: (*dur,* "bad state") Suffering, pain, alienation, disenchantment.

Epiphenomenon: A secondary phenomenon accompanying another and caused by it.

Fire Element: Energy in nature is directed by Cosmic Intelligence. The Fire Element is directed energy.

Gita: *See* Bhagavad Gita

Guna, Derived from the Indo-European base *gere,* "twirl" or "wind." The *gunas* (sattva, rajas, tamas) are the fundamental archetypal forces which underlay creation. They describe the phases of the cycle of expansion and contraction from a source that is the basis of the movement of all energy in nature.

Hal: Old English root word *hal* which means "sound, whole, and healthy."

Hamsa: Highest flying swan of *ham,* "out-breath" and *sa,* "in-breath"—symbolic of Spirit.

Hara: Fire center, center of evolutionary force, about two finger-widths below navel.

Hermetic: Refers to Hermes, the Greek messenger of the Gods and relates to teachings of a higher order which are hidden from ordinary consciousness and revealed only to those who have exalted their consciousness through the most earnest quest for insight into The Great Mystery. See Dr. Stone's analysis of "The Emerald Tablet of Hermes" in *Polarity Therapy,* Vol II, p. 229.

Hiranyargabha: Golden egg, Brahma, Self-Created.

Holomovement, Leading-edge theoretical physicists such as David Bohm have been drawn to the conclusion that there is a pattern of wholeness enfolded within all being. Bohm calls this the "implicate order," a universal pattern in which "everything is enfolded into everything." Thus "what is" can be thought of as an intrinsic hologram, and everything is to be explained in terms of forms derived from this holomovement.

Ida: Lunar, cooling prana current flowing through left nostril and centripetally around susumna in the ultrasonic core.

Implicate order: A pattern of wholeness enfolded within all being, a universal pattern in which "everything is enfolded into everything." *See also* Holomovement.

Indra: Vedic God of the atmosphere and sky.

Indriyas: Sense powers, which include five powers of cognition, *Jnanendriyas,* and five capacities for action, *Karmendriyas.*

Jagat: The world which is always changing, moving.

Jnanendriya: Senses or five organs of perception: seeing, hearing, touching, tasting, smelling.

Jivatman: (*Jiva,* "life"; *Atman,* "Self") The living, embodied Soul. The portion of the Supreme Being supporting an individual life.

Jnanmayakosa: The discriminating mind, higher in purity and vibration and capable of the levels of consciousness of intuition and yogic perception to develop spiritual insight and wisdom.

Karana sarira: Also called anandamayakosa, finest and most subtle aspect of mind is the balanced, blissful resonance.

Karma: Action, the creative force that brings beings into existence.

Karmendriyas: Organs of action: *Vach,* "Voice or Larynx"; *Pani,* "Hand," *Padas,* "Foot"; *Payus,* "Anus, organ of excretion"; *Upastha,* "Organ of generation."

Kosa: A sheath or vessel.

Kulakundalini: Goddess personifying the evolutionary force.

Kundalini: (*kund,* "to burn") The primordial cosmic energy that accelerates evolution.

Ligo: Male/protruding and female/receptive connectors.

Linga sarira: Subtle body, astral body, vehicle of the Soul— unchanging through reincarnations until liberation. The vijnanamayakosa,

manomayakosa, and prana-
mayakosa are often given the
general name of *linga sarira* or
sukshma sarira.

Macrocosm: A unity that is the
large-scale reproduction of one
of its constituents.

Mahabhuta: Great elements perceived by the senses (Ether, Air,
Fire, Water, Earth), yet dependent upon the subtle elements.

Mahat: The Great principle, Cosmic Intelligence.

Mahatattva: The first stage of
manifestation.

Mahatma: Self-realized being. A
sage who is united with the
vastness of Being.

Manas: (*man,* "to think, mind,")
Thinking faculty in general. To
think, believe, imagine, suppose, conjecture.

Manipura: Navel, directed force,
immortality, fame, dealing with
reality, will power.

Manomayakosa: The lower or
sensory mind pervades the
body to collect, organize, and
interpret sensory data. This is
conditioned mind, which is
ruled by habit and language.

Manomayapurusha: Mental consciousness of the person.

Mantra*:* ("Mind instrument")
Mantra is the scientific use of
sound to manifest states of consciousness, and is psychoactive.
The rules of Sanskrit parallel
the rules of manifestation.

Manvantara: Period of manifested
life.

Microcosm: A unity that is the
epitome of a larger unity.

Mimamsakas: Mimamsa Vedanta,
one of six orthodox schools of
Hindu philosophy.

Moksha: Liberation.

Muladhara: Root chakra at sacrococcygeal junction. Rules
Earth, boundaries, security
issues, grounding, physical
comforts, survival, shelter.

Mulaprakriti: Root nature, original root out of which all of
nature is derived. The source of
prakriti.

Nada: Mystic sound of the eternal,
the Word as the first born.

Nada Brahma: Sound is God,
sound of Brahman.

Nad yoga, nada yoga: Meditation
on primal sound vibrations.

Namaskara, namaste: (*nam,* "to
honor") A symbol of submis-
sion to the Divine. Greeting,
"honor unto thee."

Nirvana: Liberation, the
extinguishing of the painful
illusion of individuality into the
mystery of *sunyata,* the
unbounded.

Nivritti: Evolution, From *ni,*
"back"; *vrit,* "to turn," hence
to spiral centripetally.

Om: see Aum.

Omniscience: All seeing.

Omnipresence: All pervasive.

Oval fields: Energy fields which arise from and surround the chakras to function as the body.

Para: Supreme.

Pasyanti: Subtle second state of "visible sound."

Parabrahman: the Divine as Transcendent, Supracosmic Divine.

Paramatman: The Self Beyond, the Supreme Being that ensouls the universe.

Pentamirus: Harmonious reciprocal interaction of five parts.

Pingala: Solar, heating prana current flowing through right nostril and centrifugally around susumna in ultrasonic core.

Plenum: Complete in every respect.

Prakriti: Cosmic substance, primary source of all phenomena, primal nature. Prakriti consists of three constituents—sattva, rajas, and tamas.

Pralaya: Dissolution, when the cosmos merges back into the Absolute.

Prana: (*pra,* "before," "first"; *ana,* "breath") The breath of life of the universe that sustains individual life. Its movement is inward, its seat is the heart.

Prani: All breathing creatures.

Pranamayakosa: (prana, "breath"; maya, "formed) The energy sheath that sustains the physical body, "sheath built of life-force."

Pranamayapurusha: Vital and nervous consciousness or Person.

Pranava: (*pra-nu,* "to reverberate") Life field as vibrating sacred sound.

Pravritti: Involution. From *pra,* "forth"; and *vrit,* "to turn"; hence *pravrit,* "to spiral centrifugally."

Prithivi: The Earth element, is from the Sanskrit *plete,* "broad, flat, spread out, extended." It is the principle of solidity, and its function is cohesion.

Purusha: Soul of the universe, the animating principle of nature, the universal spirit. It is that which breathes life into matter; it is the source of consciousness. Purusha is postulated to account for the subjective aspect of nature. It is the universal Spirit, eternal, indestructible, and all-pervasive; it is pure Spirit without activity and attribute, without parts and form, uncaused, unqualified, and changeless. It is the Ultimate Principle of intelligence that regulates, guides, and directs the process of cosmic evolution; it accounts for the intelligent order of things—why the universe operates with such precision, why there is cosmos and not chaos. It is the efficient cause of the universe that gives the appearance of consciousness to all manifestations of matter;

it is the background that gives us the feeling of persistence; it is the static background of all manifest existence, the silent witness of nature.[6]

Rishi, rsi: (from verb root *dris,* "to see"). Seer sage; a sage who through direct experience reveals a vision of universal law. The Self-realized yogis who composed the Vedas.

Rajas guna: The force of expansion in nature. Rajas is an ascending, centrifugal current that rules all energy and motion in nature.

Sabda-Brahman: The Word, the Divine as primal sound power.

Sakti: *See* Shakti.

Samana: One of the five pranas (vital airs). It separates things. It rules digestion and is centered at the navel.

Samkhya, sankhya: A foundation of the Sanatana Dharma is the cosmology known as Samkhya, which means "to enumerate." Samkhya philosophy enumerates the *tattwas,* or fundamental forces that underlie creation. Samkhya, the oldest of the six orthodox schools of Hindu philosophy, is said to be "the philosophical foundation of all Oriental culture, the measuring rod of all Hindu literature, the basis for all knowledge of the ancient sages [*rishis*], and the key to Oriental symbolism . . ."[7]

Samskaras, sanskara: Tendencies, attachments, images, habits, recollections stored in the individual and collective mind which shape karma (action).

Sanatana Dharma: *(Sanatana,* "Timeless, eternal,") *"Dharma,* in the Indian conception is not merely the good, the right, morality and justice, ethics; it is the whole government of all the relations of man with other beings, with Nature, with God. Dharma is both that which we hold to and that which holds together our inner and outer activities, and in this, its primary sense, it means a fundamental law of our nature which secretly conditions all of our activities ... Dharma is all that helps us to grow into divine purity, largeness, light, freedom, strength, joy, love, good, unity and beauty."[8] The Sanatana Dharma refers to the formulation of natural law which has been venerated as ultimate truth and practiced as a way of life by the peoples of Bharat over the millennia.

Sanskrit: Language of the Gods, perfected language. Ancient scientifically-formed language which coordinates with universal basic principles that build and unfold creation.

Sarasvati: Goddess whose name means "Essence of Self," consort of the creator Brahma.

Sarira: (from *sri,* "to waste away") Transient form of the physical body.

Sat: Being. "The active condition of the trancendentil aspect of the Ultimate Principle."[9] "The Divine counterpart to physical substance."[10]

Satchitananda, Sachchidananda (*Sat-chit-ananda,* "Being-Consciousness-Bliss"): The three attributes of the Ultimate Principle. The Supreme Reality's very being. "Brahman is Existence, Knowledge and Bliss; but these are not attributes. Brahman cannot be said to exist. Brahman is existence itself." Brahman is absolute and infinite—the eternal, immutable, illimitable, omniscient, formless Great Mystery: "Being absolutely present, Brahman is within all creatures and objects."[11]

Sat Purusha: Spirit, the Over-Soul, the Cosmic Person. The One Life and One Consciousness of the universe which is omnipresent as the life and consciousness, the Self of all beings.

Sattva guna, (satva, satwa): The force of equilibrium in nature. The Mysterious Supreme Life Essence of a sacred living universe.

Scientism: Unfounded belief in the efficacy of science.

Shabdabrahma: Causal vibration or undifferentiated soundless sound. It is the wavelength of the experience of God. *Para shakti* is its slumbering energy. It contains the three qualities of sattva, rajas, and tamas.

Shakti, Sakti: Power, energy, strength, the power of consciousness to act. The kinetic aspect of the Ultimate Principle, personified as the consort of the God Shiva.

Shiva, Siva: The fundamental forces in creation are personified in the Hindu trinity of Brahma (Creator +), Vishnu (Sustainer 0) and Shiva (Destroyer −). Shiva personifies the centripetal force in creation, the static aspect of consciousness, tamas guna, underlying matter. Shiva the Destroyer releases Spirit from matter.

Siddhi: Accomplishment, attainment of inner power.

Siva: See Shiva.

Somatics: Somatic Therapy works with the body, conscious of its relationship to the mind, emotions, and Spirit.

Soul: The transpersonal animating force of the body.

Spanda: Pulse, movement, vibration.

Sphota: Manifestor, the seed essence of the subtle essential unmanifest sound, "significant sound," as "transcendental" or "intelligible" sound representing the Platonic ideas or *logoi*, which are eternal, ubiquitous, and noumenal.

Spiritus: (Latin, "life breath") The breath of life.

Sristi-kalpana: Creative ideation, creative unfolding, evolution.

Standing waves: Pattern of resonance that defines an energy system. Standing waves are the "natural" patterns of resonance which define phenomena.

Sthula Sarira: The physical body, "sheath built of food," same as annamayakosa.

Sukshma sarira: Subtle body, astral body, vehicle of the Soul, unchanging through reincarnations until liberation. The vijnanamayakosa, manomayakosa, and pranamayakosa are often given the general name of *linga sarira*.

Sui generis: (Latin, "of its own kind") Constituting a class by itself.

Sunyata: The unbounded, void of boundaries, the boundless void. Used by the nontheistic Buddhists to refer to the Absolute.

Surat Shabd Yoga: The practice of Soul communion through union with the Sacred Sound Current.[12]

Sushumna: With pingala and ida, form the channels of subtle energy *(nadis)* which form the nexus that steps-down energy from "the universal creative energy" (prakriti) into the physical body. Sushumna, pingala, and ida resonate with the universal vibrational quintessence of the gunas (sattva, rajas, and tamas). Sushumna is the neutral field of the ultrasonic core of the subtle body. Its current is centered in a hollow at the core of the spinal column.

Svadhishthana: Water chakra at the lumbosacral junction. It rules cohesion, cleansing, and nourishment.

Sympathetic vibration: Or resonance, "a vibration produced in one body by the vibrations of exactly the same period in another body."[13]

Synergy: Describes systems in which the organization and functioning of the parts maximally sustain the whole, and the organization of the whole maximally sustains the parts; thus the total action is greater than the sum of the parts.

Tamas guna: The force of contraction and inertia in nature.

Tanmatras: Atom, rudimentary or subtle elements. The quintessence of hearing, seeing, touching, tasting, smelling.

Theomorphic: (Latin, Greek, *Theos,* "God"; Greek, *morphe,* "form"; hence, "the form of God") In the theomorphic paradigm every aspect of creation is a reflection of the process of creation.

Transpersonal: Transcending the personal or individual. Transpersonal refers to ancient esoteric schools of Hindu and Buddhist psychology which view individuality—in a certain profound sense—as not real.

Ubiquitous: Being everywhere at the same time, constantly encountered.

Ultrasonic Core: The energy current of the Soul. It is the primary energy that builds and sustains all others. It flows through the sixth ventricle of the brain and spinal cord.

Upanishads: Upanisads: ("at the knee, listening") One of the world's oldest esoteric teachings (eighth century BC). The philosophical portion of the *Vedas*. For millennia the Sanatana Dharma has been transmitted orally through teachers' discourses on the Upanishads to earnest disciples sitting at their feet.

Vayu: The Air element is the breath of Brahma. All movement of energy is as an expression of universal law and is ruled by Air.

Ved: (Sanskrit, *vid*, "to know") The holographic knowledge which manifests The Universe.

Vedas: The oldest known Indo-European religious and philosophical writings.

Vijnamayakosa: Sheath built of discrimination.

Water element: The attractive principle in nature that underlies contraction and form.

Yang: Centrifugal force in all creation, the masculine, active, solar force that combines with yin to produce all of nature.

Yin: Centripetal force in all creation, the feminine, passive, lunar force that combines with yang to produce all of nature.

Yoga: (to yoke or join) Union with the Self. The science of yoga describes practices that facilitate Soul communion and yogic perception to foster knowledge of the Self.

Yogin, yogini (female): One who practices yoga.

Bibliography

Polarity Therapy

Arroyo, Stephen. *Astrology, Psychology, and the Four Elements*. Sebastopol, California: CRCS Publications, 1975.

Beaulieu, John. *Music and Sound In The Healing Arts*. Berrytown, New York: Station Hill Press, 1987.

————. *Polarity Therapy Workbook*. New York: Biosonic Enterprises, Ltd., 1994.

Bodary, John. *Index to the Polarity Writings of Dr. Randolph Stone*. Dearborn, Michigan: Polarity Center, 1992.

Chitty, John and Anna. *Relationships and the Human Energy Field*. Boulder, Colorado: Polarity Press, 1991.

Chitty, John, and M. L. Muller. *Energy Exercises For Health and Vitality*. Boulder, Colorado: Polarity Press, 1990.

Francis, John. *Polarity Self-Help Exercises*. Boulder, Colorado: APTA , 1985.

Gordon, Richard. *Your Healing Hands*. Berkeley, California: Wingbow, 1985.

Muller, Mary Louise. *Self-help for Cranial Integration*. Boulder, Colorado: APTA

Seidman, Maruti. *Guide to Polarity Therapy*. Boulder, Colorado: Box 3175, 1986.

Siegal, Alan. *Polarity Therapy the Power That Heals*. Garden City Park, New York: Avery, 1987.

Sills, Franklin. *The Polarity Process*. Longmead, England: Element Books, 1989.

Stone, Randolph. *Polarity Therapy*, Vol I. Sebastopol, California: CRCS Publications, 1986.

———. *Polarity Therapy*, Vol II. Sebastopol, California: CRCS Publications, 1987.

———. *Health Building*. Sebastopol, California: CRCS Publications, 1985.

———. *The Mystic Bible*. Chicago,Illinois: Radha Swami Satsang, 1956.

Young, Phil. *The Art of Polarity Therapy*. Garden City Park, New York: Avery, 1990.

Energy: Newsletter of the American Polarity Therapy Association. The American Polarity Therapy Association, 2888 Bluff St. #149, Boulder, CO 80301, (303) 545–2080, Fax (303) 545–2161.

Polarity Diet

Aihara, Herman. *Acid and Alkaline*. Oroville, California: George Ohsawa Macrobiotic Foundation, 1986.

Ballentine, Rudolph. *Diet and Nutrition: A Holistic Approach*. Honesdale, Pennsylvania: Himalayan Institute, 1978.

Chopra, Deepak, et. al. *Perfect Health*. New York: Harmony Books, 1990.

Cousens, Gabriel. *Spiritual Nutrition: The Rainbow Diet*. Boulder, Colorado: Cassandra Press, 1986.

Murietta Foundation. *Murrietta Hot Springs Vegetarian Cookbook*. Summertown, Tennessee: The Book Publishing, 1987. This is *the* Polarity diet cookbook, which was formulated by a large community of individuals applying the Polarity principles over more than a decade of healthy living.

Pitchford, Paul. *Healing With Whole Foods*. Berkeley, California: North Atlantic, 1993. This is an encyclopedia of nutrition with an oriental energy perspective, developed in teaching and healing at the Heartwood community.

Vanamali Ashram. *The Taste Divine Cookbook*. Buffalo, New York: SUNY Press, 1992. An excellent introduction to sattvic cooking.

General

Avalon, Arthur. *The Serpent Power*. New York: Dover Publications: 1974. (Arthur Avalon was a pen name of Sir John Woodroffe.)

———. *Tantra of the Great Liberation*. New York: Dover Publications, 1972.

Babbitt, Edwin, D. *The Principles of Light and Color*, Malaga, New Jersey: Spectro-Chrome Institute, 1925.

Balsekar, Ramesh S. *The Final Truth: A Guide to Ultimate Understanding*, Los Angeles: Advaita Press, 1989.

Consciousness Speaks. Los Angeles, California: Advaita Press, 1989.

Beaulieu, John. *Music and Sound in the Healing Arts*. Barreytown: New York: Station Hill Press, 1987.

Beinfield, H., and E. Korngold. *Between Heaven and Earth*. New York: Ballantine, 1991.

Bentov, Itzhak. *Stalking the Wild Pendulum*. New York: Dutton, 1977.

Besant, Annie, and Charles Leadbeater. *Occult Chemistry*, London: Theosophical Publishing, 1908.

Brendt, J. E. *Nada Brahma :The World As Sound*. London: East West Publications, 1983.

Capra, F. *The Turning Point*. New York: Bantam, 1982.

———. *The Tao of Physics*. Boston, Massachusetts: Shambala, 1976.

Chatterji, J. C. *The Wisdom of the Vedas*. Wheaton, Illinois: Quest, 1992.

Combs, A., and M. Holland. *Synchronicity: Science, Myth, and the Trickster*. New York: Paragon House, 1995.

Coomaraswamy, Ananda K. *The Dance Of Shiva*, New Delhi, India: Sagar Publications, 1991.

Cousto, H. *The Cosmic Octave*. Mendocino, California: Life Rhythm, 1988.

Dass, Baba Hari. *Between Pleasure and Pain*. Victoria, British Colombia: Dharma Sara Publications, 1976.

Deussen, Paul. *The Philosophy of the Upanishads*. New York: Dover, 1966.

Dhondhen, Dr. Yeshi, and Jhampa Kelsang. *Ambrosia Heart Tantra*. Upper Dharamsala, India: Dharamsala: Library of Tibetan Works and Archives, 1977.

Doczi, Gyorgy. *The Power of Limits*, Boston, Massachusetts: Shambala, 1985.

Easwaran, Eknath. *Dialogue With Death*. Petaluma, California: Nilgiri Press, 1981.

———, *The Bhagavad Gita for Daily Living*, Vol I. Petaluma, California: Nilgiri Press, 1979.

Frawley, David. *Ayurvedic Healing*. Salt Lake City, Utah: Passage Press, 1989.

Godwin, Joscelyn.*The Harmonies of Heaven and Earth*. Rochester, Vermont: Inner Traditions, 1987.

Hahshananda, Swami. *Hindu Gods and Goddesses*. Mysore, India: Ramakrishna Ashram, 1982.

Hamel, P. M. *Through Music To The Self.* Longmead, England: Element Books, 1976.

Hart, William. *Vipassana Meditation.* New York: Harper, 1987.

Hartmann, Franz. *Paracelsus: Life and Prophecies.* Blauvelt, NY: Rudolph Steiner Publications, 1973.

Hendricks, Gay, and Barry Weinhold. *Transpersonal Approaches to Counseling and Psychotherapy.* Denver, Colorado: Love Publishing, 1982.

Johari, Harish. *Tools for Tantra.* Rochester, Vermont: Destiny Books, 1986.

Lama Foundation. *Be Here Now: Cookbook for a Sacred Life.* Lama, New Mexico: Lama Foundation, 1970.

Lawlor, Robert. *Sacred Geometry.* London: Thames and Hudson, 1982,

Leadbeater, C. W. *The Chakras,* Madras, India: Theosophical Publishing, 1973.

Le Mee, Jean. *Hymns from the Rig Veda.* New York: Knopf, 1975.

Maharshi, Sri Ramana. *The Spiritual Teachings of Ramana Maharshi,* Boston, Massachusetts: Shambala, 1972.

———. *Who Am I ?* Tiruvnnamalai, India: Sri Ramanasramam, 1992.

McClain, E. G. *The Myth of Invariance: The Origin of the Gods, Mathematics, and Music from the Rig Veda to Plato.* York Beach, Maine: Nicolas-Hays, 1976.

Nisargadatta, Maharaj, and M. Frydman, ed., *I Am That.* Bombay, India: Chitanya, 1974.

———, and Robert Powell, ed., *The Ultimate Medicine.* San Diego, California: Blue Dove Press, 1994.

Ohsawa, George. *The Art of Peace.* Oroville, California: George Ohsawa Microbiotic Foundation, 1990.

Ouspensky, P. D. *In Search of the Miraculous.* New York: Harcourt, Brace and Jovanovich, 1949.

Poonja, H.W.L. *Wake Up and Roar.* Lucknow, India: Universal Booksellers, 1992.

Prabhavananda, Swami. *The Spiritual Heritage of India.* Los Angeles, California: Vedanta, 1979.

———, and Christopher Isherwood. *Shankara's Crest Jewel of Discrimination.* Hollywood, California: Vedanta Press, 1947.

———, and Christopher Isherwood. *The Song of God: The Bhagavad Gita.* New York: Mentor Books, 1972.

Pradhan, V. G. *Jnaneshvari.* Albany, New York: SUNY Press, 1987.

Prem, Sri Krishna, and Sri Madhava Ashish, *Man the Measure of All Things.* Wheaton, Illinois: Theosophical, 1969.

Purce, Jill. *The Mystic Spiral.* London: Thames and Hudson, 1974.

Quigley, David. *Alchemical Hypnotherapy*. Redway, California: Lost Coast Press, 1984.

Radhakrishnan, S. *The Principal Upanisads*. New York: Humanities Press, 1974.

Ramacharaka, *The Bhagavad Gita*. Chicago, Illinois: Yogi Publication Society, 1935.

Schneider, Michael S. *A Beginner's Guide to Constructing the Universe*. New York: Harper, 1994.

Seal, Brajendranath. *The Positive Sciences of the Ancient Hindus*. Delhi, India: Motilal Banarsidass, 1985.

Sogyal, Rimpoche, *The Tibetan Book of Living and Dying*. New York: Harper, 1992.

Talbot, Michael. *Beyond the Quantum*. New York: Bantam Books, 1986.

Tame, David. *The Secret Power of Music*. Rochester, Vermont: Destiny Books, 1984.

Three Initiates, *The Kybalion*. Chicago, Illinois: Yogi Publication Society, 1940.

Tyberg, J. *The Language of the Gods*. Los Angeles, California: East-West Cultural Center, 1970.

Vanamali, Mathaji. *The Play of God*. San Diego, California: Blue Dove Press (800) 691-1008, 1995.

———. *Nitya Yoga: The Path of Constant Communion*. San Diego, California: Blue Dove Press, 1998.

———. *Nitya Yoga*. San Diego, California: Blue Dove Press, 1997. Illuminating audio discourse (13 tapes) on *The Bhagavad Gita*.

———. *The Taste Divine Cookbook*. Buffalo, New York: SUNY Press, 1992.

———. *The Bhagavad Gita*. New Delhi, India: Aryan Press, 1997.

———. *The Bhagavad Gita*, San Diego, California: Blue Dove Press, 1998, (800) 691–1008. Audio tape chanting in Sanskrit and English.

Wallace, R. K. *The Physiology of Consciousness*. Fairfield, Iowa: MIU Press, 1993.

Watts, Alan W. *The Two Hands of God*. New York: Collier, 1969.

Wilber, Ken. *The Spectrum of Consciousness*. Wheaton, Illinois: Theosophical Publishing, 1977.

———. *No Boundary*. Boston, Massachusetts: Shambala, 1979.

———. *The Atman Project*. Wheaton, Illinois: Theosophical Publishing, 1980.

———. *Grace and Grit*. Boston, Massachusetts: Shambala, 1993.

———, ed., *The Holographic Paradigm*. Boston, Massachusetts: Shambala, 1982.

Woodroffe, Sir John. *The World as Power*. Madras, India: Ganesh, 1966.

————. *The Garland of Letters*. Madras, India: Ganesh, 1971.

————. *Principles of Tantra,* Part II. Madras, India: Ganesh, 1970.

Yogi, Maharishi Mahesh. *The Science of Being and Art of Living*. Fairfield, Iowa: MIU Press, 1966.

Zaehner, R. C. *The Bhagavad Gita*. New York: Oxford University Press, 1969.

Zukav, G. *The Dancing Wu Li Masters*. New York: Bantam Books, 1979.

End Notes

Throughout this work, we refer to Dr. Stone's works variously by title, book number, and abbreviation. Following the order of publication, Book I, then, is *Energy;* Book II, *The Wireless Anatomy of Man;* Book III, *Polarity Therapy;* Book IV, *The Mysterious Sacrum;* Book V, *Vitality Balance;* EEC, *Evolutionary Energy Charts;* ET, *Energy Tracing;* PN, *Private Notes;* HB, *Health Building.* Where *Polarity Therapy: The Complete Collected Works of Dr. Randolph Stone* is referenced by volume, Volume I contains Books I, II, and III. Volume II contains Books IV, V, EEC, ET, and PN.

Chapter 1

1. Stone, Book II, chart 3.
2. Transpersonal refers to ancient schools of psychology which teach that individuality (in a certain profound sense) is not real.
3. Stone, Book II, p. 3, the first principle of Polarity Therapy, Summary of Principles, in the Introduction to *The Wireless Anatomy.*
4. Ibid., p. 2.
5. Stone. Book III, p. 33.

Chapter 2

1. Stone, Book III, p. 33.
2. Stone, HB.
3. Stone, Book I, p. 72.
4. Ibid., p. 75

5. Stone, Book III, pp. 40–41.
6. Stone, Book I, p. 72.
7. Stone, Book III, p. 38.
8. Stone, Book I, p. 71.

Chapter 4

1. Stone, HB, pp. 55–60; Book II, pp. 65–66, charts 3, 5–8, 43, 44; Book III, chart 3; Book V, chart 7; EES, charts 1, 12, 17.
2. Stone, EES, chart 12.
3. Stone, Book II, p. 65.
4. Ibid., charts 9, 10.
5. Stone, Book II, charts 3, 5, 6, 7, 8; Book III, chart 3; Book V, chart 7; EES, chart 1.
6. Stone, HB, pp.111–112; Book I, pp. 22–23.

7. Stone, Book V, chart 7.
8. Stone, EES, chart 17, p. 195.
9. Stone, Book II, p. 66.
10. Stone, HB, p. 55.
11. Stone, EES, chart 2.
12. Ibid., chart 17.
13. Stone, EES, charts 2, 3, 4; Book III, chart 7.
14. Stone, Book II, charts 5–7, 17, 33.
15. Stone, Book III, chart 7.
16. Stone, Book II, chart 37; Book III, chart 3; Book V, charts 4, 14; EES, chart 6.
17. Stone, Book V, chart 4.
18. Stone, Book II, chart 54, fig. 5.
19. Stone, Book III, charts 7, 8.
20. Stone, Book V, chart 3.
21. Stone, EES, charts 2, 3, 18.
22. Ibid., chart 4.
23. Stone, Book II, chart 32; Book IV, pp. 4, 24, 41.
24. EES, chart 13.
25. Stone, Book II, chart 11.
26. Stone, EES, charts 19, 20.
27. Stone, Book III, chart 7; EES, chart 4.
28. Stone, EES, charts 2, 3, 18.
29. Stone, Book II, chart 4.
30. Stone, Book II, chart 27.
31. Stone, Book V, charts 7, 8.
32. Stone, Book I, p. 80.
33. Stone, Book I, p. 31.
34. Stone, Book II, p. 4.
35. Ibid., chart 22.
36. Ibid., p. 66.
37. Ibid., p. 65.
38. Ibid., p. 66.
39. Stone, Book I, p. 80.
40. Stone, Book II, chart 18.
41. Stone, Book II, chart 17, 31, 33; Book V, chart 4.
42. Stone, Book II, chart 2; Book III, chart 3; Book V, chart 4; EES, chart 6.
43. Stone, Book II, chart 33.
44. Ibid., chart 4.
45. Stone, Book II, chart 31; EES, chart 20.
46. Stone, Book III, chart 3; EES, charts 5, 6.
47. Stone, Book V, chart 4.
48. Ibid., chart 19.
49. Stone, Book II, charts 5, 6; EES, charts 1, 17.
50. Stone, Book I, p. 80.
51. Stone, Book II, chart 11, p. 18.
52. Stone, Book V, chart 7.
53. Chitty and Muller, Energy Exercises, (Boulder, Colorado: Polarity Press, 1990), p.114.
54. Stone, Book V, p. 36.
55. Stone, Book II, chart 38, p. 45.
56. Stone, Book III, pp. 40–41.
57. Stone, Book V, p. 39.
58. Stone, Book II, chart 17, 31, 33; Book V, chart 4.
59. Stone, Book II, chart 2; Book III, chart 3; Book V, chart 4; EES, chart 6.
60. Stone, Book III, chart 3; EES, charts 5, 6.
61. Stone, Book V, chart 4.
62. Ibid., chart 19.
63. Stone, Book II, charts 5, 6; EES, charts 1, 17.
64. Stone, Book I, p. 80.
65. Stone, Book II, chart 11.
66. Ibid., chart 2, p. 9.
67. Ibid.
68. Stone, Book II, chart 3, p. 10.
69. Stone, Book III, chart 1, p. 27.
70. Stone, HB, p. 57.
71. Stone, Book I, chart 2.
72. Stone. Book V, chart 2, 19.
73. Stone, EES, chart 20.
74. Stone, Book V, chart 19.
75. Stone, EES, chart 8, 16, 19, 20.
76. Ibid., chart 19.
87. Stone, Book II, chart 11.
78. Ibid., chart 30
79. Stone, Book V, p. 21.
80. Stone, Book II, chart 2, p. 9.
81. Ibid., chart 3, p. 10.
82. Stone, Book III, chart 1, p. 30.
83. Stone, Book V, chart 19.
84. Stone, Book II, chart 22, p. 19.

85. Stone, EES, charts 7, 8, 19, 20.
86. Ibid., chart 9.
87. Stone., Book V, chart 2.
88. Stone, Book II, pp. 90–102.
89. Ibid., chart 19.
90. Ibid., p. 85.
91. Stone, EES, chart 19.
92. Stone, Book II, chart 11.
93. Ibid., chart 30.
94. American Polarity Therapy Association, *Standards for Practice and Code of Ethics* (Boulder, Colorado: American Polarity Therapy Association, 1997), p. 26.
95. Stone, Book I, p. 38.
96. Stone, EES, chart 20.
97. Stone, Book V, p. 23.
98. Stone, Book I, p. 75.
99. Ibid., p. 80.
100. Stone, Book V, chart 17, p. 195.
101. Stone, EES, chart 20.
102. Ibid., chart 17.
103. Stone, Book V, p. 87.
104. Stone, EES, chart 7, p. 48.
105. Stone, Book V, chart 17.
106. Stone, Book 1, p. 80.
107. Stone, EES, chart 7, p. 185.
108. Stone, Book I, p. 75.
109. Stone, Book V, p. 59.
110. Ibid., p. 64.
111. Ibid., p. 87.
112. Stone, Book I, p. 86.
113. Stone, Book I, chart 8, p. 81–89; Book II, chart 30, 31; EES, chart 20.
114. Ibid., p. 51, charts 4, 5.
115. Stone, EES, charts 19, 20, 21.
116. Stone, Book II, chart 31.
117. Stone, EES, 19. 20.
118. Ibid.
119. Stone, Book II, chart 31.
120. Stone, EES, chart 19.
121. Stone, Book II, chart 11.
122. Ibid., chart 30.
123. Stone, Book II, charts 9, 10.
124. Stone, Book IV, p. 3.
125. Ibid., p. 6.
126. Ibid., p. 2.
127. Stone, Book III, chart 13, p. 71.
128. Stone, Book V, p. 65.
129. Stone, Book III, charts, 10–15; Book IV, charts 1–5.
130. Stone, Book IV, p. 19.
131. Stone, Book III, charts 10–12; Book IV, charts 1, 3; Book V, p. 95.
132. Stone, Book II, chart 18.
133. Stone, Book II, chart 33.
134. Stone, Book II, charts 31, 33; Book V, p. 95.
135. Stone, Book II, p 37.
136. Stone, Book III, chart 9.
137. Stone, Book V, p. 65; Book II, chart 25.
138. Stone, Book II, charts 25, 26.
139. American Polarity Therapy Association, *Standards for Practice and Code of Ethics*, Appendix D.
140. Stone, Book IV, chart 3.
141. Stone, Book IV, p. 3.
142. Ibid., p. 6.
143. Ibid.
144. Ibid., chart 2.
145. Stone, Book V, p. 84. See: Book I, Chart 7; Book II, charts 13, 21; Book III, charts 10–15; Book IV, charts 1–5.
146. Stone, Book IV, p. 19.
147. Stone, Book III, charts 10–12; Book IV, charts 1, 3; American Polarity Therapy Association, *Standards for Practice and Code of Ethics*, Part Three, pp. 13–15.
148. Stone, Book IV, chart 11.
149. Stone, Book II, chart 18.
150. Ibid., chart 33.
151. Stone, Book II, charts 31, 33.
152. Stone, Book I, p. 79.
153. Stone, Book II, charts 30, 31, p. 37.
154. Stone, Book IV, chart 2.
155. Stone, Book I, p. 84.
156. Stone, Book II, chart 11, p. 60; Book III, charts 9–11.

157. Stone, Book I, p. 80.
158. Stone, Book I, p. 80; Book I, chart 11; Book III, chart 4.
159. Stone, Book IV, chart 2.
160. Ibid., p. 13.
161. Ibid., charts 1, 2.
162. Stone, Book IV, p. 46.
163. Stone, Book II, chart 28.
164. Stone, Book II, chart 28, p. 35.
165. Stone, Book I, p. 70.
166. Stone, Book II, chart 29, p. 36.
167. Richard C. Miller, *Yoga Journal* (May/June, 1994,) pp. 84–85.
168. Stone, EES, charts 5, 6.
169. Stone, Book IV, chart 6.
170. Stone, Book III, chart 7.
171. Stone, Book IV, chart 6.
172. Stone, Book I, p. 44.
173. Stone, Book II, charts 28, 29.
174. Stone, Book II, chart 36; Book IV, chart 7.
175. Book IV, chart 8, p. 39.
176. Ibid., charts 8, 9.
177. Stone, Book II, chart 28.
178. Ibid., chart 59, p. 80.
179. Stone, Book V, p. 74.
180. Stone, Book II, chart 59; Book III, chart 5; EES chart 3.
181. Stone, Book II, chart 59.

Chapter 5
1. Stone, Book II, p. 3. The first principle of Polarity Therapy, "Summary of Principles."
2. Ken Wilber, *Spectrum of Consciousness* (Wheaton, Illinois: Theosophical Publishing, 1982), p. 27.
3. Will Durant, *The Case for India* (New York: Simon and Schuster, 1930).
4. Ibid.
5. Ramesh S. Balsekar, *The Final Truth: A Guide to Ultimate Understanding* (Los Angeles, California: Advaita Press, 1989).
6. Dr. Stone writes: "The Vedas . . . have probably the most lengthy and accurate of all systems on the account of creation . . . The whole Vedic system is very comprehensive." Randolph Stone, *Mystic Bible* (Beas, India: Radha Soami Satsang, 1977), p. 70.
7. Jean Le Mee. *Hymns from the Rig Veda* (New York: Knopf, 1975), pp. ix-x.
8. Bloomfield, *Religion of the Veda* (1908), p. 55, in S. Radhakrishnan, *The Principal Upanisads* (New York: Humanities Press, 1974), p. 17.
9. Prabhavananda and Christopher Isherwood, *The Song of God: Bhagavad-Gita* (New York: Mentor Books, 1972), pp. 11–13.
10. A. C. Bhaktivedanta Swami Prabhupada, *Bhagavad-Gita: As It Is* (Los Angeles, California: The Bhaktivedanta Book Trust, 1991), p. back cover. The Bhagavad Gita offers a synthesis that represents the essence of the Sanatana Dharma. Dr. Stone, the formulator of Polarity Therapy, kept the *Gita* at his bedside. An excellent discussion of *The Bhagavad Gita* is by Mathaji Vanamali, *Nitya Yoga: The Path of Constant Communion* (San Diego, California: Blue Dove Press, 1997), also an excellent presentation of Bhakti Yoga and the life of Sri Krishna is by Mathaji Vanamali, *The Play of God* (San Diego, California: Blue Dove Press, 1995).
11. Swami Prabhavananda and Christopher Isherwood, *Shankara's Crest Jewel of Discrimination* (Hollywood, California: Vedanta Press, 1947), p. 10. Shankara's teachings are the foundation of contemporary Hinduism.
12. Sir John Woodroffe, *The World as Power* (Madras, India: Ganesh, 1966), p. 158. Atman comes from

the Sanskrit verb root *at,* "to breathe." J. Tyberg, *The Language of the Gods* (Los Angeles, California, East-West Cultural Center, 1970), p. 4.

13. Prabhavananda and Isherwood, *Shankara's Crest Jewel of Discrimination,* pp. 86–87.

14. *The Upanishads,* literally, "at the knee, listening," are one of the world's oldest esoteric teachings (8th century B.C.). For millennia the Sanatana Dharma has been transmitted orally through teachers' discourses on *The Upanishads* to earnest disciples sitting at their feet. See Eknath Easwaran, *Dialogue With Death* (Petaluma, California: Nilgiri Press, 1981).

15. Prabhavananda and Isherwood, Shankara's *Crest Jewel of Discrimination,* p. 87.

16. Ibid., p. 104.

17. Ibid., p. 115.

18. Brahman is the Sanskrit term for the Absolute, The Supreme Being. R. C. Zaehner, *The Bhagavad Gita* (New York: Oxford University Press, 1969), p. 8.

19. Prabhavananda and Isherwood, *Shankara's Crest Jewel of Discrimination,* p. 104.

20. Ibid., p. 131.

21. Prabhavananda and Isherwood, *The Song of God: Bhagavad - Gita,* p. 70.

22. N. Vasudevan, "Man and Pure Consciousness," *The Mountain Path* (Tiruvannamalai, India: 1991), p. 124.

23. The Sanatana Dharma is not simply pantheistic, because it recognizes that God is beyond creation as well as within creation.

24. Paul Deussen, *The Philosophy of the Upanishads* (New York: Dover, 1966), p. 195.

25. Prabhavananda and Isherwood, *The Song of God: Bhagavad-Gita,* p. 136.

26. Tyberg, *The Language of the Gods,* p. 101.

27. Spirit is infinite (*Aparicchinna*) and formless (*Arupa*). Arthur Avalon, *The Serpent Power* (New York: Dover Publications, 1974), p. 26.

28. Robert Powell, Ph.D., Ed, *The Ultimate Medicine* (San Diego, California: Blue Dove Press, 1994).

29. The word epiphenomenon means a secondary phenomenon accompanying another and caused by it.

30. Ramcharaka, *The Bhagavad Gita* (Chicago, Illinois: Yoga Publication Society, 1935) (brackets mine).

31. Vanamali, *Nitya Yoga,* Thirteen audio tape series (San Diego, California: Blue Dove Press, 1995) Also published in book form (Tapovan Sari P.O., via Rishikesh, UP, 249192, India, Vanamali Ashram, 1995).

32. Dr. Deepak Chopra, "Ageless Body, Timeless Mind," *Yoga Journal,* Sep/Oct 1993, p. 64.

33. Prabhavananda and Isherwood, *Shankara's Crest Jewel of Discrimination,* p. 110.

34. Stone, Book II, p. 3.

35. "Physical sciences inform us that the whole of creation is built up of layers of energy, one inside the other. The subtlest constitutes the innermost stratum of creation. Underneath the subtlest layer of all that exists in the relative field is the abstract, absolute field of pure Being. It is said that Being is the ultimate reality of creation. It is present in all forms. We may say that existence is life itself,

while that which exists is the ever-changing phenomenal phase of the never changing reality of existence. Existence is the abstract basis of life on which is built the concrete structure of life." Maharishi Mahesh Yogi, *The Science of Being and Art of Living* (Livingston Manor, New York: MIU Press, 1976), pp. 25–28.

36. Avalon, *The Serpent Power*, p. 26. "Atman, the Supreme Knower in each and every one of us that is none other than Brahman, the sole and basic reality of the Universe .." Ken Wilber, *The Spectrum of Consciousness*, p. 87.

37. Prabhavananda and Isherwood, *The Song of God: Bhagavad-Gita*, p. 71.

38. Radhakrishnan, *The Principle Upanisads*, p. 82. Also, Eknath Easwaran, *The Bhagavad Gita for Daily Living*, Vol. I (Petaluma, California: Nilgiri Press, 1979), Introduction. In this explication we tend to emphasize the profound eminence of God yet we will at the same time constantly bow to the transcendent as well.

39. Please appreciate that we are not being pantheistic. While we emphasize that Brahman is the world, or the sum total of the universe, we do not overlook the equally important fact that Brahman is radically prior to the universe. See Ken. Wilber, ed., *The Holographic Paradigm* (Boston, Massachusetts: New Science Library, 1982), p. 255; Wilber, *The Spectrum of Consciousness,* p. 88.

40. Prabhavananda and Isherwood, *The Song of God: Bhagavad-Gita*, p. 131.

41. Ibid., pp. 37–38.

42. "The Greek word phusis—the root of our modern terms physics, physiology, and physician—as well as the Sanskrit Brahman, both of which denoted the essential nature of all things, derive from the same Indo-European root Bheu [to grow]." F. Capra, *The Turning Point* (New York: Bantam, 1982), p. 214.

43. For the ancients the Creator was an all-pervasive consciousness that was as near to us as our own life breath. Thus we define prana as the intelligent force which is the living breathing presence of the living intelligence of the one living being of universe in the subsystems of nature and man.

44. Prabhavananda and Isherwood, *The Song of God: Bhagavad-Gita*, p. 70.

45. Stone, HB, p. 7.

46. Ken Wilber, *No Boundary* (Boston, Massachusetts: Shambala, 1985), pp.128–129.

Chapter 6

1. Stone, Vol II, p. 207.

2. See the discussion of Maharishi Mahesh Yogi's teachings in R. K. Wallace, Ph.D, *The Physiology of Consciousness* (Fairfield, Iowa: MIU Press, 1993), pp. 214–216.

3. See the work of David Bohm, in Ken.Wilber, ed., *The Holographic Paradigm*. Also D. Bohm, *Wholeness and The Implicate Order* (London: Routledge and Kegan Paul, 1980).

4. Wallace, *The Physiology of Consciousness*, p. 8.

5. Ibid., pp. 214–216.

6. Ibid., p. 8.

7. Stone, HB, p. 7.

8. Alan W. Watts, *The Two Hands of God* (New York: Collier, 1969), p. 68.

9. E. G. McClain, *The Myth of Invariance: The Origin of the*

Gods, Mathematics and Music from the Rig Veda to Plato (York Beach, Maine: Nicolas-Hays, 1976).

10. Stone, HB, p. 85.

11. Swami Prabhavananda and Fredrick Manchester, *The Upanishads* (New York: Mentor Books, 1948). Also Avalon, *The Serpent Power*, p .47.

12. Stone, Book I, p. 22.

13. Perhaps the title of *The Bhagavad Gita*, which means "The Song of God," points to this ultimate truth. This article is developed from insights explicated in *The Bhagavad Gita*, which represents a major synthesis of Vedic thought. An excellent introduction to the *Gita* and its explication of the Sanatana Dharma is available on Audio Tape from Vanamali Gita Yoga Ashram, Tapovan Sari PO, via Rishikesh, 249192, India; or Blue Dove Press, San Diego, California, (800) 691–1008.

14. See the work of the Maharishi Mahesh Yogi, for example, *The Science of Being and the Art of Living*.

15. Lord Shiva personifies the changeless, transcendental aspect of creation, in the trinity of the Sanatana Dharma. Its power or consort is Shakti, the changing realm of nature.

16. "What is here is everywhere ... what is not here is nowhere." Visvasara-Tantra, Sir John Woodroffe, *The World as Power* (Madras, India: Ganesh, 1966), p. 291.

17. J. C. Chatterji, *The Wisdom of the Vedas* (Wheaton, Illinois: Quest, 1992), p. 23.

18. V. G. Pradhan, *Jnaneshvari* (Buffalo, New York: SUNY Press, 1987), p. 170.

19. "The Self knows and loves the Self." It is this love that is Bliss or "resting in the Self" for as it is elsewhere said, "Supreme love is bliss." Avalon, *The Serpent Power*, p. 32.

20. In the Advaita, God is often referred to as the Self. The science of Advaita is an inner process, a science (methodical and replicable) of self-realization.

21. Deepak Chopra, *Timeless Body, Ageless Mind*, audiotape.

22. Crossculturally and transhistorically the fundamental axiom of Hermetic research has been "As above, so below." See the writings of Sri Krishna Prem or Yogi Ramacharaka.

23. Stone, Book IV, p. 44.

24. Avalon, *The World as Power*, "Visvasara-Tantra," p. 22.

25. Stone, Book I, p. 23.

26. Ibid., p. 27.

27. Franz Hartmann, *Paracelsus: Life and Prophecies*, Blauvelt, NY, Rudolph Steiner Publications, 1973.

28. Vedic sutra quoted by Deepak Chopra, M.D., in a lecture to T.M. Meditators, Oxnard, California. This sutra represents the actual vision, the realization of the ancient seers.

29. K. Wilber, *Grace and Grit* (Boston, Massachusetts: Shambala, 1993), p. 17.

30. Bohm, *Wholeness and The Implicate Order*, p. 178.

31. Three Initiates, *The Kyballion* (Chicago, Illinois: Yoga Publishing, 1940).

32. *Webster's New Collegiate Dictionary* (Springfield, Massachusetts: G. and C. Merriam, 1975).

33. Brajendranath,*The Positive Sciences of the Ancient Hindus* (Delhi, India: Motilal Banarsidass, 1985), p. 17.

Chapter 7

1. Prabhavananda and Isherwood, *The Song of God: Bhagavad-Gita*, p. 80.
2. Seal, *The Positive Sciences of the Ancient Hindus*, (Delhi, India: Motilal Banarsidass, 1985), p. 121.
3. Stone, HB.
4. Watts, *The Two Hands of God*, see *The Bhagavad Gita*, Chapter XVI.
5. David Frawley, *Ayurvedic Healing* (Salt Lake City, Utah: Passage Press, 1989), p. 3.
6. Ibid., p. 3.
7. Ibid., p. 4.
8. "The Way" of living life in accordance with universal law.
9. Theos. Bernard, *Hindu Philosophy* (New York: Philosophical Library, 1947), p. 74.
10. Stone, Book I, p. 16.
11. G. Zukav, *The Dancing Wu Li Masters* (New York: Bantam Books, 1979), p. 206.
12. Stone, Book IV, p. 2.
13. Vanamali, *Nitya Yoga*. To my heart and mind, the most revealing discussion of *The Bhagavad Gita* and the wisdom of the Sanatana Dharma.
14. Wilber, *The Spectrum of Consciousness*, p. 184.
15. At every point of intersection of *rajas guna* (centrifugal force) and *tamas guna* (centripetal force) a stillpoint of sattva guna equilibrium is sustained which is the bindu ... the seed, resonating with the Unitive Field to step-down to the next octave of manifestation.
16. Stone, Book I, p. 46.
17. Ibid., p. 82.
18. Radhakrishnan, *The Principal Upanisads*, p. 73.
19. Phil Nurenberger, "Overcoming Fear," *Yoga International*, Nov. 94, p. 25.
20. Tyberg, *The Language of the Gods*, p. 103.
21. Ibid., p. 111.
22. Nurenberger, "Overcoming Fear," pp. 24–25.
23. Bernard, *Hindu Philosophy*, p. 93.
24. Ramcharaka, *The Bhagavad Gita*, p. 87.
25. Vanamali, *Nitya Yoga*, Chapter 15.

Chapter 8

1. Bruce Rawles, http://www.summum.org/phi.htm: Summum, 707 Genesee Avenue, Salt Lake City, UT, 84104 (801) 322–3738.
2. David Canright, http://www.math.nps.navy.mil/~dcanrig: Mathematics Department, Code MA/Ca, Naval Postgraduate School, Monterey, CA 93943.
3. Robert Lawlor, *Sacred Geometry* (London: Thames and Hudson,1982), p. 58.
4. David Tame, *The Secret Power of Music* (Rochester, Vermont: Destiny Books, 1984), p. 220–221.
5. Annie Besant and Charles Leadbeater, early Theosophists, studied with yogis in India and developed microclairvoyant vision. They reported the findings of this research in *Occult Chemistry* (London: Theosophical Publishing,1908). (A reprint of the 1919 edition of *Occult Chemistry* is available from, Health Research, Mokelumne Hill, CA 95245) In this clairvoyant research, which took place between 1895 and 1933, a group of Theosophists using yogic powers developed a "Periodic Table of the Elements." From carbon to uranium, they described and diagrammed fifty-

seven chemical elements—the basic constituents of matter—and developed a formula for ascertaining their atomic weights. They found ninety-nine distinct elements, including a number that were yet to be discovered. The book, *Extra Sensory Perception of Quarks* (Wheaton, Illinois: Theosophical Publishing, 1980), demonstrates that the micro-clairvoyant research of Dr. Besant and Bishop Leadbeater had indeed been accurate and has been largely confirmed by a half-century of scientific subatomic research.

6. Based on the illustrations in Gyorgy Doczi, *The Power of Limits,* (Boston: Massachusetts: Shambala, 1985), p. 104.

7. Ibid.

8. Michael S. Schneider, *A Beginner's Guide to Constructing the Universe* (New York: Harper, 1994), p. 254.

9. Tame, *The Secret Power of Music,* p. 221.

10. Lawlor, *Sacred Geometry,* p. 31.

11. Ibid., p. 229.

12. Ibid., p. 58.

13. Ibid.

14. Tame, *The Secret Power of Music,* p. 4.

15. Ibid., p. 220.

16. Stone, HB.

17. Stone, Book I, p. 18.

18. Avalon, *The Serpent Power,* p. 70.

19. Baba Hari Dass, *Between Pleasure and Pain* (Suma, Washington: Dharma Sara Publications, 1976), p. 13.

20. Shivani Arjuna, "Balancing Vata Dosha," *Energy: The Newsletter of the American Polarity Therapy Association,* Fall 1996.

21. Dr. Yeshi Dhondhen and Jhampa Kelsang, *Ambrosia Heart Tantra,* (Dharamsala: Upper Dharmsala, India: Library of Tibetan Works and Archives, 1977), p. 33.

22. William Hart, *Vipassana Meditation* (New York: Harper, 1987), p. 26. Also see the clairvoyant research of Besant, et. al., cited herein, and their description of the ultimate atom as a vortex-sustaining space.

23. Stone, PN, p. 226.

24. Radhakrishnan, *The Principal Upanisads,* p. 56.

25. Franklin Sills, *The Polarity Process* (Dorset, England: Element Books, 1989), p. 57.

26. Stone, Book V, p. 25.

27. Dhondhen and Kelsang, *Ambrosia Heart Tantra,* p. 33.

28. Doczi, *The Power of Limits,* p. 104.

29. Wilber, ed., *The Holographic Paradigm,* p. 255.

30. Stone, Book I, chart 2.

31. Stone, *Mystic Bible* (Beas, India: Radha Soami, 1977), p. 70.

32. Stone, Book II, chart 3, p. 10.

33. "In Mahayana Buddhism this 'within which is beyond' is called tathagatagarbha, or Matrix of Reality. The word 'matrix' suggests the universal field-like nature of reality, and thus is reminiscent of the Dharmadhatu or Universal Field. In fact, the tathagatagarbha is actually identical to the Dharmadhatu as centered on the individual, just as in Hinduism the Atman is identical to Brahman as centered in the individual. But the tathagatagarbha (as well as the Atman) has a more psychological and 'personal' ring, as evidenced by the fact that it also means the Womb of Reality .." Wilber, *Spectrum of Consciousness,* p. 87.

34. C. W. Leadbeater, *The Chakras*

(Madras, India: Theosophical Publishing, 1973), p. 32.

35. American Polarity Therapy Association, *Standards for Practice*, p. 8.

36. Leadbeater, *The Chakras*, pp. 4–5.

37. Ibid., p. 34.

38. Ibid., p. 8.

39. Chitty and Muller, *Energy Exercises*. This excellent chart by Mark Allison, R.P.P., is based on Stone, Book II, charts 2, 3.

40. APTA, *Standards for Practice*, p. 9.

41. Woodroffe, *The World as Power*, pp. 103–105.

42. Ibid., p. 100.

43. Ibid., p. 208.

44. Tyberg, *The Language of the Gods*, p. 29.

45. Radhakrishnan, *The Principal Upanisads*, Barhad-aranyaka Upanisad, p. 243.

46. Ibid., p. 176.

47. Stone, Book III, p. 41.

48. Avalon, *The Serpent Power*, pp. 77–78; Seal, *The Positive Sciences of The Ancient Hindus*, p. 229.

49. Eleanora Lipton, R.P.P., Polarity Wholeness Certification Program, Atlanta, Georgia.

50. Stone, Book II, p. 62.

51. Stone, Book III, p. 41.

52. Stone, Book II, p. 72.

53. Stone, Book V, pp. 67–68.

54. Harish. Johari, *Tools for Tantra* (Rochester, Vermont: Destiny Books, 1986), p. 5; Stone, Book III, chart 1.

55. Stone, HB, p. 114–116. We have substituted "currents" for "senses" to clarify what appears to be a typographical error in the original.

56. Stone, Book II, chart 3.

57. Stone, Book III, p 7.

58. Johari, *Tools for Tantra*, p. 5.

59. Dr. Stone emphasized "oval fields" ruled by the dynamic of polarity. The heart-shaped fields of our model are made up of intersecting oval fields of centrifugal and centripetal forces.

60. APTA, *Standards For Practice*, p. 8.

61. Stone, Book IV, p. 51.

62. Ibid.

63. Based on illustrations in Doczi, *The Power of Limits*, p. 10.

Chapter 9

1. George Ohsawa, *The Art of Peace* (Oroville, California: George Ohsawa Microbiotic Foundation, 1990), p. 65.

2. Theos. Bernard, Ph.D., *Hindu Philosophy*, p. 69. Bernard was a Sanskrit scholar and Professor of Philosophy at Columbia University. He was recognized as a *tulku*, a reincarnation of a Tibetan lama, and was the first Western scholar to be initiated into monastic training in Tibetan Buddhism. He spent a period of time in the 1930s studying the Sanatana Dharma in Tibet.

3. Ibid., p. 66. See also Seal's classic, *The Positive Sciences of the Ancient Hindus*.

4. Bernard, *Hindu Philosophy*, p. 70.

5. Stone, Book I, p. 19.

6. Tyberg, *The Language of the Gods*, p. 67.

7. Seal, *The Positive Sciences of the Ancient Hindus*.

8. Bernard, *Hindu Philosophy*, p. 67.

9. Tyberg, *The Language of the Gods*, pp. 117–119.

10. "The Three Gunas," from Prabhavananda and Isherwood, *The Song of God: Bhagavad-Gita*, pp. 106–110.

11. E. Partridge, *Origins* (New York: Macmillan, 1958), p. 452.

12. Prabhavananda and Isherwood, *The Song of God: Bhagavad-Gita,* p. 134.

13. Stone, Book III, p. 13.

14. Stone, HB, p. 57.

15. From Zarathustra to Einstein the helix has been pictured as the symbol of the great mystery. See the video, *Nature Was My Teacher: The Vision of Viktor Schauberger,* Borderland Research, Bayside, California.

16. Bruce Berger, *The Holographic Body,* self-published audiotape; Garberville, California, 1999.

17. See the concept of "Dinergy " in Doczi, *The Power of Limits,* which is developed in the following discussion.

18. Stone, Book I, p. 22–23.

19. Stone, Book II, p. 64.

20. Thus giving us the twelve root archetypes of the zodiac, which embody the three principles of motion expressed through the four elemental phases of resonance.

21. Jill Purce, *The Mystic Spiral* (London: Thames and Hudson, 1974), p. 8.

22. Avalon, *The Serpent Power,* p. 70.

23. Barnard, *Hindu Philosophy,* Glossary, pp. 157–207.

24. Lawlor, *Sacred Geometry,* p. 7.

25. See the discussion of standing waves which follows.

26. Lawlor, *Sacred Geometry,* p. 7.

27. Capra, *The Tao of Physics* (Boston, Massachusetts: Shambala, 1976), p. 59.

28. I once asked Baba Hari Dass of Mount Madonna, Watsonville, California, why he called energy *sound* and why not *Shakti, force, light,* etc. He wrote on his chalk board (he has been silent for decades) that *"energy IS sound."*

29. In the cosmologies of the Sanatana Dharma a distinction is made between subtle inner universal elements and the gross elements which give rise to phenomena. In the following discussion, for the sake of simplicity, we will be ignoring this distinction.

30. Sri Krishna Prem and Sri Madhava Ashish " *Man the Measure of all Things* (Wheaton, Illinois: Theosophical, 1969). p. 35.

31. Stone, PN, p. 226–227.

32. Stone, EES, chart 2.

33. Stone, Book II, p. 66.

34. Ibid., p.71.

35. Ibid., p.103.

36. Itzhak Bentov, *Stalking the Wild Pendulum* (New York: Dutton, 1977), p. 22.

37. Stone, Book III, p. 20.

38. Ibid., p. 28–29.

39. Ibid., p. 16.

40. Sri Ramana Maharshi, *Who Am I?* (Tiruvnnamalai, India,: Sri Ramanasramam, 1992).

Chapter 10

1. Stone, HB, p. 15.

2. Prabhavananda and Manchester, *The Upanishads,* pp. 50–53.

3. Tyborg, *The Language of the Gods,* p. 18.

4. J. E. Brendt, *Nada Brahma: The World As Sound* (London: East-West Pub, 1983), p. 17–18.

5. Woodroffe, *The World as Power.*

6. Seal, *The Positive Sciences of the Ancient Hindus,* p. 153.

7. Sir John Woodroffe, *The Garland of Letters* (Madras, India: Ganesh, 1971), pp. 214–227.

8. David-Neel, *Tibetan Journey,* in Capra, *The Tao of Physics,* pp. 186–187.

9. Ibid., pp. 230–231.

10. Tame, *The Secret Power of Music,* p. 228.

11. The following section on sacred

sound leans heavily on the lucid explication from John Beaulieu, R.P.P., N.D., *Music and Sound in the Healing Arts* (Barreytown, New York: Station Hill Press, 1987), pp. 35–38.

12. Bentov, *Stalking the Wild Pendulum*, p. 23.

13. Beaulieu, *Music and Sound in the Healing Arts*, p. 36.

14. Ibid., p. 36.

15. P. D. Ouspensky, *In Search of the Miraculous* (New York: Harcourt, Brace, Jovanovich, 1949), p. 122.

16. Beaulieu, *Music and Sound in the Healing Arts*.

17. Ibid., p. 36–37.

18. Hans Jenny, *Cymatics: The Structure and Dynamics of Waves and Vibrations*, quoted in Beaulieu, *Music and Sound in the Healing Arts*, pp. 38–39.

19. Ibid.

20. D. Bohm and B. Hiley quoted in Capra, *The Tao of Physics*, pp. 124–125.

21. Capra, *The Tao of Physics*. p. 56.

22. Winter and Friends, *Sacred Geometry: Alphabet of the Heart*, p. 97.

23. Capra, *The Tao of Physics*. p. 138.

24. Ouspensky, *In Search of the Miraculous*, pp. 124–125.

25. Doczi, *The Power of Limits*.

26. Winter and Friends, *Sacred Geometry: Alphabet of the Heart*, p. 92.

27. Bentov, *Stalking The Wild Pendulum*, p. 10.

28. Doczi, *The Power of Limits*, p. 104.

29. Ibid.

30. Ibid., p. 97.

31. Ibid., p. 49.

32. Bohm, *Wholeness and the Implicate Order*, p. 178.

33. Hamel, *Through Music to the Self* (Dorset, England: Element, 1976), p. 93.

34. Beaulieu, *Music and Sound in The Healing Arts*, p. 43.

35. Berendt, *Nada Brahma: The World As Sound*, p. 60.

36. Joscelyn Godwin, *The Harmonies of Heaven and Earth* (Rochester, Vermont: Inner Traditions, 1987), pp. 143–149.

37. Ibid., p. 127.

38. Ibid., pp. 127–128.

39. H. Cousto, *The Cosmic Octave* (Mendocino, California, Life Rhythm: 1988), p. 25.

40. Ibid., p. 19

41. Ibid., pp. 19–20.

42. Hamel, *Through Music to the Self*, p. 105.

43. Ibid., p. 105.

44. Based on an illustration from Doczi, *The Power of Limits.*, p.104.

45. Stone, Book I, p. 18.

46. Lawlor, *Sacred Geometry*, p. 65.

47. Doczi, *The Power of Limits*.

48. Lawlor, *Sacred Geometry*, p. 42.

49. Stone, Vol II, p. 232.

50. Kirpal Singh, *Naam or Word* (Delhi, India: Ruhani Satsang, 1960), p. 188. For more information about Surat Shabd Yoga contact Radha Soami Books, P.O. Box 242, Gardena, CA 90247, (310) 329–5635.

51. Lama Foundation, *Be Here Now: Cookbook for a Sacred Life*, (Lama, New Mexico: Lama Foundation, 1970), pp. 83–84.

52. Excellent ear plugs for meditation are available from: Gateway Books, Pacific Ave., Santa Cruz, CA 95060.

53. Lama Foundation, *Be Here Now: Cookbook for a Sacred Life*, p. 83.

Chapter 11

1. Los Angeles, California: Philosophical Research Society, 1978.
2. E. D. Babbitt, *The Philosophy of Cure* (Los Angeles, California: College of Fine Forces, 1887).
3. E. D. Babbitt, *The Principles of Light and Color* (Los Angeles, California: College of Fine Forces, 1878. Reprint, Malaga, New Jersey: Spectro-Chrome Institute, 1925), pp. 95–96. An unedited reprint of The Principles of Light and Color with color plates is available from Borderland Sciences, Bayside, California. Beware of "edited" editions. One recent edition by a myopic color-psychologist, cuts out whole chapters of Babbitt's profound insights as "Victorian fluff."
4. Babbitt, *The Principles of Light and Color*, p. 96.
5. Ibid., p. 103.
6. Ibid., p. 119.
7. Ibid., p. 117.
8. Ibid., p. 97.
9. Ibid., p. 107.
10. Babbitt, *The Principles of Light and Color*; Stone, EES, chart 12, p. 190.
11. *Scientific American* (May 1957).
12. Purce, *The Mystic Spiral*, p. 109.
13. Randolf Stone, *Mystic Bible,* Radha Soami Satsang Beas, Punjab, India, 1989.
14. Besant and Leadbeater, *Occult Chemistry.*
15. Stone, Book V, chart 1.
16. Ibid., p. 2.
17. Stone, Book III, chart 1.
18. Some occultists point out that the evolutionary currents are also driven by Jupiter (*Brihaspati*, the Guru), which would certainly account for the red, orange, and yellow fiery colors which characterize the evolutionary return currents of the Earth, Water, and Fire chakras. Both Babbitt and Gurdjieff allude to this force.
19. Doczi, *The Power of Limits*, p. 1.
20. On the etheric plane the ovoid *Hiranyagarbha*, or self-creating egg, is the fundamental gestalt. But as energy fields come into a more material resonance the negative pole predominates and the field becomes more heart-shaped, as in the torso.
21. Does this mandala look familiar? It's the Heartwood logo.
22. "Babbitt's Atom," *The Principles of Light and Color.*
23. Stone, Book I, pp. 33–34.
24. Stone, Book III, p. 9.
25. Stone, Book I, chart 2.
26. Stone, PN, p. 1.
27. Baba Hari Dass, *Between Pleasure and Pain*, pp. 10–35.
28. Stone, EES, chart 2
29. Chitty and Muller, *Energy Exercises.* This excellent chart by Mark Allison, R.P.P., is based on Stone, Book II, charts 2, 3.
30. Stone, Book II, pp. 10–15.
31. Stone, Book IV, p. 38.
32. Capra, *The Tao of Physics*, p. 78.
33. Stone, Book V, chart 7, "Primordial Mind Patterns."
34. "By the single sun, This whole world is illumined: By its one Knower, The field is illumined." Swami Prabhavananda and Christopher Isherwood, *The Song of God: Bhagavad-Gita*, p. 104.
35. Ibid., Chapter 7, verses 7–9.
36. Doczi, *The Power of Limits*, p. 4.

Chapter 12

1. Avalon, *The Serpent Power*, p. 77. Arthur Avalon is a pen name of Sir John Woodroffe.
2. Lawlor, *Sacred Geometry*, p. 7.
3. Tyberg, *The Language of the Gods*, p. 102.
4. Avalon, *The Serpent Power*, p. 77.

5. Stone, Book I, p. 23.

6. Capra, *The Turning Point*, p. 234.

7. Ken Wilber, *The Atman Project* (Wheaton, Illinois: Theosophical Publishing, 1980), p. 1.

8. Avalon, *The Serpent Power*, p. 70.

9. Stone, Book V, p. 14.

10. Sufi teachings picture a core whose out-breath is compassion, lending its essence as a vehicle to all that seeks manifestation and whose in-breath is mercy, receptive to all that seeks return.

11. Stone, Book II, p 9.

12. See the work of Viktor Schauberger in the video, *Nature Was My Teacher: The Vision of Viktor Schauberger*.

13. Stone, Book II, pp. 16–18.

14. Stone, Book II, chart 9, p. 16.

15. Purce, *The Mystic Spiral*.

16. Stone, Book I, p. 2.

17. Hart, *Vipassana Meditation*, pp. 72–73.

Chapter 13

1. Swami Hahshananda, *Hindu Gods and Goddesses* (Mysore, India: Ramakrishna Ashram, 1982), p. 36.

2. In the Sanatana Dharma, Purusha, the universal Soul, is understood as omniscient, all-knowing, and does not evolve. Prakriti, nature, has as its fundamental organizing principle evolution, or growth.

3. Radhakrishnan, *The Principal Upanisads*, p. 285 [italics mine].

4. Ibid., pp. 781–782.

5. Ibid., p. 91.

6. Prabhavananda and Isherwood, *Shankara's Crest Jewel of Discrimination*, p 70.

7. Swami Prabhavananda *The Spiritual Heritage of India* (Los Angeles, California: Vedanta, 1979), p. 52.

8. Ibid., p. 70.

9. Bernard, *Hindu Philosophy*, p. 176.

10. Three Initiates, *The Kyballion*, p. 65.

11. Sogyal Rimpoche, *The Tibetan Book of Living and Dying* (New York: Harper, 1992), p. 90.

12. P. D. Ouspensky, *Teritum Organum* (London: Rutledge, Kegan, and Paul, 1970, in Capra, *The Turning Point*, p. 86).

13. Wilber, *No Boundary*, p. 39.

14. Ibid., p. 40.

15. Capra, *The Turning Point*, p. 86.

16. A. S. Eddington, *Nature of the Physical World* (London: Cambridge University Press, 1928), p. 338.

17. Michael Talbot, *Beyond the Quantum* (New York: Bantam Books, 1986), p. 4.

18. Radhakrishnan, *The Principal Upanisads*, p. 200.

19. Stone, Book II, p. 3.

20. Stone, Book I, pp. 18–20.

21. Stone, EES, chart 1.

22. Stone, Book I, p. 54.

23. Ibid., p. 55.

24. Stone, Book III, p. 8

25. Ibid., p. 14.

26. Stone, Book IV, p. 11.

27. Ibid., p. 51.

28. Stone, Book II, p. 70.

29. Stone, HB, pp. 114–115.

30. Ibid., pp. 114–116.

31. Stone, Book III, p. 5.

Chapter 14

1. Lawlor, *Sacred Geometry*, p. 90.

2. Hartmann, *Paracelsus: Life and Prophecies*.

3. Quoted in Capra, *The Turning Point*, p. 166.

4. Prem, *The Yoga of The Bhagavat Gita*, p. 96.

5. I once experienced "letterland," an archetypal realm, where I witnessed a phantasmagoria of letters emanating in every imaginable font and color of psychedelic

expression. Ouspensky describes a similar experience in which he experienced the room archetype and witnessed a splendor of interior decorating.

6. Tame, *The Secret Power of Music*, pp. 52–53.
7. Stone, Book I, p. 16.
8. *Culpepper's Complete Herbal* (New York: Sterling Publishing).
9. Babbitt, *Principles of Light and Color.*
10. Frawley, *Ayurvedic Healing.*
11. Stephen Arroyo, *Astrology, Psychology, and the Four Elements* (Sebastopol, California: CRCS, 1975), p. 95.
12. Ibid., p. 96.
13. Ibid.
14. Ibid.
15. Barnard, *Hindu Philosophy,* 92.
16. Discourse at Siddha Yoga Dham, Oakland, California, 1977.
17. H. Beinfield and E. Korngold, *Between Heaven and Earth* (New York: Ballantine, 1991), p. 56.
18. John and Anne Chitty, *Relationships And The Human Energy Field* (Boulder, Colorado: Polarity Press, 1991) p. 7.
19. Ibid., p. 7.
20. Stone, Book IV, p. 44.

Chapter 15
1. Stone, Book V, p. 3.
2. Jelalaluddin Rumi, *The Mathnawi.*
3. Lawlor, *Sacred Geometry,* p. 89.
4. Capra, *The Turning Point,* p. 304.
5. Vanamali, *Nitya Yoga.*
6. Ken Wilber, *No Boundary,* p. 1. See R. M. Bucke, *Cosmic Consciousness* (New York, New York: Dutton, 1969)
7. Wilber, *Spectrum of Consciousness,* p. 88.
8. Wilber, *Grace and Grit,* p. 370.
9. Ramana Maharshi, *The Spiritual Teachings of Ramana Maharshi* (Boston, Massachusetts: Shambala, 1972); Sri Nisargadatta Maharaj, *I Am That* (Bombay, India: Chetana, 1973); H.W.L. Poonja, *Wake Up and Roar* (Lucknow, India: Universal Booksellers, 1992); Ramesh Balsekar, *The Final Truth: A Guide to Ultimate Understanding, Consciousness Speaks* (Los Angeles, California: Advaita Press).
10. Beinfield and Korngold, *Between Heaven and Earth,* p. 30.
11. Stone, PN, p. 227.
12. In this teaching Dr. Stone uses the word Sky, a symbol of vast space, for Ether, and Wind, the essence of movement for Air.

Chapter 16
1. Vanamali, *Nitya Yoga.*
2. In the Sanatana Dharma Purusha, the universal Soul, is understood as immutable and omniscient, all knowing. Purusha, the Cosmic Being, does not evolve. Prakriti, or nature, has as its fundamental organizing principle evolution or growth.
3. John Woodroffe, *Principles of Tantra, Part II* (Madras, India: Ganesh, 1970), p. 35 "The Divine Mother (call her Father or Father-Mother as you choose) is in every molecule, in every atom, in all things which constitute the world."
4. Polarity Press, 2410 Jasper Ct., Boulder, CO 80304.
5. Stone, Book V, chart 17.
6. Ibid., p. 23.
7. Stone, Book I, p. 75.
8. Stone, Book V, chart 7.
9. Stone, EES, chart 20.
10. Ibid.
11. Stone, Book I, p. 80.
12. Stone, Book V, chart 17.
13. Ibid., p. 23.
14. Stone, Book I, p. 75.

15. Stone, Book V, chart 7.

16. Stone, EES, chart 20.

17. Ibid.

18. David Quigley, *Alchemical Hypnotherapy*, (Redway, California: Lost Coast Press, 1984). Quigley offers a brilliant approach to effective therapy.

19. Gay Hendricks and Barry Weinhold, *Transpersonal Approaches to Counseling and Psychotherapy* (Denver, Colorado: Love Publishing, 1982). Gay and his wife Kate have written more than twenty valuable books on body-centered psychotherapy and transpersonal approaches to counseling. Hendricks Institute, Santa Barbara, CA (800) 688–0772.

20. Stone, Book II, p. 65

Chapter 17

1. This chapter relies heavily on the excellent summary of Polarity principles in Franklin Sill's book, *The Polarity Process;* and the work of John and Anna Chitty in *Relationships and the Human Energy Field.* I recommend both of these books to individuals with a serious interest in somatic psychology.

2. Ether predominates in the plasma generated in the bone marrow, then steps down through the Ether/Air-predominant joints, through the Fire-predominant muscles, and to the Water-predominant tendons, for completion on the Earth.

3. Sills, *The Polarity Process*, p. 76.

Chapter 18

1. Prabhavananda and Isherwood, *The Bhagavad Gita,* p. 55.

2. Wilber, *Grace and Grit,* pp. 370–371.

3. Prabhavananda and Isherwood, *The Song of God: Bhagavad-Gita,* p. 70.

4. Ibid., p. 38, p.70.

5. Robert Powell, Ph.D., Ed, *The Ultimate Medicine* (San Diego, California: Blue Dove Press, 1994).

6. Ananda K. Coomaraswamy, *The Dance of Shiva* (New Delhi, India: Sagar Publications, 1991), pp. 3–4.

7. Radhakrishnan, *The Principal Upanisads,* p. 117.

8. Coomaraswamy, *The Dance of Shiva*, p. 5.

9. Bernard, *Hindu Philosophy*, p. 2.

10. Nisargadatta Maharaj, *I Am That.*

11. Sogyal Rinpoche, *The Tibetan Book of Living and Dying,* p. 188.

12. Wilber, *No Boundary,* p. 128

13. Arthur Avalon, *Tantra of the Great Liberation* (New York: Dover Publications, 1972); *Mahanirvana,* Chapter VIII, verse 215; Woodroffe, *The World as Power,* p. 212.

14. Wilber, *No Boundary,* p. 128.

15. Radhakrishnan, *The Principal Upanisads,* p. 120.

16. Wilber, *No Boundary,* p. 130.

17. Vanamali, *Nitya Yoga,* audio tape 13.

18. Radhakrishnan, *The Principal Upanisads,* p. 121.

19. Prabhavananda and Isherwood, *The Song of God: Bhagavad-Gita,* p. 34.

20. M. Ghandi, "Commentary on the Gita," in *Vidya* (Santa Barbara, California: U.T.F.) Vol I, No 1.

21. Prabhavananda and Isherwood, *The Song of God: Bhagavad-Gita,* p. 65.

22. Vanamali, *Nitya Yoga,* Chapter 8, audio tape 6, side A.

23. Ibid., side B.

24. Prabhavananda and Isherwood, *The Song of God: The Bhagavad Gita,* p. 40.

25. Ibid., p. 41.
26. Ibid., p. 70.
27. Tyborg, *The Language of the Gods,* p. 2.
28. Radhakrishnan, *The Principal Upanisads,* III, 28.

Appendix A
1. Le Mee, *Hymns from the Rig Veda,* p. xii.
2. Tyborg, *The Language of the Gods,* p. 15.
3. Le Mee, *HymnsfFromtThe Rig Veda,* p. xii.
4. Johari, *Tools for Tantra,* p. 8.
5. Ibid., p. 36.
6. Avalon, *The Serpent Power,* p. 99.
7. Woodroffe, *The World As Power.*
8. Seal, *The Positive Sciences of The Ancient Hindus,* p. 153.
9. Sir John Woodroffe, *Principles of Tantra* (Madras, India: Ganesh, 1970), pp. 172–173.
10. Johari, *Tools for Tantra,* p. 8.
11. Ibid.
12. Frawley, *Ayurvedic Healing,* pp. 314–315.
13. Ibid.

Appendix A
1. Wilber, *The Holographic Paradigm;* Talbot, *The Holographic Universe.*
2. Stone, Book II, charts 3, 4; Book III, chart 3; Book V, charts 7, 8, 19; EES, charts 5, 6.

Appendix C
1. Stone, HB, "A Purifying Diet," pp. 87–95.
2. Elizabeth Shaw,"Polarity Diet Self Test" (Garberville, California: Heartwood Wellness Clinic, 1996).
3. Ibid., p. 24.
4. Ibid., p. 25.
5. Elizabeth Shaw, "Alkaline and Acid Foods" (Garberville, California: Heartwood Wellness Clinic, 1996).

Appendix E
1. American Polarity Therapy Association, *Standards for Practice and Code of Ethics* (Boulder, Colorado: American Polarity Therapy Association, 1997). Reprinted with permission, American Polarity Therapy Association, 2888 Bluff St, #149, Boulder, CO 80301.

Glossary
1. Avalon, *The Serpent Power,* p. 32.
2. Woodroffe, *The World As Power,* Ganesh, p. 158. Atman comes from the verb root *at* "to breathe." Also, Tyberg, *The Language of the Gods,* p. 4.
3. Prabhavananda and Manchester, *The Upanishads,* pp. 50–53.
4. Tyborg, The *Language of the Gods,* p. 2.
5. Ibid. p. 70.
6. Bernard, *Hindu Philosophy,* p. 69.
7. Ibid., p. 193.
8. Tyborg, *The Language of the Gods,* p. 2.
9. Bernard, *Hindu Philosophy,* p. 193.
10. Tyberg, *The Language of the Gods,* p. 71.
11. Prabhavananda and Isherwood, *The Song of God: Bhagavad-Gita,* p. 131.
12. Radha Soami Books, P.O. Box 242, Gardena, CA 90247 (310) 329–5635.
13. *Webster's New Collegiate Dictionary.*

Index